VOLUME ONE HUNDRED AND FORTY SEVEN

PROGRESS IN MOLECULAR BIOLOGY AND TRANSLATIONAL SCIENCE

Matrix Metalloproteinases and Tissue Remodeling in Health and Disease: Cardiovascular Remodeling

VOLUME ONE HUNDRED AND FORTY SEVEN

PROGRESS IN MOLECULAR BIOLOGY AND TRANSLATIONAL SCIENCE

Matrix Metalloproteinases and Tissue Remodeling in Health and Disease: Cardiovascular Remodeling

Edited by

RAOUF A. KHALIL
Harvard Medical School,
Brigham and Women's Hospital,
Division of Vascular Surgery,
Boston, MA, United States

ELSEVIER

ACADEMIC PRESS
An imprint of Elsevier

Academic Press is an imprint of Elsevier
50 Hampshire Street, 5th Floor, Cambridge, MA 02139, United States
525 B Street, Suite 1800, San Diego, CA 92101-4495, United States
The Boulevard, Langford Lane, Kidlington, Oxford OX5 1GB, United Kingdom
125 London Wall, London EC2Y 5AS, United Kingdom

First edition 2017

ISBN: 978-0-12-811637-1
ISSN: 1877-1173

For information on all Academic Press publications
visit our website at https://www.elsevier.com/books-and-journals

Publisher: Zoe Kruze
Acquisition Editor: Alex White
Editorial Project Manager: Helene Kabes
Production Project Manager: James Selvam
Cover Designer: Victoria Pearson

Typeset by SPi Global, India

CONTENTS

CONTRIBUTORS

Valerie Arpino
Centre for Critical Illness Research, Lawson Health Research Institute; Schulich School of Medicine and Dentistry, Western University, London, ON, Canada

Bethan A. Brown
School of Clinical Sciences, University of Bristol, Bristol, United Kingdom

Yunfei Chen
Vascular Surgery Research Laboratories, Brigham and Women's Hospital, Harvard Medical School, Boston, MA, United States

Ning Cui
Vascular Surgery Research Laboratories, Brigham and Women's Hospital, Harvard Medical School, Boston, MA, United States

Kristine Y. DeLeon-Pennell
Mississippi Center for Heart Research, UMMC; Research Service, G.V. (Sonny) Montgomery Veterans Affairs Medical Center, Jackson, MS, United States

Emanuela Falcinelli
Section of Internal and Cardiovascular Medicine, University of Perugia, Perugia, Italy

Sarah J. George
School of Clinical Sciences, University of Bristol, Bristol, United Kingdom

Sean E. Gill
Centre for Critical Illness Research, Lawson Health Research Institute; Schulich School of Medicine and Dentistry, Western University, London, ON, Canada

Paolo Gresele
Section of Internal and Cardiovascular Medicine, University of Perugia, Perugia, Italy

Min Hu
Vascular Surgery Research Laboratories, Brigham and Women's Hospital, Harvard Medical School, Boston, MA, United States

Mira Jung
Mississippi Center for Heart Research, UMMC, Jackson, MS, United States

Raouf A. Khalil
Vascular Surgery Research Laboratories, Brigham and Women's Hospital, Harvard Medical School, Boston, MA, United States

Christopher K.S. Lee
Centre for Critical Illness Research, Lawson Health Research Institute; Schulich School of Medicine and Dentistry, Western University, London, ON, Canada

Merry L. Lindsey
Mississippi Center for Heart Research, UMMC; Research Service, G.V. (Sonny) Montgomery Veterans Affairs Medical Center, Jackson, MS, United States

Marcello G. Masciantonio
Centre for Critical Illness Research, Lawson Health Research Institute; Schulich School of Medicine and Dentistry, Western University, London, ON, Canada

Sanjay Mehta
Centre for Critical Illness Research, Lawson Health Research Institute, London, ON, Canada

Cesar A. Meschiari
Mississippi Center for Heart Research, UMMC, Jackson, MS, United States

Stefania Momi
Section of Internal and Cardiovascular Medicine, University of Perugia, Perugia, Italy

William C. Parks
Women's Guild Lung Institute, Cedars-Sinai Medical Center, Los Angeles, CA, United States

Wei Peng
Vascular Surgery Research Laboratories, Brigham and Women's Hospital, Harvard Medical School, Boston, MA, United States

Simon W. Rabkin
University of British Columbia, Vancouver, BC, Canada

Joseph D. Raffetto
Vascular Surgery Research Laboratories, Brigham and Women's Hospital, Harvard Medical School, Boston, MA, United States

Manuela Sebastiano
Section of Internal and Cardiovascular Medicine, University of Perugia, Perugia, Italy

Kate S. Smigiel
Women's Guild Lung Institute, Cedars-Sinai Medical Center, Los Angeles, CA, United States

Helen Williams
School of Clinical Sciences, University of Bristol, Bristol, United Kingdom

PREFACE

Matrix metalloproteinases (MMPs) are a family of structurally related zinc-containing proteolytic enzymes that degrade various components of the extracellular matrix and connective tissue proteins. The MMP family includes collagenases, gelatinases, stromelysins, matrilysins and membrane-type MMPs, and other MMPs. MMPs are important regulators of tissue remodeling, cell migration, and adhesion molecules. In addition to their proteolytic effects on the extracellular matrix, recent studies suggest novel effects of MMPs on transmembrane and intracellular signaling in many cell types including the vascular endothelium and smooth muscle. MMPs are tightly regulated at the transcription level and can be activated by other MMPs or proteases into their proteolytic active forms that interact with various substrates and signaling pathways. MMP activity is also regulated by endogenous tissue inhibitors of metalloproteinases (TIMPs), which provide a balancing mechanism to prevent excessive degradation of the extracellular matrix. MMP/TIMP imbalance could lead to pathological conditions and major cardiovascular, metabolic, and musculoskeletal disorders as well as cancer. Modulation of MMP activity using genetic manipulations of endogenous TIMPs or synthetic pharmacological inhibitors could control MMP activity and may provide new approaches in the management of MMP-related diseases.

This volume of *Progress in Molecular Biology and Translational Science* provides insights into MMPs and Tissue Remodeling in Health and Disease. Because MMPs play a role in a large number of biological processes and could be involved in numerous pathological conditions, we divided this volume into two parts. Part I focuses on Cardiovascular Remodeling, and Part II covers the role of MMPs in other target tissues and diseases and the potential benefits of MMP inhibitors. Renown scientists and researchers have agreed to share their expertise and advanced knowledge on MMPs. In Part I on the role of MMPs on Cardiovascular Remodeling we will cover several important topics regarding the basic biochemical and biological properties of MMPs; the role of MMPs in myocardial infarction and heart failure; the MMP/TIMP balance and microvascular endothelial function and dysfunction; MMPs in blood platelets and leukocyte activation; the role of MMPs in atherosclerosis, aneurysm, and hepatic ischemia/reperfusion injury; and the role of MMPs in remodeling of lower extremity veins and chronic venous disease. These important reviews were written by research investigators and clinician–scientists from different parts of the world, thus promoting different viewpoints in the pathogenesis of cardiovascular disease and highlighting different approaches in the diagnosis and management of cardiovascular disorders. Thanks to the good work of the contributing authors, and the careful review of our dedicated Reviewers and Editors, we were able to put together these important topics, and present them to our readers in a clear, concise, and informative fashion. I encourage every researcher, clinician, medical, graduate, and undergraduate student with aspiration to work in the cardiovascular field to read this state-of-the-art synopsis on MMPs.

I would like to dedicate this volume to the late Dr. P. Michael Conn, the past Series Editor of *Progress in Molecular Biology and Translational Science*. Dr. Conn gave me the great opportunity to be the Editor of this special and timely volume, and for this I will always be very grateful. I also would like to thank our outstanding Senior Editorial Project Manager Mrs. Helene Kabes and our hardworking Editorial Staff who spared no effort to ensure the highest quality of the articles. I also would like to acknowledge our contributing authors not only for their excellent articles but also for sharing some of the reviewers' duties, and for being very generous with their time and effort in providing other authors with helpful comments and constructive criticism. I particularly wish to thank our readers for their interest in MMPs and cardiovascular remodeling. I encourage all of you to provide feedback and contact me directly if you have any questions, comments, suggestions, criticism, or ideas that could further enhance our knowledge and help us achieve our goals and meet the highest expectations of our readers.

Raouf A. Khalil
Harvard Medical School, Brigham and Women's Hospital,
Boston, MA, United States

CHAPTER ONE

Biochemical and Biological Attributes of Matrix Metalloproteinases

Ning Cui, Min Hu, Raouf A. Khalil[1]
Vascular Surgery Research Laboratories, Brigham and Women's Hospital, Harvard Medical School, Boston, MA, United States
[1]Corresponding author: e-mail address: raouf_khalil@hms.harvard.edu

Contents

Abstract

Matrix metalloproteinases (MMPs) are a family of zinc-dependent endopeptidases that are involved in the degradation of various proteins in the extracellular matrix (ECM). Typically, MMPs have a propeptide sequence, a catalytic metalloproteinase domain with catalytic zinc, a hinge region or linker peptide, and a hemopexin domain. MMPs are commonly classified on the basis of their substrates and the organization of their structural domains into collagenases, gelatinases, stromelysins, matrilysins, membrane-type (MT)-MMPs, and other MMPs. MMPs are secreted by many cells including fibroblasts,

Progress in Molecular Biology and Translational Science, Volume 147
ISSN 1877-1173
http://dx.doi.org/10.1016/bs.pmbts.2017.02.005

vascular smooth muscle (VSM), and leukocytes. MMPs are regulated at the level of mRNA expression and by activation of their latent zymogen form. MMPs are often secreted as inactive pro-MMP form which is cleaved to the active form by various proteinases including other MMPs. MMPs cause degradation of ECM proteins such as collagen and elastin, but could influence endothelial cell function as well as VSM cell migration, proliferation, Ca^{2+} signaling, and contraction. MMPs play a role in tissue remodeling during various physiological processes such as angiogenesis, embryogenesis, morphogenesis, and wound repair, as well as in pathological conditions such as myocardial infarction, fibrotic disorders, osteoarthritis, and cancer. Increases in specific MMPs could play a role in arterial remodeling, aneurysm formation, venous dilation, and lower extremity venous disorders. MMPs also play a major role in leukocyte infiltration and tissue inflammation. MMPs have been detected in cancer, and elevated MMP levels have been associated with tumor progression and invasiveness. MMPs can be regulated by endogenous tissue inhibitors of metalloproteinases (TIMPs), and the MMP/TIMP ratio often determines the extent of ECM protein degradation and tissue remodeling. MMPs have been proposed as biomarkers for numerous pathological conditions and are being examined as potential therapeutic targets in various cardiovascular and musculoskeletal disorders as well as cancer.

ABBREVIATIONS

BK_{Ca} large conductance Ca^{2+}-activated K^+ channel
CXCR C-X-C chemokine receptor
ECM extracellular matrix
EDHF endothelium-derived hyperpolarizing factor
EMMPRIN extracellular matrix metalloproteinase inducer
ERK extracellular signal-regulated kinase
GM-CSF granulocyte-macrophage colony-stimulating factor
GPI glycosyl phosphatidylinositol
HIF hypoxia-inducible factor
IFN interferon
IL interleukin
IVC inferior vena cava
MAPK mitogen-activated protein kinase
MI myocardial infarction
miR microRNA
MMP matrix metalloproteinase
MT-MMP membrane-type MMP
NF-κB nuclear factor-kappa-light-chain-enhancer of activated B cells
PAR protease-activated receptor
PDGF platelet-derived growth factor
PI_3K phosphoinositide 3-kinase
PMNs polymorphonuclear leukocytes
RGD Arg-Gly-Asp
siRNA small-interfering RNA

SNP single-nucleotide polymorphism
TGF-β transforming growth factor-β
TIMP tissue inhibitors of metalloproteinases
TNF-α tumor necrosis factor-α
VEGF vascular endothelial growth factor
VSM vascular smooth muscle
VSMC vascular smooth muscle cell
Zn^{2+} zinc

1. INTRODUCTION

Matrix metalloproteinases (MMPs) are a family of zinc-dependent endoproteases with multiple roles in tissue remodeling and degradation of various proteins in the extracellular matrix (ECM). MMPs promote cell proliferation, migration, and differentiation, and could play a role in cell apoptosis, angiogenesis, tissue repair, and immune response. MMPs may also affect bioactive molecules on the cell surface and modulate various cellular and signaling pathways. Alterations in MMP expression and activity occur in normal biological processes, e.g., during pregnancy and wound healing, but have also been observed in cardiovascular diseases such as atherosclerosis, aneurysms, and varicose veins, musculoskeletal disorders such as osteoarthritis and bone resorption, and in various cancers. MMPs have also been implicated in tumor progression and invasiveness.

In this chapter, we will use data reported in PubMed and other scientific databases as well as data from our laboratory to provide a general overview of the biochemical and biological properties of MMPs with emphasis on MMP structure, tissue distribution, and protein substrates. We will then describe special properties of specific classes of MMPs and provide some examples of their role in cardiovascular diseases, inflammatory, and musculoskeletal disorders, as well as cancer. We will then briefly discuss the regulation of MMP activity by endogenous tissue inhibitors of metalloproteinases (TIMPs). We will conclude the chapter by highlighting the potential benefits of MMPs as biomarkers and therapeutic targets in cardiovascular conditions, musculoskeletal disorders, and cancer. Additional information regarding specific MMP functions can be found in other reports[1–4] and are elegantly reviewed in detail in the other chapters of this book.

2. MMP STRUCTURE

In the early 1960s, MMPs were first identified as a collagen proteolytic activity that causes ECM protein degradation during resorption of the tadpole tail.[5] MMPs are now grown to a family of endopeptidases or matrixins that belong to the metzincins superfamily of proteases. MMPs are highly homologous, multidomain, zinc (Zn^{2+})-containing metalloproteinases that degrade various protein components of ECM. The MMP family shares a common core structure. Typically MMPs consist of a propeptide of about 80 amino acids, a catalytic metalloproteinase domain of about 170 amino acids, a linker peptide (hinge region) of variable length, and a hemopexin domain of about 200 amino acids (Fig. 1).[6–9]

Most MMPs also share three important characteristics. First, MMPs show homology to collagenase-1 (MMP-1). MMP-7, -23, and -26 are exceptions as they lack the linker peptide and the hemopexin domain. MMP-23 has a unique C-terminal cysteine-rich domain and an immunoglobulin-like domain immediately after the C-terminus of the catalytic domain. Second, MMPs contain a cysteine-switch motif PRCGXPD in which the cysteine sulfhydryl group chelates the active site Zn^{2+} thus keeping MMPs in their inactive pro-MMP zymogen form. Third, the catalytic domain of MMPs harbors a Zn^{2+}-binding motif to which the Zn^{2+} ion is bound by three histidines from the conserved sequence HEXXHXXGXXH, with the assistance of a conserved glutamate, and a conserved methionine sequence XBMX (Met-turn) located 8-residues down from the Zn^{2+}-binding motif that supports the structure surrounding the catalytic Zn^{2+} (Fig. 1).[10–12]

In vertebrates, the MMP family comprises 28 members, at least 23 are expressed in human tissues, and 14 of those MMPs are expressed in the vasculature (Table 1).[10] MMPs are commonly classified on the basis of their substrates and the organization of their structural domains into collagenases, gelatinases, stromelysins, matrilysins, membrane-type (MT)-MMPs, and other MMPs. Additionally, different classes of MMPs have specific structural features that distinguish them from the prototypical MMP structure (Fig. 1).[2,13,14] The topology of MMPs is well conserved, and a major difference between MMPs lies in the S1′ subsite, a well-defined hydrophobic pocket of variable depth that is critical for specific MMP–substrate interaction.[15]

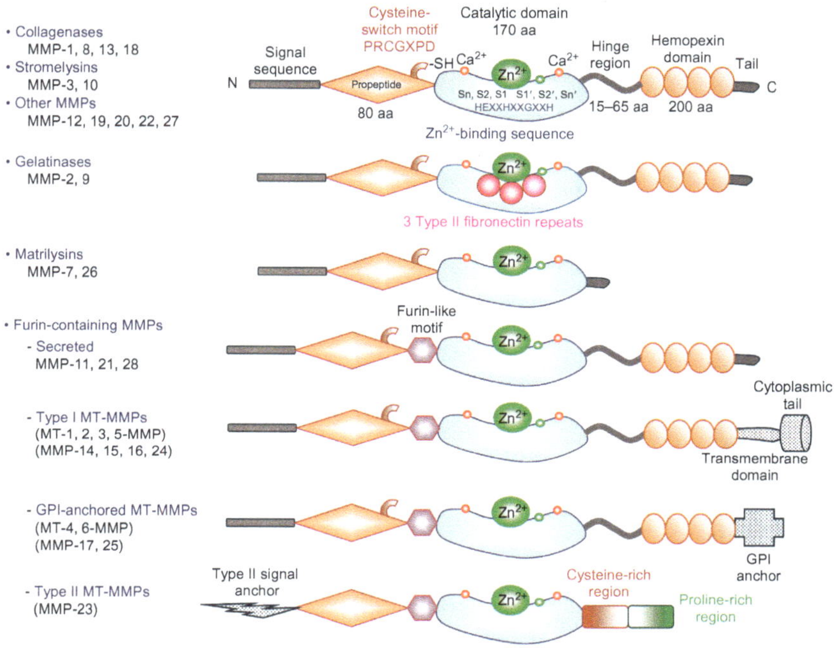

Fig. 1 Major MMPs subtypes and their structure. A typical MMP consists of a propeptide, a catalytic metalloproteinase domain, a linker peptide (hinge region), and a hemopexin domain. The propeptide has a cysteine-switch PRCGXPD whose cysteine sulfhydryl (–SH) group chelates the active site Zn^{2+}, keeping the MMP in the latent pro-MMP zymogen form. The catalytic domain contains the Zn^{2+} binding motif HEXXHXXGXXH, two Zn^{2+} ions (one catalytic and one structural), specific S1, S2,...,Sn and S1′, S2′,...,Sn′ pockets, which confer specificity, and two or three Ca^{2+} ions for stabilization. Some MMPs show exceptions in their structures. Gelatinases have three type II fibronectin repeats in the catalytic domain. Matrilysins have neither a hinge region nor a hemopexin domain. Furin-containing MMPs such as MMP-11, -21, and -28 have a furin-like pro-protein convertase recognition sequence in the propeptide C-terminus. MMP-28 has a slightly different cysteine-switch motif PRCGVTD. Membrane-type MMPs (MT-MMPs) typically have a transmembrane domain and a cytosolic domain. MMP-17 and -25 have a glycosylphosphatidylinositol (GPI) anchor. MMP-23 lacks the consensus PRCGXPD motif, has a cysteine residue located in a different sequence ALCLLPA, may remain in the latent inactive proform through its type II signal anchor, and has a cysteine-rich region and an immunoglobulin-like proline-rich region.

3. SOURCES AND TISSUE DISTRIBUTION OF MMPs

MMPs are produced by multiple tissues and cells (Table 1). MMPs are secreted by connective tissue, proinflammatory, and uteroplacental cells

Table 1 Members of the MMP Family, and Their Tissue Distribution and Substrates

MMP (Other Name) Chromosome	MW KDa Pro/Active	Distribution	Collagen Substrates	Noncollagen ECM Substrates	Other Targets and Substrates
Collagenases MMP-1 (Collagenase-1) 11q22.3	55/45	Endothelium, intima, SMCs, fibroblasts, vascular adventitia, platelets, varicose veins (interstitial/fibroblast collagenase)	I, II, III, VII, VIII, X, gelatin	Aggrecan, nidogen, perlecan, proteoglycan link protein, serpins, tenascin-C, versican	Casein, α1-antichymotrypsin, α1-antitrypsin, α1-proteinase inhibitor, IGF-BP-3 and -5, IL-1β, L-selectin, ovostatin, pro-TNF-α, SDF-1
MMP-8 (Collagenase-2) 11q22.3	75/55	Macrophages, neutrophils (PMNL or neutrophil collagenase)	I, II, III, V, VII, VIII, X, gelatin	Aggrecan, elastin, fibronectin, laminin, nidogen	α2-Antiplasmin, pro-MMP-8
MMP-13 (Collagenase-3) 11q22.3	60/48	SMCs, macrophages, varicose veins, preeclampsia, breast cancer	I, II, III, IV, gelatin	Aggrecan, fibronectin, laminin, perlecan, tenascin	Casein, plasminogen activator 2, pro-MMP-9 and -13, SDF-1
MMP-18 (Collagenase-4) 12q14	70/53	*Xenopus* (amphibian, Xenopus collagenase) heart, lung, colon	I, II, III, gelatin		α1-Antitrypsin
Gelatinases MMP-2 (Gelatinase A, Type IV Collagenase) 16q13-q21	72/63	Endothelium, VSM, adventitia, platelets, leukocytes, aortic aneurysm, varicose veins	I, II, III, IV, V, VII, X, XI, gelatin	Aggrecan, elastin, fibronectin, laminin, nidogen, proteoglycan link protein, versican	Active MMP-9 and -13, FGF-R1, IGF-BP-3, and -5, IL-1β, pro-TNF-α, TGF-β

MMP-9 (Gelatinase B, Type IV Collagenase) 20q11.2-q13.1	92/86	Endothelium, VSM, adventitia, microvessels, macrophages, aortic aneurysm, varicose veins	IV, V, VII, X, XIV, gelatin	Aggrecan, elastin, fibronectin, laminin, nidogen, proteoglycan link protein, versican	CXCL5, IL-1β, IL2-R, plasminogen, pro-TNF-α, SDF-1, TGF-β
Stromelysins MMP-3 (Stromelysin-1) 11q22.3	57/45	Endothelium, intima, VSM, platelets, coronary artery disease, hypertension, varicose veins, synovial fibroblasts, tumor invasion	II, III, IV, IX, X, XI, gelatin	Aggrecan, decorin, elastin, fibronectin, laminin, nidogen, perlecan, proteoglycan, proteoglycan link protein, versican	Casein, α1-antichymotrypsin, α1-proteinase inhibitor, antithrombin III, E-cadherin, fibrinogen, IGF-BP-3, L-selectin, ovostatin, pro-HB-EGF, pro-IL-1β, pro-MMP-1, -8, and -9, pro-TNF-α, SDF-1
MMP-10 (Stromelysin-2) 11q22.3	57/44	Atherosclerosis, uterus, preeclampsia, arthritis, carcinoma cells	III, IV, V, gelatin	Aggrecan, elastin, fibronectin, laminin, nidogen	Casein, pro-MMP-1, -8, and -10
MMP-11 (Stromelysin-3) 22q11.23	51/44	Brain, uterus, angiogenesis	Does not cleave	Aggrecan, fibronectin, laminin	α1-Antitrypsin, α1-proteinase inhibitor, IGF-BP-1
Matrilysins MMP-7 (Matrilysin-1) 11q21-q22	29/20	Endothelium, intima, VSM, uterus, varicose veins (PUMP)	IV, X, gelatin	Aggrecan, elastin, enactin, fibronectin, laminin, proteoglycan link protein	Casein, β4 integrin, decorin, defensin, E-cadherin, Fas-ligand, plasminogen, pro-MMP-2, -7, and -8, pro-TNF-α, syndecan, transferrin

Continued

Table 1 Members of the MMP Family, and Their Tissue Distribution and Substrates—cont'd

MMP (Other Name) Chromosome	MW KDa Pro/Active	Distribution	Collagen Substrates	Noncollagen ECM Substrates	Other Targets and Substrates
MMP-26 (Matrilysin-2, Endometase) 11p15	28/19	Breast cancer, endometrial tumors	IV, gelatin	Fibrinogen, fibronectin, vitronectin	Casein, β1-proteinase inhibitor, fibrin, fibronectin, pro-MMP-2
Membrane-type MMP-14 (MT1-MMP) 14q11-q12	66/56	VSM, fibroblasts, platelets, brain, uterus, angiogenesis	I, II, III, gelatin	Aggrecan, elastin, fibrin, fibronectin, laminin, nidogen, perlecan, proteoglycan, tenascin, vitronectin	$\alpha_v\beta_3$ integrin, CD44, pro-MMP-2 and -13, pro-TNF-α, SDF-1, α1-proteinase inhibitor, tissue transglutaminase
MMP-15 (MT2-MMP) 16q13	72/50	Fibroblasts, leukocytes, preeclampsia	I, gelatin	Aggrecan, fibronectin, laminin, nidogen, perlecan, tenascin, vitronectin	Pro-MMP-2 and -13, tissue transglutaminase
MMP-16 (MT3-MMP) 8q21.3	64/52	Leukocytes, angiogenesis	I	Aggrecan, fibronectin, laminin, perlecan, vitronectin	Casein, pro-MMP-2 and -13
MMP-17 (MT4-MMP) 12q24.3	57/53	Brain, breast cancer	Gelatin	Fibrin	

MMP-24 (MT5-MMP) 20q11.2	57/53	Leukocytes,lung, pancreas, kidney, brain, astrocytoma, glioblastoma	Gelatin	Chondroitin sulfate, dermatin sulfate, fibrin, fibronectin, N-cadherin	Pro-MMP-2 and -13
MMP-25 (MT6-MMP) 16p13.3	34/28	Leukocytes (Leukolysin), anaplastic astrocytomas, glioblastomas	IV, gelatin		Fibrin, fibronectin, pro-MMP-2, α1-proteinase inhibitor
Other MMPs MMP-12 (Metalloelastase) 11q22.3	54/45—22	SMCs, fibroblasts, macrophages, great saphenous vein	IV, gelatin	Elastin, fibronectin, laminin	Casein, plasminogen
MMP-19 (RASI-1) 12q14	54/45	Liver	I, IV, gelatin	Aggrecan, fibronectin, laminin, nidogen, tenascin	Casein
MMP-20 (Enamelysin) 11q22.3	54/22	Tooth enamel	V	Aggrecan, cartilage oligomeric protein, amelogenin	
MMP-21 (Xenopus-MMP) 10q26.13	62/49	Fibroblasts, macrophages, placenta			α1-Antitrypsin
MMP-22 (Chicken-MMP) 1p36.3	51	Chicken fibroblasts	Gelatin		

Continued

Table 1 Members of the MMP Family, and Their Tissue Distribution and Substrates—cont'd

MMP (Other Name) Chromosome	MW KDa Pro/Active	Distribution	Collagen Substrates	Noncollagen ECM Substrates	Other Targets and Substrates
MMP-23 (CA-MMP) 1p36.3	28/19	Ovary, testis, prostate Other (type II) MT-MMP	Gelatin		
MMP-27 (Human MMP-22 homolog) 11q24		Heart, leukocytes, macrophages, kidney, endometrium, menstruation, bone, osteoarthritis, breast cancer			
MMP-28 (Epilysin) 17q21.1	56/45	Skin, keratinocytes			Casein

CA-MMP, cysteine array MMP; *CXCL5*, chemokine (C-X-C motif) ligand 5; *FGF-R1*, fibroblast growth factor receptor 1; *IGF-BP*, insulin-like growth factor-binding protein; *IL*, interleukin; *MW*, molecular mass; *PMNL*, polymorphonuclear leukocytes; *pro-HB-EGF*, pro-heparin-binding epidermal growth factor-like growth factor; *RASI-1*, rheumatoid arthritis synovium inflamed-1; *SDF-1*, stromal cell-derived factor-1.

including fibroblasts, osteoblasts, endothelial cells, vascular smooth muscle (VSM), macrophages, neutrophils, lymphocytes, and cytotrophoblasts.

Dermal fibroblasts and leukocytes are major sources of MMPs, especially MMP-2,[16] and platelets are important sources of MMP-1, -2, -3, and -14.[17] In general, MMPs are either secreted from the cells or anchored to the plasma membrane by proteoglycans such as heparan sulfate glycosaminoglycans.[10] Membrane-type MMPs (MT-MMPs) and MMP-23 are anchored to the cell membrane by special transmembrane domains.

Because MMPs play a major role in ECM remodeling, they are highly distributed in most connective tissues. MMPs have also been localized in many cell types, suggesting other biological roles for MMPs. For example, MMP-1, -2, -3, -7, -8, -9, -12, -13, and MT1-MMP and MT3-MMP, are expressed in various vascular tissues and cells.[18] In the rat inferior vena cava (IVC), MMP-2 and -9 are localized in different layers of the venous wall including the intima, media, and adventitia, suggesting interaction with signaling pathways in endothelial cells, VSM, and ECM, respectively.[19] Other studies showed specific distribution of MMP-1, -2, -3, and -7 in endothelial cells and vascular smooth muscle cells (VSMCs), MMP-2 in the adventitia,[20] MMP-9 in endothelial cells, medial VSMCs, and adventitial microvessels, and MMP-12 in VSMCs and fibroblasts of human great saphenous vein,[21] Other studies showed intracellular localization of MMP-2 within cardiac myocytes, and colocalization of MMP-2 with troponin I within the cardiac myofilaments. MMP-2 activity has also been detected in nuclear extracts from both human heart and rat liver. Poly ADP-ribose polymerase is a nuclear matrix enzyme involved in DNA repair. Interestingly, poly ADP-ribose polymerase is susceptible to cleavage by MMP-2 in vitro, and its cleavage is blocked by MMP inhibitors. MMP-2 localization within the nucleus could play a role in degradation of poly ADP-ribose polymerase, and thereby affect DNA repair.[22]

4. MMP ACTIVATION

MMPs are regulated at multiple levels including mRNA expression, activation of the proenzyme to the active form, and the counteracting actions of endogenous TIMPs. MMPs are synthesized as pre-pro-MMPs, from which the signal peptide is removed during translation to generate pro-MMPs. In these zymogens or pro-MMPs, the cysteine from the PRCGXPD "cysteine-switch" motif coordinates with the catalytic Zn^{2+} to keep the pro-MMPs inactive.[6] In order to process and activate these zymogens or pro-MMPs, the cysteine switch is cleaved and the prodomain is detached often by other proteolytic enzymes such as serine proteases, the endopeptidase furin, plasmin, or other MMPs to produce the active MMP forms.[6] Furin-containing MMPs such as MMP-11, -21, and -28, and MT-MMPs have a furin-like pro-protein convertase recognition sequence at the C-terminus of the propeptide and are activated intracellularly by furin (Fig. 1).[23] MT-MMPs first undergo intracellular activation by furin, then proceed to the cell surface where they can cleave and activate other pro-MMPs.[13] TIMPs are also essential for the formation of noninhibitory pro-MMP/TIMP/MT-MMP complexes. Noninhibitory complexes between progelatinases and TIMPs are restricted to pro-MMP-2 and TIMP-2, -3, or -4, and to MMP-9 and TIMP-1.[24] For example, TIMP-2 first forms a complex with pro-MMP-2 by binding to its hemopexin domain, and the complex then localizes to the cell surface where it binds to the active site of a MT1-MMP molecule.[25–28] This ternary pro-MMP-2/TIMP-2/MT1–MMP complex then facilitates the cleavage and activation of its bound pro-MMP-2 to active MMP-2 by another "free" MT1-MMP molecule. This noninhibitory complex is different from the inhibitory complex of TIMP-2/active MMP-2. It is formed between the C-terminal domain of TIMP-2 and the C-terminal hemopexin of MMP-2, such that both molecules maintain their inhibitory and proteolytic properties, respectively.[24,29,30] The activation of MMP-2 on the cell surface allows it to accumulate pericellularly where it could reach marked collagenolytic activity locally in the extracellular space.[10] Similarly, the stromelysins MMP-3 and -10 are secreted from the cells as inactive pro-MMPs, but are then activated on the cell surface. MMPs can also be activated by various physicochemical agents including heat, low pH, thiol-modifying agents such as 4-aminophenylmercuric acetate, mercury chloride, *N*-ethylmaleimide, oxidized glutathione, sodium dodecyl sulfate, and chaotropic agents. Most of these activators disrupt the cysteine-Zn^{2+} coordination at the cysteine-switch motif of the MMP molecule. Other MMP activators include plasmin which activates MMP-9. Also, both MMP-3 and hypochlorous acid activate MMP-7, and MMP-7 could in turn activate MMP-1.[2]

MMP expression/activity can also be influenced by hormones, growth factors, and cytokines.[31] For example, ovarian sex hormones could affect the expression/activity of various MMPs which could in turn participate in endometrial tissue remodeling and shedding during the menstrual and estrous cycles. Also, increases in estrogen and progesterone as well as vascular endothelial growth factor (VEGF) and placental growth factor during pregnancy could promote the expression/activity of uteroplacental MMPs and in turn facilitate cytotrophoblast tissue invasion and uteroplacental growth and vascularization. MMP expression/activity also increases during the inflammatory process. MMPs are secreted by proinflammatory cells and their secretion is promoted by proinflammatory cytokines.

MMPs can be regulated by growth factors.[32] For example, overexpression of VEGFa in SNU-5 cells increases MMP-2 expression, while downregulation of VEGFa decreases MMP-2 expression.[33] Also, platelet-derived growth factor-BB (PDGF-BB) increases MMP-2 expression in rat VSMCs, possibly via Rho-associated protein kinase, extracellular signal-regulated kinases (ERK), and phosphorylation of p38 mitogen-activated protein kinase (MAPK).[34] Also, in carotid artery plaques, epidermal growth factor (EGF) upregulates MMP-1 and -9 mRNA transcripts and increases MMP-9 activity in VSMCs.[35] In contrast, transforming growth factor-β1 (TGF-β1) may downregulate MMPs via a TGF-β1 inhibitory element in the MMP promoter. Interestingly, MMP-2 does not have this element, and therefore may not be affected, or in some instances upregulated, by TGF-β1.

MMP activity is also regulated by endogenous TIMPs. Increased MMP expression/activity or decreased TIMPs could lead to MMP/TIMP imbalance and results in various pathological conditions most notably heart failure, osteoarthritis, and cancer.

5. MMP SUBSTRATES

ECM has three main components: fibers, proteoglycans, and polysaccharides. Fibers are largely glycoproteins that include collagen, which is the main ECM protein, and elastin, which is not glycosylated and provides plasticity and flexibility to certain tissues such as the arteries, lungs, and skin. Laminin is a glycoprotein localized in the basal lamina of the epithelium. Fibronectin is a glycoprotein used by cells to bind to ECM and can modulate the cytoskeleton to facilitate or hinder cell movement. Proteoglycans have more carbohydrates than proteins and attract water to keep the ECM hydrated. Proteoglycans also facilitate binding of growth factors to the ECM milieu. Syndecan-1 is a proteoglycan and integral transmembrane protein that bind chemotactic cytokines during the inflammatory process. Other ECM proteins include glycoproteins such as vitronectin, aggrecan, entactin, fibrin, and tenascin, and polysaccharides such as hyaluronic acid.[2]

MMPs play a major role in tissue remodeling by promoting turnover of various ECM proteins including collagens, elastin, gelatin, and other matrix glycoproteins, and proteoglycans. Collagen and elastin are essential for the structural integrity of the vascular wall and are important MMP substrates. MMPs break down collagen type I, II, III, IV, V, VI, VII, VIII, IX, X, and XIV with different efficacies. MMP degrades other ECM protein substrates such as aggrecan, entactin, fibronectin, tenascin, laminin, myelin basic protein, and vitronectin (Table 1). While casein is not a physiological MMP substrate, it is digested by several MMPs and, therefore, is used to measure the activity of these MMPs in zymography assays.[2]

The hemopexin domain may confer most of the MMP substrate specificity.[36,37] The hemopexin domain may be essential in the recognition and subsequent catalytic degradation of fibrillar collagen, whereas the catalytic domain may be sufficient in the degradation of noncollagen substrates.[10] MMPs catalytic activity generally requires Zn^{2+} and a water molecule flanked by three-conserved histidine residues and a conserved glutamate, with a conserved methionine acting as a hydrophobic base to support the structure surrounding the catalytic Zn^{2+} in the MMP molecule. During the initial transition states of the MMP–substrate interaction, Zn^{2+} is penta-coordinated with a substrate's carbonyl oxygen atom, one oxygen atom from the MMP glutamate-bound water, and the three-conserved histidines in the MMP molecule. The Zn^{2+}-bound water then performs a nucleophilic attack on the substrate, resulting in the breakdown of the substrate and the release of a water molecule (Fig. 2).[11,38–40] The MMP–substrate interaction may involve alternative transition states, whereby Zn^{2+} is penta-coordinated with a substrate's carbonyl oxygen atom, two oxygen atoms from the MMP conserved glutamate, and two of the three-conserved histidines. One oxygen from glutamate then performs a nucleophilic attack and causes breakdown of the substrate.[41] Peptide catalysis and substrate degradation is also influenced by specific subsites or pockets (S) within the MMP molecule that interact with corresponding substituents (P) in the substrates (Fig. 2). The most important pocket for substrate specificity and binding is the MMP S1′ pocket, which is extremely variable and could have a shallow, intermediate, or deep location.[11,38,39] MMPs with shallow S1′ pocket include MMP-1 and -7. MMP-2, -9, and -13 have intermediate S1′ pocket, while MMP-3, -8, and -12 have deep S1′ pocket.[38] S2′ and S3′ pockets are shallower than S1′ pocket and, therefore, are more exposed to solvents than S1′.[39] Second to the S1′ pocket, the S3 pocket may contribute to substrate specificity.[6]

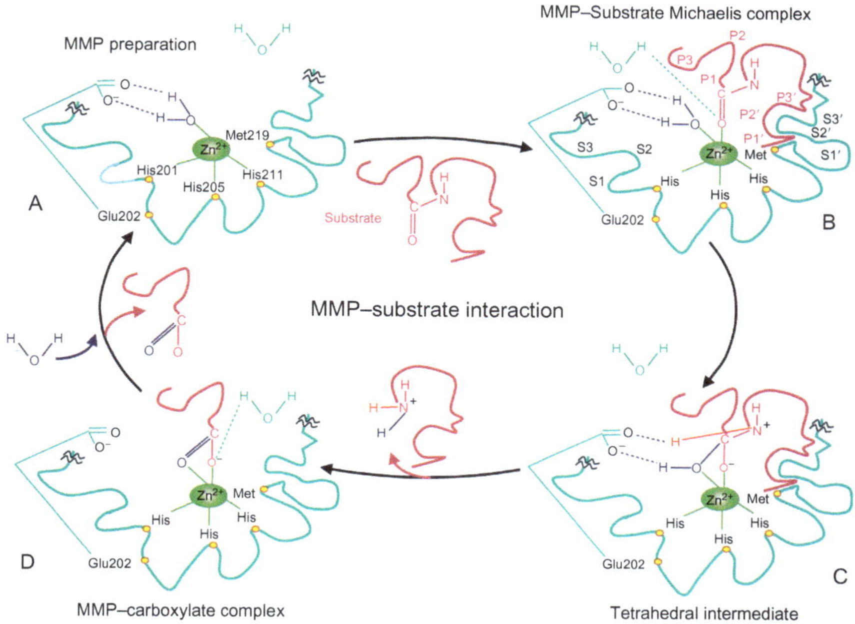

Fig. 2 MMP–substrate interaction. MMP-3 is used as an example, and slight variations in the MMP–substrate interaction and the positions of the conserved His and Glu may occur with other MMPs. Only the MMP catalytic domain is illustrated, and the remaining part of the MMP molecule is truncated by squiggles. (A) In the quiescent MMP molecule, the catalytic Zn^{2+} is supported in the HEXXHXXGXXH-motif by binding to the imidazole rings of the three histidines His201, 205, 211. Additionally, the methionine-219 (Met219) in the conserved XBMX Met-turn acts as a hydrophobic base to further support the structure surrounding the catalytic Zn^{2+}. In preparation of MMP for substrate binding, an incoming H_2O molecule is polarized between the MMP acidic Zn^{2+} and basic glutamate-202 (Glu202). (B) Using H^+ from free H_2O, the substrate carbonyl group binds to Zn^{2+}, forming a Michaelis complex. This allows the MMP S1, S2, S3,…,Sn pockets on the right side of Zn^{2+} and the primed S1′, S2′, S3′,…,Sn′ pockets on the left side of Zn^{2+} to confer specific binding to the substrate P1, P2, P3,…,Pn and the primed P1′, P2′, P3′…,Pn′ substituents, respectively. The MMP pockets are organized such that the S1 and S3 pockets are located away from the catalytic Zn^{2+}, while the S2 pocket is closer to Zn^{2+}. (C) The substrate-bound H_2O is freed, the Zn^{2+}-bound O from the Glu-bound H_2O executes a nucleophilic attack on the substrate carbon, and the Glu202 extracts a proton from the Glu-bound H_2O to form an N–H bond with the substrate N, resulting in a tetrahedral intermediate. (D) Freed H_2O is taken up again, and the second proton from Glu-bound H_2O is transferred to the substrate, forming an additional N–H bond. As a result, the substrate scissile C–N bond breaks, thus releasing the N portion of the substrate, while the carboxylate portion of the substrate remains in an MMP–carboxylate complex. Another H_2O is taken up, thus releasing the remaining carboxylate portion of the substrate, and the MMP is prepared to attack another substrate (A).

Specific MMPs degrade specific protein substrates. Stromelysin-1 and -2 (MMP-3 and -10) do not cleave interstitial collagen, but degrade other ECM protein substrates and may participate in cleaving certain pro-MMPs to their active form. Although MMP-3 and -10 have similar substrate specificity, MMP-3 has greater proteolytic efficiency than MMP-10. Stromelysin-3 (MMP-11) is distantly related to stromelysin-1 and -2. MMP-11 does not cleave interstitial collagen and shows very weak proteolytic activity toward other ECM protein substrates.[23] Importantly, different MMPs may cooperate in order to completely degrade a protein substrate. For example, the collagenases MMP-1, -13, and -18 first unwind triple-helical collagen and hydrolyze the peptide bonds of fibrillar collagen type I, II, and III into 3/4 and 1/4 fragments.[6,42] The resulting single α-chain gelatins are further degraded by the gelatinases MMP-2 and -9 into smaller oligopeptides.[36] Gelatinases have three type II fibronectin repeats in their catalytic domain that allow them to bind not only gelatin but also collagen and laminin (Fig. 1). Therefore, while MMP-2 is primarily a gelatinase, it can function much like the collagenase MMP-1, albeit in a weaker manner.[6] MMP-2 can degrade collagen in two steps: first by inducing a weak interstitial collagenase-like collagen degradation into 3/4 and 1/4 fragments, then second by promoting gelatinolysis using the fibronectin-like domain.[43] MMP-9 could also act as a collagenase and gelatinase. As a collagenase, MMP-9 binds the α2 chains of collagen IV with high affinity even when it is inactive, making the substrate readily available.[44]

6. MMPs, ECM DEGRADATION, AND TISSUE REMODELING

MMPs are important in many biological processes including cell proliferation, migration, and differentiation, remodeling of ECM, and tissue invasion and vascularization (Fig. 3). These biological processes take place multiple times during normal development and organogenesis, but, if not properly balanced, could also contribute to harmful pathological conditions such as cancer, tumor progression, and tissue invasion (Fig. 4). MMPs can participate in these processes by several mechanisms including proteolytic cleavage of growth factors so that they become available to cells that are not in direct physical contact, degradation of ECM so that founder cells can move across the tissues into nearby stroma, and regulated receptor cleavage in order to terminate migratory signaling and cell migration.[45]

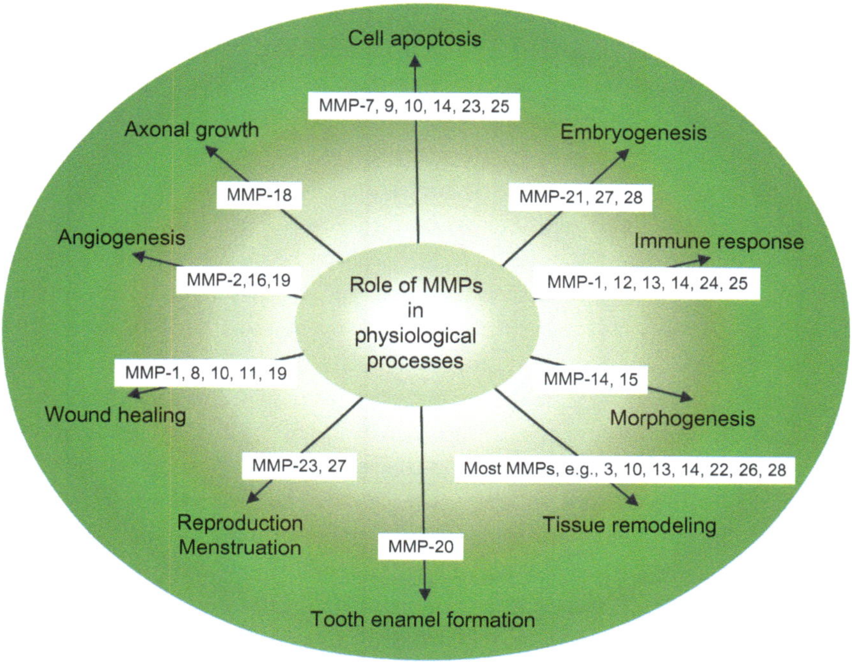

Fig. 3 Representative roles of MMPs in physiological processes.

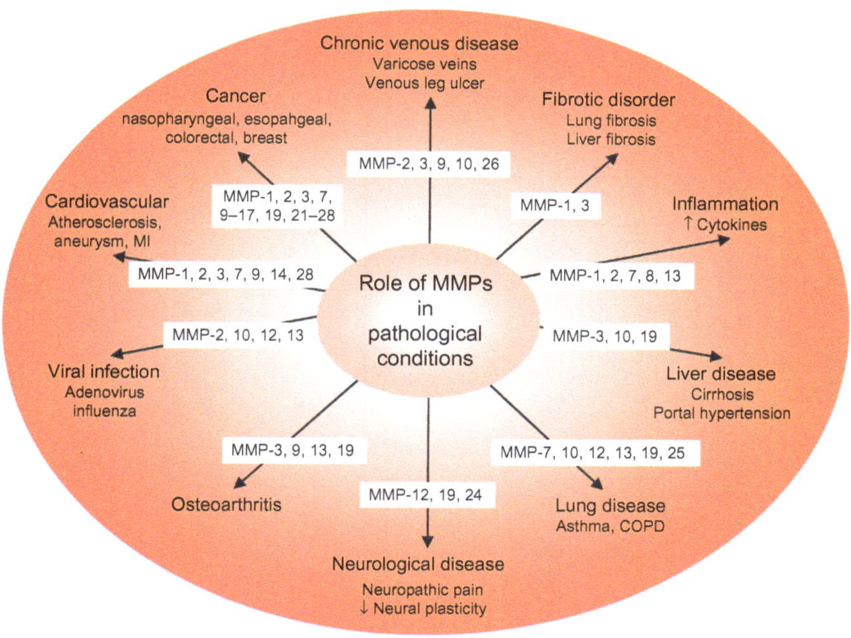

Fig. 4 Representative roles of MMPs in pathological conditions. *COPD*, chronic obstructive pulmonary disease; *MI*, myocardial infarction.

Dynamic modulation of the physical contacts between neighboring cells is integral to epithelial processes such as tissue repair. MMPs participate in tissue repair after acute injury.[46] Induction of MMP activity contributes to the disassembly of intercellular junctions and the degradation of ECM, thus overcoming the physical constraint to cell movement.[47]

MMPs may affect VSMC growth, proliferation, and migration. MMPs induce the release of growth factors by cleaving the growth factor-binding proteins and matrix molecules.[48] MMPs can facilitate VSMC proliferation by promoting permissive interactions between VSMCs and components of ECM, possibly via integrin-mediated pathways.[49] MMP-1 and -9 increase human aortic SMC migration.[50,51] MMP-induced ECM proteolysis can modulate cell-ECM adhesion either by removal of sites of adhesion or by exposing a binding site and in turn facilitate VSMC migration.

Alterations in MMPs expression/activity may be associated with cardiovascular disease. Evidence suggests associations between polymorphisms in MMP-1, -2, -3, -9, and -12 with ischemic stroke incidence, pathophysiology, and clinical outcome. Polymorphisms in the MMP genes can be influenced by racial and ethnic background, and could ultimately affect the presentation of ischemic stroke.[52] MMPs also play key roles in the spread of viral infection, inflammation, and remodeling of the respiratory airways and tissue fibrosis.[53] MMPs may also participate in cancer development, progression, invasiveness, and dissemination by promoting a protumorigenic microenvironment and modulating the cell-ECM and cell-to-cell contacts.[46] MMPs could break the cell-to-cell and cell-to-ECM adhesion, degrade ECM proteins, and promote angiogenesis, and thereby facilitate cancer invasion and metastasis.[54]

7. MMPs AND CELL SIGNALING

In addition to their role in ECM degradation, immunohistochemical studies have localized MMPs in many cell types. Localization of MMPs in certain cells not only supports that these cells could be a source of the MMPs released in ECM but also suggests a role of MMP in cell signaling and intracellular pathways. Evidence for MMP-induced signaling pathways has been demonstrated in several tissues including blood vessels.[55,56]

7.1 MMPs and VSM Function

MMPs may affect VSM contraction mechanisms. VSM contraction is triggered by increases in Ca^{2+} release from the intracellular stores in the sarcoplasmic reticulum and Ca^{2+} entry from the extracellular space through different types of Ca^{2+} channels. We have shown that MMP-2 and -9 do not inhibit phenylephrine-induced contraction of isolated aortic segments incubated in Ca^{2+}-free physiological solution, suggesting that these MMPs do not inhibit the Ca^{2+} release mechanism from the intracellular stores.[57] However, MMP-2 and -9 cause relaxation of phenylephrine-precontracted aortic segments and inhibit phenylephrine-induced Ca^{2+} influx.[57] Similarly, MMP-2 inhibits Ca^{2+}-dependent contraction mechanisms in isolated segments of rat IVC.[56] It has been proposed that during substrate degradation MMPs may produce Arg-Gly-Asp (RGD)-containing peptides, which could bind to $\alpha_v\beta_3$ integrin receptors and inhibit Ca^{2+} entry into VSM.[58] This is unlikely as RGD peptides do not affect IVC contraction.[56] The mechanism by which MMPs inhibit Ca^{2+} entry could involve direct effects on Ca^{2+} or K^+ channels. In rat IVC, MMP-2-induced relaxation is abolished in high KCl-depolarizing solution, which prevents K^+ ion from moving out of the cell via K^+ channels. Importantly, blockade of large conductance Ca^{2+}-activated K^+ channels (BK_{Ca}) by iberiotoxin inhibits MMP-2 induced IVC relaxation, suggesting that MMP-2 actions may involve activation of BK_{Ca} and membrane hyperpolarization, which in turn decreases Ca^{2+} influx through voltage-gated Ca^{2+} channels.[59] The MMP-induced inhibition of venous tissue Ca^{2+} influx and contraction may lead to prolonged venous dilation and varicose veins.

While MMP-2 and -9 reduce Ca^{2+} influx in both arteries and veins,[56,57] veins differ from arteries in their structure and function, and the effects of MMPs on the veins should not always be generalized to the arteries. Veins have fewer layers of VSMCs compared to the several layers of VSMCs in the arteries. Also, venous and arterial VSMCs originate from distinct embryonic locations and are exposed to different pressures and hemodynamic effects in the circulation.[60] Studies have also shown that MMP-2 expression is higher in cultured VSMCs from human saphenous veins compared with those from human coronary artery. In contrast, MMP-3, -10, -20, and -26 expression is less in saphenous vein than coronary artery VSMCs.[60] Interestingly, while some studies suggest that MMP-2 and -9 levels could be similar in cultured saphenous vein and internal mammary artery VSMCs, venous VSMCs exhibit more proliferation, migration, and invasion compared to arterial VSMCs.[61] These observations make it important to further study the differences in the expression and activity of MMPs in veins vs arteries and in venous vs arterial disease.

7.2 MMPs and Endothelial Cell Function

The endothelium controls vascular tone by releasing relaxing factors including nitric oxide and prostacyclin, and through hyperpolarization of the underlying VSMCs by endothelium-derived hyperpolarizing factor (EDHF).[62] MMPs may stimulate protease-activated receptors (PARs). PARs 1–4 are G-protein coupled receptors that have been identified in humans and other species. PAR-1 is expressed in VSMCs,[63] endothelial cells, and platelets,[64] and is coupled to increased nitric oxide production,[65] and in turn contributes to vasodilation. MMP-1 has been shown to activate PAR-1.[66]

EDHF-mediated relaxation may involve the opening of small and intermediate conductance Ca^{2+}-activated K^+ channels and hyperpolarization of endothelial cells. Endothelial cell hyperpolarization may spread via myoendothelial gap junctions and cause relaxation of VSMCs. EDHF could also cause hyperpolarization through opening of BK_{Ca} in VSM.[62] MMP-2 may increase EDHF release and enhance K^+ efflux via BK_{Ca}, leading to venous tissue hyperpolarization and relaxation.[59] In contrast, MMP-3 may impair endothelium-dependent vasodilation,[67] making it important to further examine the effects of MMPs on EDHF.

8. SPECIAL ATTRIBUTES OF SPECIFIC MMPs

8.1 Collagenases

Collagenases include MMP-1 (interstitial collagenase), -8 (neutrophil collagenase), -13, and -18. These MMPs play an important role in cleaving fibrillar collagen type I, II, and III into characteristic 3/4 and 1/4 fragments. They first unwind triple-helical collagen, then hydrolyze the peptide bonds. The MMPs hemopexin domains are essential for cleaving native fibrillar collagen while the catalytic domains are needed for cleaving noncollagen substrates.[42,68]

8.1.1 MMP-1

MMP-1, also termed collagenase-1 or interstitial collagenase, has a gene locus on chromosome 11q22.3, i.e., MMP-1 is coded on the q arm of chromosome 11. MMP-1 degrades collagen and gelatin. MMP-1 also cleaves pro-MMP-9 into its active form. As with many other MMPs, the levels of MMP-1 are very low in most cells under physiological conditions, but are upregulated in inflammatory conditions and autoimmune disease.[1]

Increased levels and activities of MMP-1, -8, and -9 with relatively low levels of TIMP have been identified in slow-to-heal wounds and venous wounds.[69] MMP-1 expression is augmented by inflammatory cytokines such as tumor necrosis factor-α (TNF-α) and interleukin-1 (IL-1).[70] In cultured human vocal fold fibroblasts, TNF-α inhibits cell proliferation, downregulates TIMP-3 and the mRNA transcript levels for collagen III and fibronectin, and upregulates MMP-1 and -2 expression, resulting in increased MMP/TIMP-3 ratio, which may accelerate wound healing following vocal fold injury.[71] MMP-1 may also play a role in the circulatory disturbance and inflammation associated with sudden deafness. In a Korean population, a single-nucleotide polymorphism (SNP) of MMP-1 at the promotor region −1607G/2G is associated with increased risk of sudden deafness when compared with the G/2G and G/G genotypes.[70]

Localized controlled release of antifibrogenic factors can prevent tissue fibrosis surrounding biomedical prostheses such as breast implants and vascular stents. In a rabbit ear fibrotic model, topically applied stratifin prevents dermal fibrosis and promotes normal tissue repair by regulating ECM deposition. Studies have tested the antifibrogenic effect of a controlled release form of stratifin in the prevention of fibrosis induced by dermal poly(lactic-*co*-glycolic acid) (PLGA) microsphere/poly(vinyl alcohol) (PVA) hydrogel implants. Controlled release of stratifin from PLGA microsphere/PVA hydrogel implants increased MMP-1 expression in the surrounding tissue, resulted in less collagen deposition, moderated dermal fibrosis and inflammation by reducing collagen deposition, total tissue cellularity, and infiltrated CD3(+) immune cells in the surrounding tissue. These stratifin-eluting PLGA/PVA composites may be used as coatings to decrease fibrosis around implanted biomedical prostheses such as breast implants and vascular stents.[72] Kynurenic acid is a downstream end product of kynurenine that has antiscarring properties and is unlikely to pass the blood–brain barrier or cause central side effects. Studies showed that kynurenic acid did not cause adverse effects on dermal cell viability, and markedly increased the expression of MMP-1 and -3, and suppressed the production of type I collagen and fibronectin by fibroblasts. The findings suggest that kynurenic acid could be a candidate antifibrogenic agent to improve healing outcome in patients at risk of hypertrophic scarring.[73] Kynurenine treatment appears to increase the levels of MMP-1 and -3 expression through activation of the (MAPK)/extracellular signal-regulated kinase (ERK1/2) signaling pathway.[74] In human primary chondrocytes, IL-1β-induced activation of p38 MAPK may increase MMP-1 and -13 production and glycosaminoglycan release. Thus, activated p38 could accelerate cartilage breakdown by enhancing the expression of MMP-1 and -13 which promote collagen cleavage, and therefore p38 inhibitors may have chondroprotective effects in osteoarthritis.[75]

MMP-1 may play a role in cancer development and metastasis. Studies have suggested an association between SNP of MMP-1 −1607 2G/2G and poor prognosis in malignant tumors such as tongue squamous cell carcinoma.[76] Also, in patients with invasive well-differentiated thyroid carcinoma, MMP-1 expression correlates with tumor aggressiveness manifested as laryngotracheal invasion, multifocality of the tumor, and the presence of metastases. MMP-1 expression is associated with poor prognosis in esophageal cancer[77] and may serve as a prognostic marker and an indicator for the need for more aggressive surgical intervention.[78]

8.1.2 MMP-8

MMP-8, also termed collagenase-2 or neutrophil collagenase, has a gene locus on chromosome 11q22.3. MMP-8 was discovered in cDNA library constructed from mRNA extracted from peripheral leukocytes of a patient with chronic granulocytic leukemia. The library was screened with an oligonucleotide probe constructed from the putative Zn^{2+}-binding region of fibroblast collagenase. Eleven positive clones were identified, of which the one bearing the largest insert (2.2 kb) was sequenced. From the nucleotide sequence of the 2.2-kb cDNA clone, a 467-amino acid sequence representing the entire coding sequence of the enzyme was deduced.[79] Being a collagenase, MMP-8 can cleave interstitial collagens I, II, and III at a site within the triple-helical domain about 3/4 down from the N-terminus.[10] While some pro-MMPs are secreted then form heterodimeric complexes bound to TIMPs, e.g., the MMP-2/TIMP-2 complex, secreted pro-MMP-8 remains in its free form. The pro-MMP-8 activity is then regulated by proteolytic cleavage by other MMPs such as MMP-3 and -10.[80]

MMP-8 is the first collagenase to appear during dermal wound healing and its levels peak earlier than that of MMP-1, supporting time-dependent expression of different MMPs during wound healing.[81] Mice deficient in MMP-8 show delayed healing of cutaneous wounds and increased inflammatory responses, supporting that MMP-8 is a necessary component in dermal wound healing and the regulation of the inflammatory process.[82] In a study to assess the temporal relationship between periodontal tissue destruction and activity of collagenase, exudate from inflamed periodontal tissues was collected, and latent and active collagenase activities were measured.

It was found that the collagenase activity was derived from neutrophils, and there was an overall 40% increase of pooled active collagenase activity in all subjects with progressive loss of connective tissue. These findings suggest a role of neutrophil collagenase or MMP-8 in the destruction of periodontal connective tissue, and MMP-8 expression in the saliva may be used as a marker of diseases involving connective tissue breakdown and advanced periodontitis.[83] MMP-8 can also be detected and analyzed in gingival crevicular fluid using time-resolved immunofluorometric assay, a MMP-8 specific chair-side dip-stick test, a dentoAnalyzer device, and an ELISA kit. Western immunoblots confirmed that immunofluorometric assay and dentoAnalyzer can detect activated 55 kDa MMP-8 species in periodontitis-affected gingival crevicular fluid.[84]

8.1.3 MMP-13

MMP-13, also termed collagenase-3, has a gene locus on chromosome 11q22.3. MMP-13 is very efficient in degrading type II collagen. MMP-13 was first thought to be expressed in connective tissue particularly cartilage and developing bone. However, MMP-13 has also been detected in epithelial and neuronal cells. MMP-13 is overexpressed in cartilage tissues of osteoarthritis patients, and increased expression of MMP-13 in chondrocytes may contribute to the development of osteoarthritis.[85] MMP-13 has been suggested as a direct target gene of micoRNA (miR)-411 in chondrocytes. Overexpression of miR-411 inhibits MMP-13 expression, and increases the expression of type II and IV collagen in chondrocytes. In comparison with normal cartilage, osteoarthritis cartilage shows downregulation of miR-411 and increased MMP-13 expression. These findings suggest that miR-411 may regulate MMP-13 expression and ECM remodeling in human chondrocytes, and may be a therapeutic target in treatment of osteoarthritis.[86] Low ratio of linoleic acid (n-6)/α-linolenic acid ($n-3$) polyunsaturated fatty acids reduces MMP-13 expression in inflammatory chondrocytes in vitro and in vivo, and may be a means to control or reduce the symptoms of osteoarthritis. In cultured human chondrocytes, low 1:1 and 2:1 n-6/n-3 ratios decreased the mRNA expression and protein levels of MMP-13 without affecting chondrocytes proliferation. In rat model of arthritis produced by injection of Freund's complete adjuvant, low 1:1 and 2:1 n-6/n-3 dietary ratio reduced paw swelling rate, decreased serum MMP-13 and IL-1 levels, and alleviated cartilage damage.[87]

MMP-13 may be involved in lung diseases such as acute lung injury, viral infections, and chronic obstructive pulmonary disease. In human small airway epithelial cells, polyinosinic-polycytidylic acid stimulated the secretion of MMP-13, and MMP-13 secretion was abolished by p38 MAPK inhibitor SB304680, phosphoinositide 3-kinase (PI_3K) inhibitor LY294002, Janus kinase (JAK) inhibitor I, RNA-activated protein kinase inhibitor, and nuclear factor-κB (NF-κB) inhibitor Bay 11-7082. Interferon-β (IFN-β) also caused stimulation of MMP-13 secretion that was inhibited by all modulators except Bay 11-7082. It was suggested that MMP-13 secretion was mediated through IFN receptor pathways independently of NF-κB and that polyinosinic-polycytidylic acid stimulated IFN secretion in an NF-κB-dependent manner, leading to IFN-stimulated MMP-13 secretion from human small airway epithelial cells. MMP-13 inhibitors and MMP-13 siRNA inhibited IFN-stimulated secretion of IFNγ-inducible protein 10 and regulated on activation normal T-cell expressed and secreted (RANTES), suggesting that MMP-13 is involved in the secretion of these virus-induced proinflammatory chemokines. Also, a novel polymorphism was identified in the promoter region of the MMP-13 gene. These observations support that MMP-13 plays a role in defense mechanisms of airway epithelial cells.[88]

MMP-13 may be involved in ECM degradation in brain astrocytes. Conditioned medium collected from activated microglia increased IL-18 production and enhanced MMP-13 expression in astrocytes. Treatment with recombinant IL-18 increased MMP-13 protein and mRNA levels in astrocytes. Recombinant IL-18 stimulation also increased the enzymatic activity of MMP-13 and the migratory activity of astrocytes, and MMP-13 or pan-MMP inhibitors antagonized IL-18-induced migratory activity of astrocytes. Treatment of astrocytes with recombinant IL-18 led to the phosphorylation of JNK, Akt, or PKC-δ, and treatment of astrocytes with JNK, PI_3K/Akt, or PKC-δ inhibitors decreased IL-18-induced migratory activity. These findings suggest that IL-18 is an important regulator of MMP-13 expression and cell migration in astrocytes, likely via JNK, PI_3K/Akt, and PKC-δ signaling pathways.[89]

Liver fibrosis is the final stage of liver diseases that lead to liver failure and cancer and studies have tested whether overexpressing MMP-13 gene in rat liver could prevent liver fibrosis progression. In a rat model of liver fibrosis model established by ligating the bile duct, liver-targeted hydrodynamic gene delivery of a MMP-13 expression vector, containing a CAG promoter-MMP-13-IRES-tdTomato-polyA cassette caused a peak in serum level of MMP-13 after 14 days that was sustained for the next 60 days.

Hyaluronic acid levels were lower in the treated vs nontreated rats, suggesting therapeutic effect of MMP-13 overexpression. Quantitative analysis of tissues stained with the collagen stain sirius red showed a statistically smaller volume of fibrotic tissue in MMP-13-treated vs nontreated rats. Liver-targeted hydrodynamic delivery of MMP-13 gene could be useful in prevention of liver fibrosis.[90]

MMP-13 is often overexpressed in tumors and may increase the risk of tumor progression and metastasis. MMP-13 is overexpressed in nasopharyngeal cancer cells and exosomes purified from conditioned medium, as well as plasma of nasopharyngeal cancer patients. Transwell analysis revealed that MMP-13-containing exosomes facilitated the metastasis of nasopharyngeal cancer cells. MMP-13 siRNA reduced the effect of MMP-13-containing exosomes on tumor cell metastasis and angiogenesis.[91]

8.1.4 MMP-18

MMP-18, also termed collagenase-4, has a gene locus on chromosome 12q14. In the 1990s, sequence similarity searching of databases containing expressed sequence tags identified a partial cDNA encoding the 3′ end of a putative novel human MMP. The remaining 5′ end of the MMP cDNA was amplified by PCR from human mammary gland cDNA. The predicted protein product displayed all the structural features characteristic of the MMP family and showed closest identity with MMP-1, -3, -10, and 11, and was designated MMP-18. MMP-18 differs structurally from other MMPs in that its amino acid sequence contains two cleavage sites for activation. MMP-18 mRNA is expressed in several normal human tissues, but is not detected in the brain, skeletal muscle, kidney, liver, or leukocytes.[92] MMP-18 is expressed in migrating macrophages.[93] Growth of peripheral axons is strongly attracted toward limb buds and skin explants in vitro. Directed axonal growth toward skin explants of *Xenopus laevis* in matrigel is associated with expression of MMP-18 and other MMPs, and is inhibited by the MMP inhibitors BB-94 and GM6001. Also, forced expression of MMP-18 in COS-7 cell aggregates enhances axonal growth from Xenopus dorsal root ganglia explants. Nidogen is the target of MMPs released by cultured skin in matrigel, whereas other components remain intact. These findings suggest a link between MMP-18 and ECM breakdown in the control of axonal growth.[94] Despite its diverse tissue distribution and function, MMP-18 has not been directly linked to a specific pathological condition.

8.2 Gelatinases

Gelatinases include gelatinase A (MMP-2) and gelatinase B (MMP-9). MMP-2 and -9 are structurally similar to other proteinases in the MMP family, but differ in that they have a distinct collagen-binding domain composed of three fibronectin type II tandem repeats in the N-terminus of the catalytic domain, which is needed for gelatin binding.[95,96] MMP-2 and -9 have been long recognized as major contributors to proteolytic degradation of ECM. In recent years, a plethora of nonmatrix proteins have been identified as gelatinase substrates thus broadening our understanding of these enzymes as proteolytic executors and regulators in various physiological and pathological states including embryonic growth and development, angiogenesis, vascular diseases, inflammation, infective diseases, and degenerative diseases of the brain and tumor progression.

MMP-2 and MMP-9 are particularly involved in cancer invasion and metastasis. Gelatin zymography in situ showed increased gelatinolytic activity of MMP-2 and -9 in esophageal squamous cell carcinomas, with different intensities of localization in the tumor nest itself and the stromal cells adjacent to tumor nests.[97] Although the effect of broad-spectrum MMP inhibitors in the treatment of cancer has been disappointing in clinical trials, novel mechanisms of gelatinase inhibition have been identified. Inhibition of the association of gelatinases with cell-surface integrins appears to offer highly specific means to target these enzymes without inhibiting their catalytic activity in multiple cell types including endothelial cells, leukocytes, and tumor cells.[98]

8.2.1 MMP-2

MMP-2, also termed gelatinase A or type IV collagenase, has a gene locus on chromosome 16q13-q21. MMP-2 cleaves collagen in two phases, the first resembling that of interstitial collagenase, followed by gelatinolysis, which is promoted by the fibronectin-like domain.[36,43] The collagenolytic activity of MMP-2 is much weaker than collagenases. However, pro-MMP-2 is recruited to the cell surface and undergoes autocatalytic cleavage at the cell surface with the support of MT1-MMP/TIMP-2 complex, and therefore accumulates pericellularly and causes marked local collagenolytic activity.[6,99] MMP-2 is ubiquitous in many cells and tissues, and is involved in a variety of physiological and pathological processes, including angiogenesis, tissue repair, and inflammation. MMP-2 and its inhibitors TIMP-1 and -2, also play a role in tumor invasion and metastasis, and MMP-2/TIMPs imbalance may contribute to tumor progression. The involvement of MMP-2 in cancer has been studied in different malignancies including esophageal cancer.[77,100] MMP-2 activity was correlated with lymph node metastasis, and lymphatic and vascular invasion, supporting an important role of MMP-2 in the invasion of esophageal carcinoma.[97] MMP-2 levels also correlate with invasiveness of cancer cells and shortened survival independent of major prognostic indicators in patients with primary breast carcinoma.[101] MMP-2 may play a role in malignant tumors of the central nervous system, and because of the highly proliferative and aggressive nature of these tumors, current treatments are not been very successful, and new lines of therapy to target MMP-2 have been explored. An adenoviral vector expressing small-interfering RNA (siRNA) against the MMP-2 gene was constructed to specifically inhibit MMP-2 expression, and to test its effects on invasion, angiogenesis, tumor growth, and metastasis of A549 lung cancer cells. Adenoviral-mediated MMP-2 siRNA infection of A549 lung cancer cells caused downregulation of MMP-2, mitigated lung cancer invasion and migration, and reduced tumor cell-induced angiogenesis in vitro. In a mouse model of metastatic lung tumor, treatment of established tumors with adenoviral-mediated MMP-2 siRNA inhibited subcutaneous tumor growth and formation of lung nodules in mice. Adenoviral-mediated MMP-2 siRNA may have a therapeutic potential for lung cancer in part by inhibiting angiogenesis.[102]

Integrins control a variety of signal transduction pathways central to cell survival, proliferation, and differentiation, and their functions and expression levels are altered in many types of cancer. In a study to examine the mechanisms underlying the involvement of α5β1 integrin in tumor invasion, its expression in MCF-7Dox human breast carcinoma cells was depleted using siRNA. Concomitant to α5β1 integrin depletion, there was a sharp decrease in MMP-2 expression and inhibition of the invasiveness of these cells in vitro. Similar reduction of invasion potential was observed upon siRNA-mediated silencing of the MMP-2 gene. Downregulation of α5β1 integrin was associated with decrease in the amounts of active phosphorylated forms of Akt, ERK1/2 kinases, and c-Jun oncoprotein. Also, in MCF-7Dox cells, inhibition of these kinases reduced expression of MMP-2 and c-Jun, and suppressed invasion of the cells in vitro. Coimmunoprecipitation experiments demonstrated that α5β1 integrin interacted with MMP-2 on the surface of MCF-7Dox breast carcinoma cells. These findings suggest that α5β1 integrin controls invasion of breast cancer cells via regulation of MMP-2 expression through signaling pathways involving PI_3K, Akt, and ERK kinases and the c-Jun or via direct recruitment of MMP-2 to the cell surface.[103]

MMP-2 is markedly upregulated in glioblastomas.[104] Knockdown of MMP-2 using MMP-2 siRNA in human glioma xenograft cell lines 4910 and 5310 decreased cell proliferation. Cytokine array and Western blotting using tumor-conditioned media displayed modulated secretory levels of various cytokines including granulocyte-macrophage colony-stimulating factor (GM-CSF), IL-6, IL-8, IL-10, TMF-α, angiogenin, VEGF, and PDGF-BB in MMP-2 knockdown cells. Further, cDNA PCR array suggested potential negative regulation of Janus kinase/Stat3 pathway in MMP-2 knockdown cells. Mechanistically, MMP-2 is involved in complex formation with α5β1 integrin and MMP-2 downregulation inhibited α5β1 integrin-mediated Stat3 phosphorylation and nuclear translocation. Electrophoretic mobility shift assay and chromatin immunoprecipitation assays showed inhibited Stat3 DNA-binding activity and recruitment at CyclinD1 and c-Myc promoters in MMP-2 siRNA-treated cells. MMP-2/α5β1 binding is enhanced in human recombinant MMP-2 treatments, resulting in elevated Stat3 DNA-binding activity and recruitment on CyclinD1 and c-Myc promoters. In vivo experiments in orthotropic tumor model revealed decreased tumor size upon treatment with MMP-2 siRNA. Immunofluorescence studies in tumor sections showed high expression and colocalization of MMP-2/α5β1, which is decreased along with reduced IL-6, phospho-Stat3, CyclinD1, and c-Myc expression levels upon treatment with MMP-2 siRNA. These observations suggest a role of MMP-2/α5β1 interaction in the regulation of α5β1-mediated IL-6/Stat3 signaling and highlight the therapeutic potential of blocking MMP-2/α5β1 interaction in glioma treatment.[105]

8.2.2 MMP-9

MMP-9 or gelatinase B is also a type IV collagenase that has a gene locus on chromosome 20q11.2-q13.1. MMP-9 is produced by a variety of cells including epithelial cells, fibroblasts, keratinocytes, osteoblasts, dendritic cells, macrophages, granulocytes, and T-cells. In the house ear institute-organ of Corti 1 choclear cells, IL-1β induces expression of MMP-9 in a dose- and time-dependent manner, and dexamethasone and p38 MAPK inhibitor SB203580 inhibit IL-1β-induced MMP-9 expression/activity.[106] MMP-9 plays a key role in inflammatory cell migration and in the destructive behavior of cholesteatoma. However, serum levels of MMP-9 might not correctly reflect the extent of localized tissue inflammation. In a study of patients with cholesteatoma, MMP-9 and TIMP-1 serum levels were similar with those in control group. In contrast, the levels of MMP-9 and TIMP-1 were higher in cholesteatoma tissues than normal skin specimens. These findings suggest better clinical usefulness of MMP-9 and TIMP-1 expression in cholesteatoma tissues than either serum or plasma levels of these proteins and that the higher the expression of MMP-9 the stronger the inflammation-accompanied cholesteatoma.[107]

Chronic sinonasal inflammation is associated with tissue remodeling and sinonasal osteitis, which could be a marker of refractory disease. Bone real-time polymerase chain reaction (RT-PCR) revealed upregulation of MMP-9 in all patients with chronic rhinosinusitis, but the magnitude of MMP-9 upregulation decreased with severity of osteitis. Mucosa RT-PCR showed upregulation of MMP-9 in moderate/severe osteitis only. The pattern of expression suggests a time- and tissue-dependent role for MMP-9 in the pathophysiology of osteitis.[108]

In the cornea, galectin-3 is a carbohydrate-binding protein that promotes cell–cell detachment and redistribution of the tight-junction protein occludin through its N-terminal-polymerizing domain. Galectin-3 initiates cell–cell disassembly by inducing MMP-9 expression in a manner that is dependent on the interaction with and clustering of the extracellular MMP inducer EMMPRIN (CD147, basigin) on the cell surface. Corneas of control mice expressing galectin-3 had a substantial amount of MMP-9 in the migrating epithelia of healing corneas. In contrast, corneas of galectin-3-knockout mice show impairment in MMP-9 expression. These findings suggest a galectin-3-mediated regulatory mechanism for induction of MMP-9 expression and disruption of cell–cell contacts required for cell motility in migrating epithelia.[47]

MMP-9 is also expressed in migrating macrophages.[93] MMP-9 has also been detected in esophageal cancer,[77] and gelatin zymography showed a correlation between MMP-9 activity and vascular invasion of esophageal carcinoma.[97]

8.3 Stromelysins

Stromelysins 1, 2, and 3, also known as MMP-3, -10, and -11, respectively, have the same domain arrangement as collagenases, but do not cleave interstitial collagen. MMP-3 and -10 are similar in structure and substrate specificity, while MMP-11 is distantly related. MMP-3 and MMP-10 digest a number of ECM molecules and participate in pro-MMP activation, but

MMP-11 has very weak activity toward ECM molecules. Also, MMP-3 and -10 are secreted from the cells as inactive pro-MMP, but MMP-11 is activated intracellularly by furin and secreted from the cells as an active enzyme.[23]

8.3.1 MMP-3

MMP-3, also known as stromelysin-1, has a gene locus on chromosome 11q22.3. Structurally, MMP-3 possesses some unique characteristics. First, MMP-3 retains protease capability even if the zinc moiety is replaced with cobalt, manganese, cadmium, or nickel ions, but depending on the moiety, the protease activity becomes sensitive to different substrates. Second, MMP-3 has a unique deep active site that transverses the length of the enzyme.[31] MMP-3 is well known as a secretory endopeptidase that degrades ECM.[109] MMP-3 preferentially cleaves proteins at sites where the first three amino acids following the cleavage site are hydrophobic.[31] MMP-3 degrades collagen type II, IV, and IX as well as a variety of proteoglycans, elastin, fibronectin, and laminin. MMP-3 may activate other MMPs necessary for tissue remodeling including MMP-1, -7, and -9.[31] MMP-3 has been detected in the nucleus, and human nuclear MMP-3 may function as a trans-regulator of connective tissue growth factor. MMP-3 has also been detected in the nuclei of hepatocytes and may be involved in apoptosis.[110] MMP-3 was detected in the nuclei of cultured chondrocytic cells and in normal and osteoarthritic chondrocytes in vivo. Nuclear translocation of externally added recombinant MMP-3, and six putative nuclear localization signals in MMP-3 have been identified. Heterochromatin protein-γ regulates connective tissue growth factor by interacting with MMP-3, and MMP-3 knockdown suppresses connective tissue growth factor expression. These observations suggest that MMP-3 may be involved in the development, tissue remodeling, and pathology of arthritic diseases through regulation of connective tissue growth factor.[109]

Posttraumatic osteoarthritis is characterized by progressive cartilage degeneration in injured joints, and fibronectin fragments may degrade cartilage through upregulating MMPs. Studies have profiled the catabolic events, fibronectin fragmentation, and proteinase expression in bovine osteochondral explants following a single blunt impact on cartilage. Impacted cartilage released higher amount of chondrolytic fibronectin fragments and proteoglycan than nonimpacted controls. Those increases coincided with upregulation of MMP-3 in impacted cartilage, suggesting that posttraumatic osteoarthritis may be propelled by fibronectin fragments which act as catabolic mediators through upregulating cartilage-damaging proteinases such as MMP-3.[111]

In addition to its role in arthritis, MMP-3 may be involved in the development of atherosclerosis, and tumor growth and metastasis.[112,113] Serum levels of MMP-3 and VEGF are higher in patients with malignant adrenal incidentalomas than in those with benign ones, and therefore can be used as markers of malignancy of incidentalomas. Also, MMP-3 and VEGF levels decreased after tumor resection in patients with malignant tumors and increased in patients with recurrence, and therefore, could be of prognostic value in these patients.[114] MMP-3 activation may also be a key upstream event that leads to induction of mitochondrial reactive oxygen species and NADPH oxidase 1 (Nox1) and eventual dopaminergic neuronal death.[115]

8.3.2 MMP-10

MMP-10 or stromelysin-2 has a gene locus on chromosome 11q22.3. MMP-10 is a secreted protein that may play a role in pulmonary fibrosis. In patients with idiopathic pulmonary fibrosis, serum levels of MMP-7 and -10 correlate with both the percentage of predicted forced vital capacity and the percentage of predicted diffusing capacity of the lung for carbon. MMP-7 and -10 levels in bronchoalveolar lavage fluid correlate with their corresponding serum levels. Serum MMP-10 predicted clinical deterioration within 6 months and overall survival. In idiopathic pulmonary fibrosis lungs, the expression of MMP-10 was enhanced and localized to the alveolar epithelial cells, macrophages, and peripheral bronchiolar epithelial cells. These findings suggest that MMP-10 may be a useful biomarker of disease severity and prognosis in patients with idiopathic pulmonary fibrosis.[116]

Respiratory syncytial virus is an important pathogen of bronchiolitis, asthma, and severe lower respiratory tract disease in infants and young children. Studies have investigated the regulation of MMP in respiratory syncytial virus-infected human nasal epithelial cells in vitro. MMP-10 mRNA expression was increased in human nasal epithelial cells after respiratory syncytial virus infection, together with induction of mRNAs of MMP-1, -7, -9, and -19. The amount of MMP-10 released from human nasal epithelial cells was also increased in a time-dependent manner after respiratory syncytial virus infection as is that of chemokine RANTES. The upregulation of MMP-10 was prevented by inhibitors of NF-κB and pan-PKC with inhibition of respiratory syncytial virus replication. Upregulation of MMP-10 was prevented by inhibitors of JAK/STAT, MAPK, and EGF receptors without inhibition of respiratory syncytial virus replication. In lung tissue of an infant with severe respiratory syncytial virus infection in which a few respiratory syncytial virus antibody-positive macrophages were observed, MMP-10 was expressed at the apical side of the bronchial epithelial cells and alveolar epithelial cells. These findings suggest that MMP-10 induced by respiratory syncytial virus infection in human nasal epithelial cells is regulated via distinct signal transduction pathways with or without relation to virus replication. MMP-10 may play an important role in the pathogenesis of respiratory syncytial virus diseases and may have the potential to be a marker and therapeutic target for respiratory syncytial virus infection.[53]

MMP-10 may be associated with peripheral arterial disease. Studies have analyzed MMP-10 levels in patients with peripheral arterial disease according to disease severity and cardiovascular risk factors and evaluated the prognostic value of MMP-10 for cardiovascular events and mortality in lower limb arterial disease after a follow-up period of 2 years. Patients with peripheral arterial disease showed increased levels of MMP-10 and decreased levels of TIMP-1 compared with controls. Among patients with peripheral arterial disease, those with critical limb ischemia showed higher levels of MMP-10 compared with those with intermittent claudication, whereas the MMP-10/TIMP-1 ratio remained similar. The univariate analysis showed an association between MMP-10 levels, age, hypertension, and ankle-brachial index in peripheral arterial disease patients. Patients with the highest MMP-10 tertile had an increased incidence of all-cause mortality and cardiovascular mortality. These observations suggest that MMP-10 is associated with severity and poor outcome in peripheral arterial disease.[117]

MMP-10 is expressed by macrophages and epithelium in response to injury, and its role in wound repair has been examined. In wounds of MMP-10 KO mice, collagen deposition and skin stiffness is increased, with no change in collagen expression or reepithelialization. Increased collagen deposition in MMP-10 KO wounds was accompanied by less collagenolytic activity and reduced expression of specific metallocollagenases, particularly MMP-8 and -13, where MMP-13 was the key collagenase. Ablation and adoptive transfer approaches and cell-based models demonstrated that the MMP-10-dependent collagenolytic activity was a product of alternatively activated (M2) resident macrophages. These observations suggest a role for MMP-10 in controlling the tissue-remodeling activity of macrophages and moderating scar formation during wound repair.[118]

MMP-10 may be involved in pelvic organ prolapse. In a study exploring the correlation between genetic mutations in MMP-10 and susceptibility to pelvic organ prolapse, serum MMP-10 levels were higher in the pelvic organ prolapse group than in the control group. Also, there was a marked difference between the two groups in the distribution frequency of the MMP-10 rs17435959G/C genotype. Patients with parity > 2 and postmenopausal women had elevated serum MMP-10 levels, and patients with parity > 2 and postmenopausal women who carried the G/C + C/C genotype in the MMP-10 gene had an increased risk of pelvic organ prolapse. These observations suggest that the rs17435959 polymorphism of the MMP-10 gene may be associated with an increased risk of pelvic organ prolapse.[119]

MMP-10 is often expressed in human cancers and could play a role in tumor progression, migration, and invasion. Nonneoplastic oral epithelium does not show MMP-10 expression. MMP-10 may be involved in the transformation of normal oral epithelium to oral verrucous carcinoma and squamous cell carcinoma. MMP-10 expression levels are higher in oral squamous cell carcinoma than verrucous carcinoma, and therefore can be used in the differential diagnosis of the two tumors.[120] MMP-10 is limited to epithelial cells and may facilitate tumor cell invasion by targeting collagen, elastin, and laminin. Increased MMP-10 expression has been linked to poor clinical prognosis in some cancers. MMP-10 expression is positively correlated with the invasiveness of human cervical and bladder cancers. MMP-10 can regulate tumor cell migration and invasion, and endothelial cell tube formation, and these effects are associated with resistance to apoptosis. Increasing MMP-10 expression stimulates the expression of hypoxia-inducible factor (HIF-1α) and MMP-2 (proangiogenic factors) and plasminogen activator inhibitor type 1 (PAI-1) and C-X-C chemokine receptor CXCR2 (prometastatic factors). Targeting MMP-10 with siRNA in vivo results in decreased xenograft tumor growth, reduced angiogenesis, and apoptosis. These findings suggest that MMP-10 can play a role in tumor growth and progression, and MMP-10 inhibition may represent a rational strategy for cancer treatment.[54]

MMP-10 plays a role in liver regeneration. Studies have examined MMP-10 expression and function in human hepatocellular carcinoma and diethylnitrosamine-induced mouse hepatocarcinogenesis. MMP-10 was induced in human and murine hepatocellular carcinoma tissues and cells. MMP-10-deficient mice showed less hepatocellular carcinoma incidence, smaller histological lesions, reduced tumor vascularization, and less lung metastases. Importantly, expression of the protumorigenic, C-X-C chemokine receptor-4 (CXCR4), was reduced in diethylnitrosamine-induced hepatocarcinogenesis in MMP-10-deficient mice livers. Human hepatocellular carcinoma cells stably expressing MMP-10 had increased CXCR4 expression and migratory capacity. Pharmacological inhibition of CXCR4 reduced MMP-10-stimulated hepatocellular carcinoma cell migration. MMP-10 expression in hepatocellular carcinoma cells was induced by hypoxia and the CXCR4 ligand, stromal-derived factor-1 (SDF-1), through the ERK1/2 pathway, involving an activator protein 1 site in MMP-10 gene promoter. These findings suggest that MMP-10 contributes to hepatocellular carcinoma development and participates in tumor angiogenesis, growth, and dissemination. Reciprocal crosstalk between MMP-10 and the CXCR4/SDF-1 axis may contribute to hepatocellular carcinoma progression and metastasis.[46]

8.3.3 MMP-11

MMP-11 or stromelysin-3 has a gene locus on chromosome 22q11.23. MMP-11 was first identified in stromal cells surrounding invasive breast carcinoma and has been proposed as one of the stroma-derived factors that play a role in the progression of epithelial malignancies.[121] Like all other members of the MMP gene family, stromelysin-3 is synthesized as an inactive precursor that must be processed to its mature form in order to express enzymatic activity. However, compared to other MMPs which require activation extracellularly, MMP-11 is secreted in its active form. MMP-11 can be processed directly to its enzymatically active form by an obligate intracellular proteolytic event that occurs within the constitutive secretory pathway. Like other furin-containing MMPs, intracellular activation of MMP-11 is regulated by a 10-amino acid insert sandwiched between the pro- and catalytic domains of MMP-11, which is encrypted with an Arg-X-Arg-X-Lys-Arg recognition motif for the Golgi-associated proteinase furin, a mammalian homologue of the yeast Kex2 pheromone convertase. A furin-MMP-11 processing axis not only differentiates the regulation of this enzyme from other nonfurin-containing MMPs but also identifies proprotein convertases as potential targets for therapeutic intervention in matrix-destructive disease states.[23]

Some of the MMP-11 substrates include laminin receptor and α-1-proteinase inhibitor.[1,122] MMP-11 is expressed in tissues undergoing the active remodeling associated with embryonic development, wound healing, and tumor invasion.[23] MMP-11 may promote tumorigenicity. In breast cancer, MMP-11 is a bad prognosis marker and its expression is associated with a poor clinical outcome.[15] In a study investigating the influence of genetic polymorphisms of MMP-11 gene on the susceptibility to oral squamous cell carcinoma in a Taiwanese population,

MMP-11 gene polymorphisms exhibited synergistic effects with the environmental factors betel nut chewing and tobacco use on the susceptibility to oral squamous cell carcinoma. Among patients with oral squamous cell carcinoma with betel nut consumption, those who have at least one polymorphic C allele of MMP-11 rs738792 have an increased incidence of lymph node metastasis when compared with patients homozygous for T/T. These observations suggest combined effects of MMP-11 gene polymorphisms and environmental carcinogens in the increased risk for oral squamous cell carcinoma and may be a predictive factor for tumor lymph node metastasis in Taiwanese with oral squamous cell carcinoma.[123]

MMP-11 levels are elevated in specimens from patients with esophageal squamous cell carcinoma. Patents with esophageal dysplasia also show elevated MMP-11, suggesting that these alterations are early events in esophageal tumorigenesis. In postesophagectomy follow-up, patients with MMP-11 positive TIMP-2 negative carcinoma had shorter disease-free survival compared with patients with other MMP/TIMP profiles. These findings suggest that MMP-11 positive TIMP-2 negative phenotype may be associated with adverse prognosis in patients with esophageal cancer.[124] MMP-11 is also overexpressed in sera of cancer patients compared with normal control group, and in tumor tissue specimens from patients with laryngeal, gastric, pancreatic, and breast cancer. The presence of MMP-11 in tumor tissues has suggested that it could promote cancer development by inhibiting apoptosis as well as enhancing migration and invasion of cancer cells. However, studies in animal models suggest that MMP-11 may play a negative role against cancer progression by suppressing metastasis.[125]

In patients with laryngeal squamous cell carcinoma, the expression of MMP-11 mRNA expression and the tumor suppressor gene p14ARF was different in tumor tissues compared with their corresponding adjacent tissues and was associated with several clinical characteristics. Patients with low MMP-11 and high p14ARF expression had better survival compared with those with high MMP-11 and low p14ARF expression. It was suggested that altered expression of MMP-11 and p14ARF in tumors may individually, or in combination, predict poor prognosis of laryngeal squamous cell carcinoma.[126]

8.4 Matrilysins

Matrilysins include MMP-7 and -26, and they both lack the hemopexin domain and the hinge region.

8.4.1 MMP-7

MMP-7 or matrilysin-1 has a gene locus on chromosome 11q21-q22. Structurally, MMP-7 is one of the smallest MMPs. MMP-7 is expressed by Xenopus embryonic macrophages.[93] Common substrates of MMP-7 include proteoglycans, fibronectin, casein, and gelatin types I, II, IV, and V. MMP-7 plays a role in remodeling of tissues involved in development and reproduction such as the uterus, and could play a role in remodeling following tissue injury.[31] MMP-7 degrades ECM components and cleaves cell-surface molecules such as Fas–ligand, pro-TNF-α, syndecan-1, and E-cadherin to generate soluble forms.[127] MMP-7 can have dual effects on apoptosis, whereby it can induce apoptosis by releasing Fas–ligand or inhibit apoptosis by producing heparin-binding epidermal growth factor.[31] MMP-7 acts intracellularly in the intestine to process procryptidins to bactericidal forms.

Studies have examined MMP-2, MMP-7, MMP-9, and TIMP-1 in dysregulated turnover of connective tissue matrices in tonsillar specimens from children with recurrent tonsillitis and undergoing tonsillectomy. MMP-7 level of the superficial part and MMP-9 level at both the superficial and core regions were higher in patients with grade III and IV than patients with grade I and II tonsillar hypertrophy. The balance between MMP-7 and -9 and TIMP-1 activities tended to shift toward the MMP-7 and -9 side with increased tonsillar grade. The presence of MMPs in tonsil tissue suggested a role of degraded ECM proteins in the pathophysiology of chronic tonsillitis. The specific increases in MMP-7 and -9 activities suggest that they are the main promoters of ECM degradation that responded to inflammatory changes in the tonsillar tissue.[128] MMP-7 has also been described as a useful biomarker for idiopathic pulmonary fibrosis.[116]

MMP-7 may play a role in cancer development and metastasis. Serum levels of anti-MMP-7 antibody are higher in patients with oral squamous cell carcinoma, and those with poorly differentiated tumors have more MMP-7 antibody than those with well to moderate tumor. Patients with oral squamous cell carcinoma at late tumor lymph node metastasis (TNM) stages (III, IV) and lymph node metastases have higher serum MMP-7 antibody levels than those at earlier stages (I, II). Serum MMP-7 antibody positivity independently predicted poor overall survival in patients with oral squamous cell carcinoma. MMP-7 mRNA and protein expression increased in tumor tissues from patients with oral squamous cell carcinoma and high serum MMP-7 antibody. These findings suggested that serum anti-MMP-7 antibody might be useful as a diagnostic and prognostic biomarker for oral squamous cell carcinoma.[129]

8.4.2 MMP-26

MMP-26, also known as matrilysin-2 or endometase, has a gene locus on chromosome 11p15. The chromosomal location of the MMP-26 gene shows that it maps to the short arm of chromosome 11, a location distinct from that of other MMP genes.[130] The cDNA-encoding MMP-26 was cloned from fetal cDNA. The deduced 261-amino acid sequence is homologous to macrophage metalloelastase. It includes only the minimal characteristic features of the MMP family: a signal peptide, a prodomain, and a catalytic domain.[131] As with MMP-7, MMP-26 lacks the hemopexin domain, believed to be involved in substrate recognition, and also the hinge region.[130] The amino acid sequence of MMP-26 also contains a threonine residue adjacent to the Zn^{2+}-binding site that is a specific feature of matrilysin.[130] MMP-26 mRNA is specifically expressed in the placenta and uterus. Recombinant MMP-26 demonstrates proteolytic activity toward several substrates including type IV collagen, β-casein, fibrinogen, fibronectin, gelatin, and vitronectin.[130–132] MMP-26 also activates pro-MMP-9 (gelatinase B).[130] MMP-26 mRNA is also detected in human cell lines such as HEK 293 kidney cells and HFB1 lymphoma cells, and is widely expressed in malignant tumors from different sources as well as in multiple tumor cell lines. MMP-26 is also expressed in cancer cells of epithelial origin, including carcinomas of the lung, prostate, and breast.[132,133] The broad proteolytic activity and distribution of MMP-26 in different cell lines suggest that it may play a role in tissue-remodeling events associated with angiogenesis and tumor progression.[130,132]

MMP-26 expression may be linked to tumor invasion induced by GM-CSF. GM-CSF promotes tumor progression in different tumor models, and is associated with highly angiogenic and invasive tumors. In colon adenocarcinoma, GM-CSF overexpression and treatment reduces tumor cell proliferation and tumor growth in vitro and in vivo, but contributes to tumor progression, tumor invasion into the surrounding tissue, and induction of an activated tumor stroma. Enhanced GM-CSF expression is also associated with a discontinued basement membrane deposition likely due increased expression/activity of MMP-2, -9, and -26. Treatment with GM-CSF blocking antibodies reverses this effect. Expression of MMP-26 is predominantly located in pre- and early-invasive areas suggesting that

MMP-26 expression is an early event in promoting GM-CSF-dependent tumor invasion.[134] Pancreatic adenocarcinoma is recognized for its early aggressive local invasion and high metastatic potential. Patients with metastatic lymph nodes had increased expression of MMP-26 in actual tumor samples, and the putative role of MMP-26 as a marker of metastases warrants further studies.[135] MMP-26 is negatively regulated by TIMP-2 and -4, with TIMP-4 being more potent inhibitor of MMP-26-induced tissue remodeling.[136]

8.5 Membrane-Type MMPs

Membrane-Type MMPs (MT-MMPs) include four transmembrane MMPs, MMP-14, -15, -16, and -24, and two glycosyl phosphatidylinositol (GPI)-anchored MMPs, MMP-17 and -25 (Table 1).[8,9] MT-MMPs have a furin-like proprotein convertase recognition sequence at the C-terminus of the propeptide. They are activated intracellularly and the active enzymes are expressed on the cell surface. MT-MMPs have membrane-anchoring domains and display protease activity at the cell surface, and therefore they are optimal pericellular proteolytic machines.[137] All MT-MMPs except MT4-MMP (MMP-17) can activate pro-MMP-2.[13] MT1-MMP (MMP-14) activates pro-MMP-13 on the cell surface.[138]

8.5.1 MMP-14

MMP-14 or MT1-MMP has a gene locus on chromosome 14q11-q12. MMP-14 is one of four type I transmembrane proteins (MT1, 2, 3, and 5-MMP or MMP-14, -15, -16, and -24, respectively). Type I MT-MMPs, MT1-, MT2-, MT3-, and MT5-MMPs, have about a 20-amino acid cytoplasmic tail following the transmembrane domain.[139] MT1-MMP is ubiquitously expressed, binds TIMP-2, activates MMP-2, and stimulates cell migration in various cell types.[140] MMP-14 is best known for its collagenolytic activity, digesting type I (guinea pig), II (bovine), and III (human) collagens into characteristic 3/4 and 1/4 fragments. MT1-MMP may also degrade cartilage proteoglycan, fibronectin, laminin-1, vitronectin, α1-proteinase inhibitor, and α2-macroglobulin.[8] The activity of MT1-MMP on type I collagen is synergistically increased with coincubation with MMP-2.[8] MMP-2 is secreted as a proenzyme (pro-MMP-2, progelatinase A) which is bound and activated on the surface of normal and tumor cells. MT1-MMP induces activation of pro-MMP-2. In COS-1 cells, MT1-MMP could induce cell-surface binding of pro-MMP-2, which is consequently processed to an intermediate form.

Processing from the intermediate to the fully active form is dependent on MMP-2 concentration. Thus the MT1-MMP-induced cell-surface binding concentrates the MMP-2 intermediate form locally to allow autoproteolytic processing to the fully active MMP-2 form.[26] One difference between MT-MMPs and the other MMP family members is the insertion of eight amino acids between strands βII and III in the catalytic domain. In MT1-MMP, the best characterized of these enzymes to date, these residues consist of (163)PYAYIREG(170). Characterization of the activity of the soluble forms toward peptides and fibrinogen revealed that neither mutation nor deletion of residues 163–170 impaired catalytic function, suggesting these residues have little influence on conformation of the active site cleft. On the other hand, characterization of the kinetics of activation of pro-MMP-2 with and without its gelatin binding region by the mutants generated have shown that efficient activation of pro-MMP-2 is, at least in part, through an interaction with residues 163–170 of MT1-MMP.[13] Also, in a study using sandwich enzyme linked immunoassay systems, the levels of MMP-1, -2, -13, MT1-MMP, and TIMP-1 were higher in homogenates of human salivary gland carcinomas than non-neoplastic control salivary glands. Gelatin zymography demonstrated that the activation ratio of the MMP-2 zymogen was higher in the carcinomas than in the controls, and the pro-MMP-2 activation correlated directly with the MT1-MMP/TIMP-2 ratio. Immunohistochemistry and in situ zymography demonstrated localization of MMP-2, MT1-MMP, and TIMP-2 to carcinoma cells. These findings suggest that enhanced activation of pro-MMP-2 mediated by MT1-MMP is implicated in tumor invasion and metastasis and that TIMP-2 may regulate pro-MMP-2 activation in salivary gland carcinomas in part by inhibiting MMP-14.[141]

In another study to examine the relation between expression of MT-MMPs and MMP-2, which is one of the key proteinases in invasion and metastasis of various cancers, all head and neck squamous cell carcinoma cell lines examined consistently expressed MT1-MMP and MMP-2, but not MT2-MMP or MT3-MMP. Also, in the clinical specimens, there was a correlation in coexpression of mRNA and colocalization by immunohistochemistry for MT1-MMP and MMP-2. Relative mRNA expression levels of MT1-MMP and MMP-2 in the carcinoma tissues were higher than those of the control tissues. Both mRNA expression and immunopositivity of MT1-MMP correlated with lymph node metastasis. The localization of MMP-2 closely corresponded to that of MT1-MMP. These observations suggest that MT1-MMP possesses a role as a determinant of lymph node metastasis, and that concurrent expression of MT1-MMP and MMP-2 are involved in progression of head and neck squamous cell carcinoma.[142]

MT1-MMP could be an important molecular tool for cellular remodeling of the surrounding matrix. MT1-MMP-deficient mice show craniofacial dysmorphism, arthritis, osteopenia, dwarfism, and fibrosis of soft tissues likely due to ablation of a collagenolytic activity that is essential for modeling of skeletal and extraskeletal connective tissues. These observations demonstrate the pivotal function of MT1-MMP in connective tissue metabolism.[9] MMP-14 may promote vulnerable plaque morphology in mice, whereas TIMP-3 overexpression is protective. High MMP-14 low TIMP-3 rabbit foam cells are more invasive and more prone to apoptosis than low MMP-14 high TIMP-3 cells. Proinflammatory stimuli increase MMP-14 and decrease TIMP-3 mRNA expression and protein levels in human macrophages. Conversion to foam cells with oxidized LDL is associated with increased MMP-14 and decreased TIMP-3, independently of inflammatory mediators and partly through posttranscriptional mechanisms. Within atherosclerotic plaques, MMP-14 is prominent in foam cells with either pro- or antiinflammatory macrophage markers, whereas TIMP-3 is present in less foamy macrophages and colocalized with CD206. MMP-14 positive macrophages are more abundant, whereas TIMP-3 positive macrophages are less abundant in plaques histologically designated as rupture prone. These findings suggest that foam-cells with high MMP-14 low TIMP-3 expression are prevalent in rupture-prone atherosclerotic plaques, independent of pro- or antiinflammatory activation, and that reducing MMP-14 activity and increasing TIMP-3 could be valid therapeutic approaches to reduce plaque rupture and myocardial infarction (MI).[143]

MT1-MMP has a major impact on invasive cell migration in both physiological and pathological settings such as immune cell extravasation or metastasis of cancer cells.[144] Surface-associated MT1-MMP is able to cleave components of ECM, which is a prerequisite for proteolytic invasive migration. In a study of the mechanisms that regulate MT1-MMP trafficking to and from the cell surface, three members of the RabGTPase family, Rab5a, Rab8a, and Rab14 were found to be crucial regulators of MT1-MMP trafficking and function in primary human macrophages. Both overexpressed and endogenous forms show prominent colocalization with MT1-MMP-positive vesicles, whereas expression of mutant constructs, as well as siRNA-induced knockdown, reveal that these RabGTPases are crucial in the regulation of MT1-MMP surface exposure, contact of MT1-MMP-positive vesicles with podosomes, ECM degradation, and proteolytic invasion of macrophages. Thus, Rab5a, Rab8a, and Rab14 are major regulators of MT1-MMP trafficking and invasive migration of human macrophages, and could be potential targets for manipulation of immune cell invasion.[144] Of note, MT1-MMP is overexpressed in malignant tumor tissues, including lung and stomach carcinomas that contain activated MMP-2.[145]

8.5.2 MMP-15

MMP-15 or MT2-MMP has a gene locus on chromosome 16q13. MMP-15 is an understudied member of the MMP family. Like MT1-MMP, MT2-MMP localizes on the cell surface and mediates the activation of MMP-2,[145] which is associated with tumor invasion and metastasis.

MT-MMPs are essential for pericellular matrix remodeling in late stages of development, as well as in growth and tissue homeostasis in postnatal life. A study has examined MT1-MMP and MT2-MMP, and their roles in the process of placental morphogenesis in mice. The fetal portion of the placenta, in particular the labyrinth, displays strong overlapping expression of MT1-MMP and MT2-MMP, which is critical for syncytiotrophoblast formation and in turn for fetal vessels. Disruption of trophoblast syncytium formation leads to developmental arrest with only a few poorly branched fetal vessels entering the labyrinth causing embryonic death at day 11.5. Knockdown of either MT1-MMP or MT2-MMP is crucial during the development of the labyrinth. In contrast, knockdown of MT-MMP activity after labyrinth formation is compatible with development to term and postnatal life. These findings identify essential but interchangeable roles for MT1-MMP or MT2-MMP in placental vasculogenesis, and suggest selective temporal and spatial MMP activity during development of the mouse embryo.[146]

MMP-15 appears to be upregulated during colorectal tumorigenesis, with different expression patterns. MMP-15 expression level increases from normal mucosa to microadenomas, and immunofluorescence analysis showed a stromal localization of MMP-15 in the early phases of neoplastic transformation.[147] The mRNA and protein expression of MMP-14, -15, and -16 are increased in supraglottic carcinoma tissues compared to control adjacent nonneoplastic tissues. Expression of MMP-14, but not MMP-15 and MMP-16, is markedly increased in the T3 and neck nodal metastasis groups compared with the T1-2 group and the group without nodal metastasis. Also, MMP-14 mRNA and protein are higher in tumors at clinical stage III–IV compared with stage I–II tumors. Groups with high

MMP-14 protein expression had a poorer prognosis than patients with weak or negative expression of MMP-14. Thus while MMP-15 is expressed, MMP-14 appears to play a more dominant role in the tumor progression and may serve as prognostic factor in patients with supraglottic carcinoma.[148]

A study examined the relation of expressions of MT1, MT2, and MT3-MMP to the invasion and metastases in laryngeal cancer. The expression of MT1, MT2, and MT3-MMP was higher in laryngeal cancer tissues than those in para-tumorous tissues and had a close relationship with invasive depth. The expression of MT1-MMP was higher in patients with metastatic lymph nodes than in patients without metastatic lymph nodes. Thus MT1, MT2, and MT3-MMP play a role in the progression of laryngeal cancer, MT1-MMP may serve as a reliable marker in estimating invasive and metastatic potency of laryngeal cancer, and suppressing expressions of MT1, MT2, and MT3-MMP may inhibit the invasion and metastases of laryngeal cancer.[149]

8.5.3 MMP-16

MMP-16 or MT3-MMP has a gene locus on chromosome 8q21.3. MMP-16 is a membrane-bound protein with a cytoplasmic tail. As a type I MT-MMP, MMP-16 could transform pro-MMP-2 to active MMP-2 and thereby facilitate tumor invasion. In human cardiomyocyte progenitor cells, MMP-16 may activate MMP-2 and -9, which could in turn facilitate undesired cell migration after targeted cell transplantation and potentially limit the beneficial effects of cardiac regeneration. Treatment with MMP-16 siRNA or an MMP-16 blocking antibody blocked cell migration, suggesting that reducing MMP-16 expression/activity could have beneficial effects in progenitor cell transplantation and cardiac regeneration.[150]

In patients with melanoma, increased expression of MMP-16 is associated with poor clinical outcome, collagen bundle assembly around tumor cell nests, and lymphatic invasion. In cultured WM852 melanoma cells derived from human melanoma metastasis, silencing of MMP-16 resulted in cell-surface accumulation of the MMP-16 substrate MMP-14 (MT1-MMP) as well as L1CAM cell adhesion molecule. When limiting the activities of these transmembrane protein substrates toward pericellular collagen degradation, cell junction disassembly, and blood endothelial transmigration, MMP-16 supported nodular-type growth of adhesive collagen-surrounded melanoma cell nests, steering cell collectives into lymphatic vessels. These findings suggest that restricted collagen infiltration and limited mesenchymal invasion are unexpectedly associated with the properties of the most aggressive tumors, and reveal MMP-16 as a putative indicator of adverse melanoma prognosis.[151]

Other studies have suggested that TGF-β1 is involved in the migration and metastases of bladder cancer by inducing epithelial–mesenchymal transition and upregulation of MMP-16. These findings suggest and an association between TGF-β1, MMP-16, and epithelial–mesenchymal transition, in the setting of tumor invasion and metastasis in bladder cancer.[152]

MMP-16 enhances invasion of breast cancer cells. In MCF-7 breast cancer cells, the antitumoral and antiproliferative compound catalpol reduced MMP-16 activity and cell proliferation, promoted apoptosis, and increased the expression of miR-146a. These findings suggested that miR-146a may control the expression of MMP-16, and that catalpol suppresses proliferation and facilitates apoptosis of MCF-7 breast cancer cells through upregulating miR146a and downregulating MMP-16 expression.[153] Likewise, miR-155 may directly target MMP-16, and in turn reduce MMP-2 and -9 activities and as a result efficiently inhibit migration of human cardiomyocyte progenitor cells, suggesting that miR-155 could be used to improve local retention of progenitor cells after intramyocardial delivery.[150]

Alveolarization requires coordinated ECM remodeling, and MMPs play an important role in this process. Polymorphisms in MMP genes might affect MMP function in preterm lungs and thus influence the risk of bronchopulmonary dysplasia. In a study in neonates with bronchopulmonary dysplasia 9 SNPs were sought in the MMP-2, MMP-14, and MMP-16 genes. After adjustment for birth weight and ethnic origin, the TT genotype of MMP-16 C/T (rs2664352) and the GG genotype of MMP-16 A/G (rs2664349) were found to protect from bronchopulmonary dysplasia. These genotypes were also associated with a smaller active fraction of MMP-2 and a threefold-lower MMP-16 level in tracheal aspirates. Further evaluation of MMP-16 expression during the course of normal human and rat lung development showed relatively low expression during the canalicular and saccular stages and a clear increase in both mRNA and protein levels during the alveolar stage. In newborn rat models of arrested alveolarization the lung MMP-16 mRNA level was less than 50% of normal. These findings suggest that MMP-16 may be involved in the development of lung alveoli, and that MMP-16 polymorphisms may influence the pulmonary expression and function of MMP-16 and the risk of bronchopulmonary dysplasia in premature infants.[154]

8.5.4 MMP-17

MMP-17 or MT4-MMP has a gene locus in chromosome 12q24.3. MMP-17 is one of six human MT-MMPs, but unlike type I MT-MMPs, and as one of GPI anchor MT-MMPs (MT4-MMP and MT6-MMP, or MMP-17 and -25, respectively) it does not positively regulate pro-MMP-2 (progelatinase A). In the mid-1990s, MMP-17 was cloned from a human breast carcinoma cDNA library. The isolated cDNA contained an open-reading frame 1554 bp long, encoding a polypeptide of 518 amino acids. The predicted amino acid sequence displayed a similar domain organization as other MMPs, including a prodomain with the activation locus, a Zn^{2+}-binding site, and a hemopexin domain. In addition, it contained a C-terminal extension, rich in hydrophobic residues and similar in size to those present in other MT-MMPs. MT4-MMP also contains a nine-residue insertion between the propeptide and the catalytic domain, which is a common feature of MT-MMPs and stromelysin-3. This amino acid sequence insertion ends with the consensus sequence R-X-R/K-R, which seems to be essential for the activation of these proteinases by furin. Unlike MT1-, MT2-, MT3-, and MT5-MMPs which have about a 20-amino acid cytoplasmic tail following the transmembrane domain, and similar to MMP-25, MMP-17 lacks the cytoplasmic tail, and instead, has a GPI anchor, which confers MMP-17 (MT4-MMP) and MMP-25 (MT6-MMP) a unique set of regulatory and functional mechanisms that separates them from the rest of the MMP family.[139] MT4-MMP shedding from the cell surface appears to require an endogenous metalloproteinase.[139]

Discovered almost a decade ago, the body of work on GPI-MT-MMPs today is still limited when compared to other MT-MMPs. Accumulating biochemical and functional evidence also highlights their distinct properties.[137] MMP-17 gene is expressed in a variety of human tissues mainly leukocytes, colon, ovary, testis, and the brain. The expression of MMP-17 in leukocytes together with its membrane localization suggest that it could be involved in activation of membrane-bound precursors of growth factors or inflammatory mediators such as TNF-α. GPI-MT-MMPs are highly expressed in human cancer, where they are associated with tumor progression. MMP-17 transcripts are detected in all breast cancer cell lines, suggesting a role in tumor development/progression.[155]

8.5.5 MMP-24

MMP-24 or MT5-MMP maps to chromosome 20q11.2, a region frequently amplified in tumors from diverse sources. A cDNA-encoding

MT5-MMP was identified and cloned from a human brain cDNA library. The isolated cDNA encoded a polypeptide of 645 amino acids that displayed a similar domain organization as other MMPs, including a predomain with the activation locus, a Zn^{2+}-binding site, and a hemopexin domain. The deduced amino acid sequence contains a C-terminal extension, rich in hydrophobic residues and similar in size to the equivalent domains identified in MT-MMPs. Immunofluorescence and Western blot analysis of COS-7 cells transfected with the isolated cDNA revealed that the encoded protein is localized in the plasma membrane. Northern blot analysis demonstrated that MT5-MMP is predominantly expressed in brain, kidney, pancreas, and lung. In addition, MT5-MMP transcripts were detected at high levels compared to normal brain tissue in a series of brain tumors, including astrocytomas and glioblastomas. MMP-24 can cleave pro-MMP-2 (progelatinase A) into its active MMP-2 form. The catalytic domain of MT5-MMP, produced in *Escherichia coli* as a fusion protein with glutathione *S*-transferase, exhibits a potent proteolytic activity against pro-MMP-2, leading to the generation of the Mr 62,000 active MMP-2. MT5-MMP may contribute to the activation of pro-MMP-2 in tumor tissues, in which it is overexpressed, thereby facilitating tumor progression.[156]

MT5-MMP was also isolated from mouse brain cDNA library. It is predicted to contain a candidate signal sequence, a propeptide region with the highly conserved PRCGVPD sequence, a potential furin recognition motif RRRRNKR, a zinc-binding catalytic domain, a hemopexin-like domain, a 24-residue hydrophobic domain as a potential transmembrane domain, and a short cytosolic domain. MT5-MMP is expressed in a brain-specific manner. It is also highly expressed during embryonic development. In contrast to other MT-MMPs, MT5-MMP tends to shed from cell surface as soluble proteinases, thus offering flexibility as both a cell bound and soluble proteinase for ECM remodeling.[157] In relation to its location in the brain, MT5-MMP is coexpressed with N-cadherin in adult neural stem cells and ependymocytes. N-cadherin mediates anchorage of neural stem cells to ependymocytes in the adult murine subependymal zone and in turn modulates their quiescence. Importantly, MT5-MMP regulates adult neural stem cell functional quiescence by cleaving and shedding of the N-cadherin ectodomain, supporting that the proliferative status of stem cells can be dynamically modulated by regulated cleavage of cell adhesion molecules.[158]

MMP-24 is neuron-specific, and is believed to contribute to neuronal circuit formation and plasticity. MT5-MMP cleaves N-cadherin, a protein critical to synapse stabilization, and studies have shown time- and injury-dependent expression of MT5-MMP and N-cadherin during reactive synaptogenesis following neural injury.[159] MMP-24-deficient mice do not develop neuropathic pain with mechanical allodynia and do not show sprouting and invasion of Abeta-fiber after sciatic nerve injury. These findings suggest that MT5-MMP is essential for the development of mechanical allodynia and plays an important role in neuronal plasticity.[160]

MMP-24 is an essential modulator of neuroimmune interactions in thermal pain stimulation, and a mediator of peripheral thermal nociception and inflammatory hyperalgesia. MT5-MMP is expressed by CGRP-containing peptidergic nociceptors in dorsal root ganglia. MMP-24-deficient mice display enhanced sensitivity to noxious thermal stimuli under basal conditions, but do not develop thermal hyperalgesia during inflammation, a phenotype that appears associated with alterations in N-cadherin-mediated cell–cell interactions between mast cells and sensory fibers. These findings demonstrate an essential role of MT5-MMP in the development of dermal neuroimmune synapses and suggest that it may be a target for pain control.[161]

In a study investigating the expression of MMPs in different grades of human breast cancer tissues, mRNA expressions of MMP-1, -9, -11, -15, -24, and -25 were upregulated, while MMP-10 and -19 were downregulated in breast cancer compared with normal breast tissues. There was also a tumor grade-dependent increase in MMP-15 and -24 mRNA expression, supporting that MMPs are differentially regulated in breast cancer tissues and that they might play various roles in tumor invasion, metastasis, and angiogenesis.[162]

8.5.6 MMP-25

MMP-25 or MT6-MMP has a gene locus on chromosome 16p13.3. MMP-25 is one of the least studied members of the MMP family.[140] MMP-25 is a GPI-anchored MMP that is highly expressed in leukocytes and some cancer tissues. Natural MT6-MMP is expressed on the cell surface as a major reduction-sensitive form of ~120 kDa species, likely representing enzyme homodimers held by disulfide bridges. The stem region of MT6-MMP contains three cysteine residues at positions 530, 532, and 534 which may contribute to dimerization. A systematic site-directed mutagenesis study of the Cys residues in the stem region shows that Cys_{532} is involved in MT6-MMP dimerization by forming an intermolecular disulfide bond. Mutagenesis data also suggest that Cys_{530} and Cys_{534} form an intramolecular disulfide bond. Dimerization is not essential for transport of MT6-MMP to the cell surface, partitioning into lipid rafts or cleavage of α1-proteinase inhibitor. Monomeric forms of MT6-MMP exhibited enhanced autolysis and metalloprotease-dependent degradation. These findings suggest that the stem region of MT6-MMP is a dimerization interface, an event whose outcome lends protease stability to the protein.[163]

MT6-MMP is present in lipid rafts and faces inward in living human polymorphonuclear leukocytes (PMNs), but translocates to the cell surface during neutrophil apoptosis. PMNs express high levels of MT6-MMP. MT6-MMP is present in the membrane, granules, and nuclear/endoplasmic reticulum/Golgi fractions of PMNs where it is displayed as a disulfide-linked homodimer of ~120 kDa. Stimulation of PMNs results in secretion of active MT6-MMP into the supernatants. Membrane-bound MT6-MMP, conversely, is located in the lipid rafts of resting PMNs and stimulation does not alter this location. Interestingly, living PMNs do not display MT6-MMP on the cell surface. However, induction of apoptosis induces MT6-MMP relocation on PMNs' cell surface.[164] Because of its localization in PMNs, MMP-25 may play a role in respiratory burst and IL-8 secretion.[164]

To further assess the biochemical features of MT6-MMP, studies have expressed the MT6-MMP construct tagged with a FLAG tag in breast carcinoma MCF-7 and fibrosarcoma HT1080 cells. Phosphatidylinositol-specific phospholipase C was then used to release MT6-MMP from the cell surface and the solubilized MT6-MMP fractions were characterized. It was found that cellular MT6-MMP partially exists in a complex with TIMP-2. Both TIMP-1 and TIMP-2 are capable of inhibiting the proteolytic activity of MT6-MMP. MT6-MMP does not stimulate cell migration. MT6-MMP, however, generates an adequate level of gelatinolysis of fluorescein isothiocyanate-labeled gelatin and exhibits an intrinsic, albeit low, ability to activate MMP-2. As a result, it is exceedingly difficult to record the activation of MMP-2 by cellular MT6-MMP. Because of its lipid raft localization, cellular MT6-MMP is inefficiently internalized. MT6-MMP is predominantly localized in the cell-to-cell junctions. MT6-MMP has been suggested to play a role in autoimmune multiple sclerosis and cancer, but its physiologically relevant cleavage targets remain to be determined.[140] MT6-MMP mRNA expression is elevated in several human cancers including brain (anaplastic astrocytomas and glioblastomas), colon, urothelial, and prostate cancers.[137,165] MT6-MMP mRNA expression was identified in colon cancer,[165] and immunohistochemical studies confirmed the presence of MT6-MMP in samples of invasive colon cancer.[166] While MT6-MMP protein is absent in normal colonic epithelium, it is highly expressed in invasive adenocarcinomas in 50 out of 60 cases examined.[166]

8.6 Other MMPs

Other MMPs include MMP-12, -19, -20, -21, -22, -23, -27, and -28.

8.6.1 MMP-12

MMP-12 or macrophage metalloelastase has a gene locus on chromosome 11q22.3. As indicated by its name, MMP-12 degrades elastin and is highly expressed by macrophages and other stromal cells. MMP-12 is essential for macrophage migration,[167] and is also found in hypertrophic chondrocytes and osteoclasts.[168,169] Interferon-α (IFN-α) is essential for antiviral immunity, but in the absence of MMP-12 or IκBα (encoded by NFKBIA), IFN-α is retained in the cytosol of virus-infected cells and is not secreted, suggesting that the export of IFN-α from virus-infected cells require activated MMP-12 and IκBα. The inability of cells in MMP-12 KO mice to express IκBα and thus export IFN-α makes coxsackievirus type B3 infection lethal and renders respiratory syncytial virus more pathogenic. It has been suggested that after macrophage secretion, MMP-12 is transported into virus-infected cells. In HeLa cells, MMP-12 is translocated to the nucleus, where it binds to the NFKBIA promoter, driving NFKBIA transcription, and leading to IFN-α secretion and host protection. On the other hand, extracellular MMP-12 cleaves off the IFN-α receptor 2 binding site of systemic IFN-α, preventing an unchecked immune response. Consistent with a role for MMP-12 in clearing systemic IFN-α, treatment of coxsackievirus type B3-infected wild-type (WT) mice with a membrane-impermeable MMP-12 inhibitor elevates systemic IFN-α levels and reduces viral replication in the pancreas while sparing intracellular MMP-12, suggesting that inhibiting extracellular MMP-12 could be a new avenue for antiviral treatment.[170]

MMP-12 plays a role in airway inflammation and remodeling. MMP-12 expression is increased in the lungs of asthmatic patients. Compound 27 is a potent and selective inhibitor of MMP-12 that is orally efficacious in a mouse model of MMP-12 induced ear-swelling inflammation, and may be a candidate drug for treatment of asthma.[171]

MMP-12 may affect the blood–brain barrier after cerebral ischemia. In rats subjected to middle cerebral artery occlusion and reperfusion, MMP-12 was upregulated ~31-, 47-, and 66-fold in rats subjected 1-, 2-, or 4-h ischemia, respectively, followed by 1-day reperfusion. MMP-12 suppression by infusion of nanoparticles of MMP-12 shRNA-expressing plasmid protected the blood–brain barrier integrity by inhibiting the degradation of tight-junction proteins, and reduced the percent Evans blue dye extravasation and infarct size. MMP-12 suppression reduced the levels of the other endogenous proteases tissue-type plasminogen activator and MMP-9, which are key players in blood–brain barrier damage. These findings demonstrate the adverse role of MMP-12 in acute brain damage after ischemic stroke and suggest that MMP-12 suppression could be a therapeutic target for cerebral ischemia.[172]

Studies have examined possible correlation between the expression of MMPs in the primary tumor of head and neck squamous cell carcinomas and the presence of extracapsular spread in cervical nodes metastasis. MMP-2, -3, -12, and -14 were expressed in 27%, 47.5%, 55%, and 57.5% of cases, respectively. MMP-12 expression was associated with extracapsular spread and correlated with nodal metastasis. MMP-12 expressed in the primary tumor may be a molecular marker for predicting extracapsular spread in head and neck squamous cell carcinomas patients with metastatic nodal disease.[173]

8.6.2 MMP-19

MMP-19 or RASI-1 or stromelysin-4 has a gene locus on chromosome 12q14. The catalytic domain of MMP-19 can hydrolyze the basement membrane-type IV collagen, laminin, and nidogen, as well as the large tenascin-C isoform, fibronectin, and type I gelatin in vitro, suggesting that MMP-19 is a potent proteinase capable of hydrolyzing a broad range of ECM components. Neither the catalytic domain nor the full-length MMP-19 can degrade triple-helical collagen. Also, the MMP-19 catalytic domain can process pro-MMP-9 to its active form, but may not activate other latent forms of MMPs such as MMP-1, -2, -3, -13, and -14 in vitro.[174]

MMP-19 is a potent basement membrane-degrading enzyme that plays a role in tissue remodeling, wound healing, and epithelial cell migration by cleaving laminin-5γ2 chain.[175–178] Angiogenesis is the process of forming new blood vessels from existing ones and requires degradation of the vascular basement membrane and remodeling of ECM in order to allow endothelial cells to migrate and invade the surrounding tissue. Angiostatin, a proteolytic fragment of plasminogen, is a potent antagonist of angiogenesis that inhibits migration and proliferation of endothelial cells. MMP-19 may exhibit antiangiogenic effects on endothelial cells by processing human plasminogen in a characteristic cleavage pattern to generate three angiostatin-like fragments with a molecular weight of 35, 38, and 42 kDa that decrease the phosphorylation of c-met, inhibit the proliferation of human microvascular endothelial cells, and reduce formation of capillary-like structures.[179]

Idiopathic pulmonary fibrosis is a progressive interstitial lung disease characterized by aberrant activation of epithelial cells that induce the migration, proliferation, and activation of fibroblasts. The resulting distinctive fibroblastic/myofibroblastic foci are responsible for the excessive ECM production and abnormal lung remodeling. MMP-19-deficient mice develop an exaggerated bleomycin-induced lung fibrosis. Microarray analysis of MMP-19-deficient lung fibroblasts revealed the dysregulation of several profibrotic pathways, including ECM formation, migration, proliferation, and autophagy. Compared with WT mice, MMP-19-deficient lung fibroblasts show increased α1 (I) collagen gene and collagen protein levels at baseline and after TGF-β treatment and increased smooth muscle-α actin expression. MMP-19-deficient lung fibroblasts also show an increase in proliferation, transmigration, and locomotion over Boyden chambers coated with type I collagen or Matrigel. Thus, in lung fibroblasts, MMP-19 has strong regulatory effects on the synthesis of key ECM components, on fibroblast to myofibroblast differentiation, and in migration and proliferation.[180]

Bleomycin-induced lung fibrosis was evaluated in MMP-19-deficient and WT mice. Laser capture microscope followed by microarray analysis revealed MMP-19 in hyperplastic epithelial cells adjacent to fibrotic regions. MMP-19-deficient mice showed increased lung fibrotic response to bleomycin compared with WT mice. A549 alveolar epithelial cells transfected with human MMP-19 stimulated wound healing and cell migration, whereas silencing MMP-19 had the opposite effect. Gene expression microarray of transfected A549 cells showed prostaglandin-endoperoxide synthase 2 (PTGS2) as one of the highly induced genes. PTGS2 was overexpressed in idiopathic pulmonary fibrosis lungs and colocalized with MMP-19 in hyperplastic epithelial cells. PTGS2 was increased in bronchoalveolar lavage and lung tissues after bleomycin-induced fibrosis in WT mice, but not MMP-19-deficient mice. Inhibition of MMP-19 by siRNA resulted in reduction of PTGS2 mRNA and protein level. These findings suggest that during lung injury upregulation of MMP-19 may protect against fibrosis through the induction of PTGS2.[181]

Liver fibrosis is characterized by the deposition and increased turnover of ECM. MMP-19 is highly expressed in liver, and its role during the development and resolution of liver fibrosis was studied in MMP-19-deficient and WT mice exposed to chronic carbon tetrachloride intoxication. Loss of MMP-19 was beneficial during liver injury, as plasma ALT and AST levels, deposition of fibrillar collagen, and phosphorylation of SMAD3, a TGF-ss1 signaling molecule, were reduced. The ameliorated course of the disease in MMP-19-deficient mice likely results from a slower rate of basement membrane destruction and ECM remodeling as the knockout mice maintained higher levels of type IV collagen and lower expression and activation of MMP-2. Liver regeneration upon removal of the toxin was also hastened in MMP-19-deficient mice. MMP-19 deficiency may decrease the development of hepatic fibrosis through decreased replacement of physiological ECM with fibrotic deposits in the beginning of the injury.[182]

MMP-19 mRNA is widely expressed in the synovium of normal and rheumatoid arthritic patients. MMP-19 cleaves aggrecan and cartilage oligomeric matrix protein, two of the macromolecules characterizing the cartilage ECM, supporting that MMP-19 may participate in the degradation of aggrecan and cartilage oligomeric matrix protein in arthritic disease.[175]

Patients with a congenital cavitary optic disc anomaly (CODA) have profound excavation of the optic nerve resembling glaucoma. A recent study mapped the gene that causes autosomal-dominant CODA in a large pedigree to a chromosome 12q locus. Comparative genomic hybridization and quantitative PCR analysis of this pedigree identified a 6-kbp heterozygous triplication upstream of the MMP-19 gene, present in all 17 affected family members, but not normal members. The same 6-kbp triplication was identified in one of 24 unrelated CODA patients and in none of 172 glaucoma patients. Analysis with a Luciferase assay showed that the 6-kbp sequence has transcription enhancer activity. A 773-bp fragment of the 6-kbp DNA segment increased downstream gene expression eightfold, suggesting that triplication of this sequence may lead to dysregulation of the downstream MMP-19 gene in CODA patients. Immunohistochemical analysis of human donor eyes revealed strong expression of MMP-19 in optic nerve head. These findings suggest that triplication of an enhancer may lead to overexpression of MMP-19 in the optic nerve that causes CODA.[183]

MMP-19 may play a role in cancer. MMP-19-deficient mice develop diet-induced obesity due to adipocyte hypertrophy, but are less susceptible to skin cancers induced by chemical carcinogens.[184] In patients with gallbladder carcinoma loss of expression of the tumor suppressor N-myc downstream-regulated gene 2 (NDRG2) was an independent predictor of decreased survival and was associated with a more advanced T stage, higher cellular grade, and lymphatic invasion. Gallbladder carcinoma cells with loss of NDRG2 expression showed enhanced proliferation, migration, and invasiveness in vitro, and tumor growth and metastasis in vivo. Loss of NDRG2 induced the expression of MMP-19, which regulated the expression of Slug at the transcriptional level. MMP-19-induced Slug, increased the expression of a receptor tyrosine kinase, Axl, which maintained Slug expression through a positive-feedback loop, and stabilized epithelial–mesenchymal transition of gallbladder carcinoma cells. NDRG2 could be a favorable prognostic indicator and promising target for therapeutic agents against gallbladder carcinoma, and the effects of NDRG2 could be related to suppression of MMP-19.[185]

MMP-19 appears to be upregulated during colorectal tumorigenesis, with different expression patterns. Increased MMP-19 mRNA expression and protein levels were observed in the progression of colonic lesions, and MMP-19 staining increased in the normal mucosa–microadenoma–carcinoma sequence.[147]

MMP-19 may play a role in nonsmall cell lung cancer. MMP-19 gene expression and protein levels are increased in lung cancer tumors compared with adjacent normal lung tissues. Increased MMP-19 gene expression conferred a poorer prognosis in nonsmall cell lung cancer. Overexpression of MMP-19 promotes epithelial–mesenchymal transition, migration, and invasiveness in multiple nonsmall cell lung cancer cell lines. Also, miR-30 isoforms, a microRNA family predicted to target MMP-19, are downregulated in human lung cancer. Thus MMP-19 may be associated with the development and progression of nonsmall cell lung cancer and may be a potential biomarker of disease severity and outcome.[186]

On the other hand, MMP-19 may be one of the MMPs downregulated in the nasopharyngeal carcinoma cell lines. Allelic deletion and promoter hypermethylation may contribute to MMP-19 downregulation. Comparative studies of the WT and the catalytically inactive mutant MMP-19 suggest that the catalytic activity of MMP-19 may play a role in antitumor and antiangiogenesis activities. In the in vivo tumorigenicity assay, MMP-19 transfectants suppress tumor formation in only in the WT, but not mutant, nude mice. In the in vitro colony formation assay, WT MMP-19 reduced colony-forming ability of nasopharyngeal carcinoma cell lines, when compared to the inactive mutant. In the tube formation assay of human umbilical vein endothelial cells and human microvascular endothelial cells, secreted WT MMP-19, but not mutant MMP-19, caused reduction of tube-forming ability in endothelial cells, and decreased VEGF in conditioned media. The antiangiogenic activity of WT MMP-19 is correlated with suppression of tumor formation. Thus the catalytic activity of MMP-19 may be essential for its tumor suppressive and antiangiogenic effects in nasopharyngeal carcinoma.[187]

8.6.3 MMP-20

MMP-20 is also known as enamelysin. The human enamelysin gene maps to chromosome 11q22, clustered to at least seven other members of the MMP gene family. Enamelysin is a tooth-specific MMP expressed in newly formed tooth enamel.[188] MMP-20 is specifically expressed by ameloblasts and odontoblasts of dental papilla, and hence its name—enamelysin. A cDNA-encoding MMP-20 was cloned from RNA prepared from human odontoblastic cells. The open-reading frame of the cloned cDNA codes for a polypeptide of 483 amino acids. Human enamelysin has a domain organization similar to other MMPs, including a signal peptide, a prodomain with the conserved motif PRCGVPD involved in maintaining enzyme latency, a catalytic domain with a Zn^{2+}-binding site, and a C-terminal fragment similar to the sequence of hemopexin. The calculated molecular mass of human enamelysin is about 54 kDa, which is similar to that of collagenases or stromelysins. However, human MMP-20 lacks a series of structural features distinctive of subfamilies of MMPs. MMP-20 contains a very basic hinge region compared to the hydrophobic hinge region of stromelysins and the acidic hinge region of MMP-19. The full-length human enamelysin cDNA has been expressed in *E. coli*, and the purified and refolded recombinant protein degrades synthetic peptides used as substrates of MMPs. The recombinant human enamelysin also degrades amelogenin,[188] the major protein component of the enamel matrix. On the basis of its degrading activity on amelogenin, and its highly restricted expression to dental tissues, it was suggested that human enamelysin plays a central role in tooth enamel formation.[189]

Enamelysin is expressed during the early through middle stages of enamel development. The enamel matrix proteins amelogenin, ameloblastin, and enamelin are also expressed during this developmental time period, suggesting that enamelysin may be involved in their hydrolysis. Amelogenin imperfecta is a genetic disorder with defective enamel formation involving mutation at MMP-20 cleavage sites.[190] Enamelysin null mice show severe amelogenesis imperfecta tooth phenotype that does not process amelogenin properly, altered enamel matrix and rod pattern, hypoplastic enamel that delaminates from the dentin, and a deteriorating enamel organ morphology as development progresses. These findings support that enamelysin activity is essential for proper enamel development.[191]

MMP-20 also hydrolyzes aggrecan efficiently at the well-described MMP cleavage site between residues Asn(341) and Phe(342). MMP-20 also cleaves cartilage oligomeric matrix protein in a distinctive manner, generating a major proteolytic product of 60 kDa. Due to the unique expression pattern of MMP-20, it may primarily be involved in the turnover of aggrecan and cartilage oligomeric matrix protein during tooth development.[175]

8.6.4 MMP-21

MMP-21 also known as Xenopus-MMP has a genetic code on chromosome 1,[31] in contrast to the normal 11q location of other MMPs. MMP-21 is an MMP with measurable gelatinolytic activity expressed in various fetal and adult tissues, macrophages of granulomatous skin lesions, fibroblasts in dermatofibromas, and basal and squamous cell carcinomas.[192,193] MMP-21 may play a role in embryogenesis and tumor progression and could be a target of the Wnt, Pax, and Notch signaling pathways. MMP-21 mRNA was detected in mouse embryos aged 10.5, 12.5, 13.5, and 16.5 days, and in various adult murine organs. In both humans and mice, MMP-21 has been detected in the epithelial cells of developing kidney, intestine, neuroectoderm, and skin, but not in normal adult skin. MMP-21 is present in invasive cancer cells of aggressive basal and squamous cell carcinomas, but not in skin disorders characterized by mere keratinocyte hyperproliferation. TGF-β1 induced MMP-21 in HaCaTs and keratinocytes in vitro. MMP-21 expression is temporally and spatially tightly controlled during development. Unlike many classical MMPs, MMP-21 is present in various normal adult tissues. Among epithelial MMPs, MMP-21 has a unique expression pattern in cancer.[193]

MMP-21 could be an indicator of poor prognosis for certain types of cancer. Increased MMP-21 expression in metastatic lymph nodes may predict unfavorable prognosis and overall survival for oral squamous cell carcinoma patients with lymphatic metastasis.[194] MMP-21 expression is increased in esophageal squamous cell carcinoma and is associated with tumor invasion, lymph node metastasis, and distant metastasis. Patients with tumors of positive MMP-21 staining tend to have worse overall survival. Multivariate analysis showed that MMP-21 was an independent prognostic factor for overall survival in patients with esophageal squamous cell carcinoma. These findings support a role of MMP-21 in tumor progression and prognosis of human esophageal squamous cell carcinoma.[195]

MMP-21 expression is higher in colorectal cancer compared with that in normal epithelial tissue. MMP-21 expression correlates with tumor invasion, lymph node metastasis, and distant metastasis of colorectal cancer. MMP-21 may also be an independent prognostic factor in patients with stage II and III colorectal cancer.[196]

Merkel cell carcinoma is an aggressive cutaneous tumor with increasing incidence and poor outcome, and shows differential expression pattern of MMPs. MMP-28 was observed in tumor cells of 15/44 samples especially in tumors <2 cm in diameter while 21/44 specimens showed MMP-28 in the tumor stroma. Expression of MMP-21 was demonstrated in tumor cells of 13/43 samples. MMP-26 was positive in stromal cells (17/44) and its expression associated with tumors >2 cm in diameter. Stromal expression of MMP-10 was the most frequent finding of the studied samples (31/44), and MMP-10 was detected also in tumor cells (17/44). Most of the metastatic lymph nodes expressed MMP-10 and MMP-26. MMP-10, MMP-21, and MMP-28 mRNAs and corresponding proteins were basally expressed by the UISO cells. IFN-α and TNF-α downregulated MMP-21 and MMP-28 expression. These findings suggest that MMP-26 expression in stroma is associated with larger tumors with poor prognosis. Expression of MMP-21 and MMP-28 seems to associate with the tumors of less malignant potential. The study also confirms the role of MMP-10 in the pathogenesis of Merkel cell carcinoma.[197]

Pancreatic adenocarcinoma shows early aggressive local invasion and high metastatic potential, and therefore a low 5-year survival rate. MMP-21 was expressed in well-differentiated cancer cells and occasional fibroblasts, but tended to diminish in intensity from grade I to grade III tumors. All cultured cancer cell lines expressed MMP-21 basally at low levels. MMP-21 expression was induced by epidermal growth factor in PANC-1 cells. Thus MMP-21 may not be a marker of invasiveness, but rather of differentiation, in pancreatic cancer, and may be upregulated by epidermal growth factor.[135]

8.6.5 MMP-22

MMP-22 also known as chicken-MMP has a gene locus on chromosome 1p36.3. The terminal end of the short arm of human chromosome 1, 1p36.3, is frequently deleted in a number of tumors and is believed to be the location of multiple tumor suppressor genes. MMP-21 and MMP-22 genes have been identified in the Cdc2L1-2 locus, which spans approximately 120 kb on 1p36.3. These genes encode MMPs that contain prepro, catalytic, cysteine-rich, IL-1 receptor-related, and proline-rich domains. Their catalytic domains are most closely related to stromelysin-3 and contain the consensus HEXXH Zn^{2+}-binding region required for enzyme activation, while their cysteine-rich domains appear to be related to a number of human, mouse, and *Caenorhabditis elegans* MMP sequences. These MMPs lack the highly conserved cysteine residue in the proenzyme domain, the so-called "cysteine switch," which is involved in the autocatalytic activation of many MMPs. The MMP-21/22 genes express multiple mRNAs, some of which are derived by alternative splicing, in a tissue-specific manner.[198]

8.6.6 MMP-23

MMP-23 or cysteine array (CA)-MMP has a gene locus on chromosome 1p36.3. A cDNA-encoding MMP-23 has been cloned from an ovary cDNA library. This protein exhibits sequence similarity with MMPs, but displays a different domain structure. MMP-23 lacks a recognizable signal sequence and has a short prodomain, although it contains a single cysteine residue that can be part of the cysteine-switch mechanism operating for maintaining enzyme latency. Whereas all human MMPs, with the exception of matrilysin, contain four hemopexin-like repeats, the C-terminal domain of MMP-23 is considerably shortened and shows no sequence similarity to hemopexin. MMP-23 is devoid of structural features distinctive of the diverse MMP subclasses, including the specific residues located close to the Zn^{2+}-binding site in collagenases, the transmembrane domain of membrane-type MMPs, or the fibronectin-like domain of gelatinases. MMP-23 is unique among MMPs as it lacks the cysteine-switch motif in the propeptide, and the hemopexin domain is substituted by cysteine-rich immunoglobulin-like domains.[199]

MMPs are either secreted or membrane anchored via a type I transmembrane domain or a GPI linkage. Lacking either membrane-anchoring mechanism, MMP-23 was reported to be expressed as a cell-associated protein. MMP-23 is expressed as an integral membrane zymogen with an N-terminal signal anchor and secreted as a fully processed mature enzyme. MMP-23 is a type II membrane protein regulated by a single proteolytic cleavage for both its activation and secretion.[14] MMP-23 is predominantly expressed in ovary, testis, and prostate, suggesting that it may have a specialized role in reproductive processes.[199] Gene expression of MMP-23 is elevated and may promote invasiveness in MDA-MB-231 breast cancer cells.[200]

8.6.7 MMP-27

MMP-27 is a human MMP-22 homolog with a gene locus on chromosome 11q24. MMP-27 is classified as a stromelysin and holds 51.6% structural homology with MMP-10. MMP-27 is a poorly characterized and barely secreted MMP. Sequence comparison suggests that a C-terminal extension includes a potential transmembrane domain as in some MT-MMPs. Subcellular fractionation and confocal microscopy suggest retention of endogenous MMP-27 or recombinant rMMP-27 in the endoplasmic reticulum with locked exit across the intermediate compartment. Conversely, truncated rMMP-27 without C-terminal extension accessed downstream secretory compartments in endoplasmic reticulum intermediate compartment and Golgi and was constitutively secreted. Neither endogenous nor recombinant MMP-27 partitioned in the detergent phase after Triton X-114 extraction, indicating that MMP-27 is not an integral membrane protein. Due to its unique C-terminal extension, which does not lead to stable membrane insertion, MMP-27 is efficiently stored within the endoplasmic reticulum until it is ready to be released.[201]

MMP-27 is expressed in B-lymphocytes and is overexpressed in cultured human lymphocytes treated with anti-(IgG/IgM).[202] MMP-27 is expressed in CD163+/CD206+ M2 macrophages in the cycling human endometrium and in superficial endometriotic lesions. MMP-27 mRNA is detected throughout the menstrual cycle. MMP-27 mRNA levels are increased from the proliferative to the secretory phase, to peak during the menstrual phase. MMP-27 is immunolocalized in large isolated cells scattered throughout the stroma and around blood vessels: these cells are most abundant at menstruation and are identified by immunofluorescence as CD45(+), CD163(+), and CD206(+) macrophages. CD163(+) macrophages are abundant in endometriotic lesions, and colabeling for CD206 and MMP-27 is observed in ovarian or peritoneal endometriotic lesions. Thus MMP-27 is expressed in a subset of endometrial macrophages related to menstruation and in ovarian and peritoneal endometriotic lesions.[203] Several MMPs show a stronger expression in breast cancer tissue compared to normal breast tissue. Of those, expression of MMP-27 is related to tumor grade since it is higher in G3 compared to G2 tissue samples. MDA-MB-468 breast cancer cell line show the strongest mRNA and protein expression for most of the MMPs studied.

MMP-27 may be involved breast cancer development and tumor progression.[204]

8.6.8 MMP-28

MMP-28 or epilysin has a gene locus on chromosome 17q21.1. MMP-28 shows high expression in the epidermis. Epilysin was first cloned from the human keratinocyte and testis cDNA libraries.[205,206] Like most MMPs, the deduced 520-amino acid sequence of MMP-28 includes a signal peptide, a prodomain with an unusual cysteine-switch PRCGVTD motif followed by the furin activation sequence RRKKR, a Zn^{2+}-binding catalytic domain with an HEIGHTLGLTH sequence, a hinge region and a hemopexin-like domain, but no transmembrane sequence. Within the cysteine switch, MMP-28 contains a threonine residue at position 94, instead of a proline as in most MMPs. Also, compared to the 10–12-amino acid stretch in other MMPs, a longer 22 residues follows the cysteine switch before the furin cleavage region. The MMP-28 gene uniquely mapped to chromosome 17q11.2 includes eight exons and seven introns, and five exons are spliced at sites not used by other MMPs. Also, exon 4 is alternatively spliced to a transcript that does not encode the N-terminal half of the catalytic domain. Recombinant epilysin degrades casein in zymography assay, and its proteolytic activity is inhibited by EDTA and the MMP inhibitor batimastat. Immunohistochemical staining showed epilysin in the basal and suprabasal epidermis of intact skin. In injured skin, epilysin staining is seen in basal keratinocytes both at and some distance from the wound edge, a pattern distinct from that of other MMPs expressed during tissue repair. Epilysin is expressed at high levels in testis and at lower levels in lungs, heart, intestine, colon, placenta, and brain. MMP-28 may function in several tissues both in tissue homeostasis and in repair.[205,206]

The broad range of expression in normal adult and fetal tissues and in carcinomas suggests important roles for MMP-28.[207] Epilysin is expressed in a number of normal tissues, suggestive of functions in tissue homeostasis. The mRNA expression of MMP-28 was highest in healthy tissues when compared to subjects with chronic periodontitis and aggressive periodontitis. The elevated MMP-28 level in healthy tissues support that it may be involved in normal tissue homeostasis and remodeling, and its decreased levels could serve as a biomarker for periodontal health.[208]

MMP-28 transcript and protein are expressed in rhesus monkey placenta during early pregnancy. MMP-28 mRNA expression was shown by in situ hybridization after day 12 of pregnancy, and both the syncytial and the cytotrophoblastic cell layers of placental villi, the cytotrophoblast cells of the trophoblastic column, and the extravillous trophoblast cells of trophoblastic shell were primary producers of MMP-28 transcript. Expression of MMP-28 mRNA was undetectable in the endovascular trophoblast cells, decidual cells, luminal and glandular epithelium, arterioles, and myometrium. The restricted distribution pattern of MMP-28 in the villous and extravillous trophoblasts during rhesus monkey early pregnancy suggests a potential role in trophoblast invasion associated with embryo implantation.[209]

MMP-28 may regulate the inflammatory and ECM responses in cardiac aging. In a mouse model of MI of the left ventricle induced by permanent coronary artery ligation, MMP-28 expression was decreased post-MI, and its cell source shifted from myocytes to macrophages. In MMP-28 KO mice, MMP-28 deletion increased day 7 mortality because of increased cardiac rupture post-MI. MMP-28 KO mice exhibited larger left ventricular volumes, worse left ventricular dysfunction, worse left ventricular remodeling index, and increased lung edema. Plasma MMP-9 levels were unchanged in the MMP-28 KO mice but increased in WT mice at day 7 post-MI. The mRNA levels of inflammatory and ECM proteins were attenuated in the infarct regions of MMP-28 KO mice, indicating reduced inflammatory and ECM responses. M2 macrophage activation was impaired in MMP-28 KO mice. MMP-28 deletion also led to decreased collagen deposition and fewer myofibroblasts. Collagen cross-linking was impaired as a result of decreased expression and activation of lysyl oxidase in the infarcts of MMP-28 KO mice. These findings suggest that MMP-28 deletion aggravated MI-induced left ventricular dysfunction and rupture as a result of defective inflammatory response and scar formation by suppressing M2 macrophage activation.[210]

Studies have examined the cellular location and putative function of MMP-19, MMP-26 (matrilysin-2), and MMP-28 (epilysin), in normal, inflammatory, and malignant conditions of the intestine in tissue specimens from patients with ulcerative colitis and archival tissue samples of ischemic colitis, Crohn's disease, ulcerative colitis, colon cancer, and healthy intestine. Unlike many classical MMPs, MMP-19, -26, and -28 were all expressed in normal intestine. In inflammatory bowel disease, MMP-19 was expressed in nonmigrating enterocytes and shedding epithelium. MMP-26 was detected in migrating enterocytes, unlike MMP-28. In colon carcinomas, MMP-19 and MMP-28 expression was downregulated in tumor epithelium. Staining for MMP-26 revealed a meshwork-like pattern between cancer islets, which was absent from most dedifferentiated areas.

These findings suggest that MMP-19 is involved in epithelial proliferation and MMP-26 in enterocyte migration, while MMP-28 expression is not associated with inflammatory and destructive changes seen in inflammatory bowel disease. In contrast to previously characterized MMPs, MMP-19 and MMP-28 are downregulated during malignant transformation of the colon and may play a prominent role in tissue homeostasis.[211]

MMP-28 is also elevated in cartilage from patients with osteoarthritis and rheumatoid arthritis.[212,213]

9. MMP/TIMP RATIO

TIMPs are endogenous, naturally occurring MMP inhibitors that bind MMPs in a 1:1 stoichiometry.[6,11] Four homologous TIMPs, TIMP-1, -2, -3, and -4, have been identified. TIMP-1 and -3 are glycoproteins, while TIMP-2 and -4 do not contain carbohydrates. TIMPs have poor specificity for a given MMP, and each TIMP can inhibit multiple MMPs with different efficacies.[214–216] A change in either TIMP or MMP levels could alter the MMP/TIMP ratio and cause a net change in specific MMP activity.

10. CONCLUDING REMARKS

MMPs are involved in many biological processes and could be important biomarkers for cardiovascular disease, musculoskeletal disorders, and cancer. One challenge to understanding the role of specific MMPs in pathological conditions is that studies often focus on few MMPs or TIMPs, and it is important not to generalize the findings to other MMPs and TIMPs. Because tissue remodeling is a dynamic process, an increase in one MMP in a certain region may be paralleled by a decrease of other MMPs in other regions. Also, because of the differences in the proteolytic activities of MMPs toward different substrates, the activities of MMPs may vary during the course of disease. This makes it important to examine different MMPs and TIMPs in various tissue regions and at different stages of the disease. Another challenge is that the topology of MMPs is well conserved, making it difficult to design highly specific MMP inhibitors. Endogenous TIMPs are not very specific and often inhibit multiple MMPs. Likewise, synthetic MMP inhibitors have poor selectivity and many biologic actions, and therefore often cause side effects.[217] New synthetic MMP inhibitors are being developed, and their effectiveness in cardiovascular disease and cancer needs to be examined. Another strategy is to develop specific approaches to target

MMPs locally in the vicinity of a localized pathology, and thus minimize undesirable systemic effects.

ACKNOWLEDGMENTS

This work was supported by grants from National Heart, Lung, and Blood Institute (HL-65998, HL-111775). Dr. N.C. was a visiting scholar from the Department of Gastroenterology, Renmin Hospital of Wuhan University, Wuhan, Hubei Province, P.R. China, and a recipient of scholarship from the China Scholarship Council. Dr. M.H. was a visiting scholar from the Department of Cardiovascular Surgery, Tongji Hospital, Tongji Medical College, Huazhong University of Science and Technology, Wuhan, P.R. China.

Conflict of interest: None.

REFERENCES

1. Mittal R, Patel AP, Debs LH, et al. Intricate functions of matrix metalloproteinases in physiological and pathological conditions. *J Cell Physiol*. 2016;231(12):2599–2621.
2. Kucukguven A, Khalil RA. Matrix metalloproteinases as potential targets in the venous dilation associated with varicose veins. *Curr Drug Targets*. 2013;14(3):287–324.
3. MacColl E, Khalil RA. Matrix metalloproteinases as regulators of vein structure and function: implications in chronic venous disease. *J Pharmacol Exp Ther*. 2015;355(3): 410–428.
4. Raffetto JD, Khalil RA. Matrix metalloproteinases and their inhibitors in vascular remodeling and vascular disease. *Biochem Pharmacol*. 2008;75(2):346–359.
5. Gross J, Lapiere CM. Collagenolytic activity in amphibian tissues: a tissue culture assay. *Proc Natl Acad Sci USA*. 1962;48:1014–1022.
6. Nagase H, Visse R, Murphy G. Structure and function of matrix metalloproteinases and TIMPs. *Cardiovasc Res*. 2006;69(3):562–573.
7. Cauwe B, Van den Steen PE, Opdenakker G. The biochemical, biological, and pathological kaleidoscope of cell surface substrates processed by matrix metalloproteinases. *Crit Rev Biochem Mol Biol*. 2007;42(3):113–185.
8. Ohuchi E, Imai K, Fujii Y, Sato H, Seiki M, Okada Y. Membrane type 1 matrix metalloproteinase digests interstitial collagens and other extracellular matrix macromolecules. *J Biol Chem*. 1997;272(4):2446–2451.
9. Holmbeck K, Bianco P, Caterina J, et al. MT1-MMP-deficient mice develop dwarfism, osteopenia, arthritis, and connective tissue disease due to inadequate collagen turnover. *Cell*. 1999;99(1):81–92.
10. Visse R, Nagase H. Matrix metalloproteinases and tissue inhibitors of metalloproteinases: structure, function, and biochemistry. *Circ Res*. 2003;92(8): 827–839.
11. Bode W, Fernandez-Catalan C, Grams F, et al. Insights into MMP-TIMP interactions. *Ann N Y Acad Sci*. 1999;878:73–91.
12. Bode W, Gomis-Ruth FX, Stockler W. Astacins, serralysins, snake venom and matrix metalloproteinases exhibit identical zinc-binding environments (HEXXHXXGXXH and Met-turn) and topologies and should be grouped into a common family, the 'metzincins'. *FEBS Lett*. 1993;331(1–2):134–140.
13. English WR, Holtz B, Vogt G, Knauper V, Murphy G. Characterization of the role of the "MT-loop": an eight-amino acid insertion specific to progelatinase A (MMP2) activating membrane-type matrix metalloproteinases. *J Biol Chem*. 2001;276(45): 42018–42026.

14. Pei D, Kang T, Qi H. Cysteine array matrix metalloproteinase (CA-MMP)/MMP-23 is a type II transmembrane matrix metalloproteinase regulated by a single cleavage for both secretion and activation. *J Biol Chem.* 2000;275(43):33988–33997.
15. Gall AL, Ruff M, Kannan R, et al. Crystal structure of the stromelysin-3 (MMP-11) catalytic domain complexed with a phosphinic inhibitor mimicking the transition-state. *J Mol Biol.* 2001;307(2):577–586.
16. Saito S, Trovato MJ, You R, et al. Role of matrix metalloproteinases 1, 2, and 9 and tissue inhibitor of matrix metalloproteinase-1 in chronic venous insufficiency. *J Vasc Surg.* 2001;34(5):930–938.
17. Seizer P, May AE. Platelets and matrix metalloproteinases. *Thromb Haemost.* 2013;110(5):903–909.
18. Chen Q, Jin M, Yang F, Zhu J, Xiao Q, Zhang L. Matrix metalloproteinases: inflammatory regulators of cell behaviors in vascular formation and remodeling. *Mediators Inflamm.* 2013;2013:928315.
19. Raffetto JD, Qiao X, Koledova VV, Khalil RA. Prolonged increases in vein wall tension increase matrix metalloproteinases and decrease constriction in rat vena cava: potential implications in varicose veins. *J Vasc Surg.* 2008;48(2):447–456.
20. Sansilvestri-Morel P, Fioretti F, Rupin A, et al. Comparison of extracellular matrix in skin and saphenous veins from patients with varicose veins: does the skin reflect venous matrix changes? *Clin Sci (Lond).* 2007;112(4):229–239.
21. Woodside KJ, Hu M, Burke A, et al. Morphologic characteristics of varicose veins: possible role of metalloproteinases. *J Vasc Surg.* 2003;38(1):162–169.
22. Kwan JA, Schulze CJ, Wang W, et al. Matrix metalloproteinase-2 (MMP-2) is present in the nucleus of cardiac myocytes and is capable of cleaving poly (ADP-ribose) polymerase (PARP) in vitro. *FASEB J.* 2004;18(6):690–692.
23. Pei D, Weiss SJ. Furin-dependent intracellular activation of the human stromelysin-3 zymogen. *Nature.* 1995;375(6528):244–247.
24. Morgunova E, Tuuttila A, Bergmann U, Tryggvason K. Structural insight into the complex formation of latent matrix metalloproteinase 2 with tissue inhibitor of metalloproteinase 2. *Proc Natl Acad Sci USA.* 2002;99(11):7414–7419.
25. Strongin AY, Collier I, Bannikov G, Marmer BL, Grant GA, Goldberg GI. Mechanism of cell surface activation of 72-kDa type IV collagenase. Isolation of the activated form of the membrane metalloprotease. *J Biol Chem.* 1995;270(10):5331–5338.
26. Sato H, Takino T, Kinoshita T, et al. Cell surface binding and activation of gelatinase A induced by expression of membrane-type-1-matrix metalloproteinase (MT1-MMP). *FEBS Lett.* 1996;385(3):238–240.
27. Butler GS, Butler MJ, Atkinson SJ, et al. The TIMP2 membrane type 1 metalloproteinase "receptor" regulates the concentration and efficient activation of progelatinase A. A kinetic study. *J Biol Chem.* 1998;273(2):871–880.
28. Zucker S, Drews M, Conner C, et al. Tissue inhibitor of metalloproteinase-2 (TIMP-2) binds to the catalytic domain of the cell surface receptor, membrane type 1-matrix metalloproteinase 1 (MT1-MMP). *J Biol Chem.* 1998;273(2):1216–1222.
29. Kolkenbrock H, Orgel D, Hecker-Kia A, Noack W, Ulbrich N. The complex between a tissue inhibitor of metalloproteinases (TIMP-2) and 72-kDa progelatinase is a metalloproteinase inhibitor. *Eur J Biochem.* 1991;198(3):775–781.
30. Fridman R, Bird RE, Hoyhtya M, et al. Expression of human recombinant 72 kDa gelatinase and tissue inhibitor of metalloproteinase-2 (TIMP-2): characterization of complex and free enzyme. *Biochem J.* 1993;289(pt 2):411–416.
31. Verma RP, Hansch C. Matrix metalloproteinases (MMPs): chemical-biological functions and (Q)SARs. *Bioorg Med Chem.* 2007;15(6):2223–2268.
32. Hollborn M, Stathopoulos C, Steffen A, Wiedemann P, Kohen L, Bringmann A. Positive feedback regulation between MMP-9 and VEGF in human RPE cells. *Invest Ophthalmol Vis Sci.* 2007;48(9):4360–4367.

33. Mao D, Zhang Y, Lu H, Zhang H. Molecular basis underlying inhibition of metastasis of gastric cancer by anti-VEGFa treatment. *Tumour Biol.* 2014;35:8217–8223.
34. Cui Y, Sun YW, Lin HS, et al. Platelet-derived growth factor-BB induces matrix metalloproteinase-2 expression and rat vascular smooth muscle cell migration via ROCK and ERK/p38 MAPK pathways. *Mol Cell Biochem.* 2014;393(1–2): 255–263.
35. Rao VH, Kansal V, Stoupa S, Agrawal DK. MMP-1 and MMP-9 regulate epidermal growth factor-dependent collagen loss in human carotid plaque smooth muscle cells. *Physiol Rep.* 2014;2(2): e00224.
36. Patterson ML, Atkinson SJ, Knauper V, Murphy G. Specific collagenolysis by gelatinase A, MMP-2, is determined by the hemopexin domain and not the fibronectin-like domain. *FEBS Lett.* 2001;503(2–3):158–162.
37. Suenaga N, Mori H, Itoh Y, Seiki M. CD44 binding through the hemopexin-like domain is critical for its shedding by membrane-type 1 matrix metalloproteinase. *Oncogene.* 2005;24(5):859–868.
38. Park HI, Jin Y, Hurst DR, et al. The intermediate S1' pocket of the endometase/matrilysin-2 active site revealed by enzyme inhibition kinetic studies, protein sequence analyses, and homology modeling. *J Biol Chem.* 2003;278(51):51646–51653.
39. Jacobsen JA, Major Jourden JL, Miller MT, Cohen SM. To bind zinc or not to bind zinc: an examination of innovative approaches to improved metalloproteinase inhibition. *Biochim Biophys Acta.* 2010;1803(1):72–94.
40. Pelmenschikov V, Siegbahn PE. Catalytic mechanism of matrix metalloproteinases: two-layered ONIOM study. *Inorg Chem.* 2002;41(22):5659–5666.
41. Manzetti S, McCulloch DR, Herington AC, van der Spoel D. Modeling of enzyme-substrate complexes for the metalloproteases MMP-3, ADAM-9 and ADAM-10. *J Comput Aided Mol Des.* 2003;17(9):551–565.
42. Chung L, Dinakarpandian D, Yoshida N, et al. Collagenase unwinds triple-helical collagen prior to peptide bond hydrolysis. *EMBO J.* 2004;23(15):3020–3030.
43. Aimes RT, Quigley JP. Matrix metalloproteinase-2 is an interstitial collagenase. Inhibitor-free enzyme catalyzes the cleavage of collagen fibrils and soluble native type I collagen generating the specific 3/4- and 1/4-length fragments. *J Biol Chem.* 1995;270(11):5872–5876.
44. Olson MW, Toth M, Gervasi DC, Sado Y, Ninomiya Y, Fridman R. High affinity binding of latent matrix metalloproteinase-9 to the alpha2(IV) chain of collagen IV. *J Biol Chem.* 1998;273(17):10672–10681.
45. Chang C, Werb Z. The many faces of metalloproteases: cell growth, invasion, angiogenesis and metastasis. *Trends Cell Biol.* 2001;11(11):S37–S43.
46. Garcia-Irigoyen O, Latasa MU, Carotti S, et al. Matrix metalloproteinase 10 contributes to hepatocarcinogenesis in a novel crosstalk with the stromal derived factor 1/C-X-C chemokine receptor 4 axis. *Hepatology.* 2015;62(1):166–178.
47. Mauris J, Woodward AM, Cao Z, Panjwani N, Argueso P. Molecular basis for MMP9 induction and disruption of epithelial cell-cell contacts by galectin-3. *J Cell Sci.* 2014;127(pt 14):3141–3148.
48. Zhang H, Chalothorn D, Jackson LF, Lee DC, Faber JE. Transactivation of epidermal growth factor receptor mediates catecholamine-induced growth of vascular smooth muscle. *Circ Res.* 2004;95(10):989–997.
49. Morla AO, Mogford JE. Control of smooth muscle cell proliferation and phenotype by integrin signaling through focal adhesion kinase. *Biochem Biophys Res Commun.* 2000;272(1):298–302.
50. Shi ZD, Ji XY, Berardi DE, Qazi H, Tarbell JM. Interstitial flow induces MMP-1 expression and vascular SMC migration in collagen I gels via an ERK1/2-dependent and c-Jun-mediated mechanism. *Am J Physiol Heart Circ Physiol.* 2010;298(1): H127–H135.

51. Jin UH, Suh SJ, Chang HW, et al. Tanshinone IIA from Salvia miltiorrhiza BUNGE inhibits human aortic smooth muscle cell migration and MMP-9 activity through AKT signaling pathway. *J Cell Biochem*. 2008;104(1):15–26.
52. Chang JJ, Stanfill A, Pourmotabbed T. The role of matrix metalloproteinase polymorphisms in ischemic stroke. *Int J Mol Sci*. 2016;17(8). pii: E1323. http://dx.doi.org/10.3390/ijms17081323. Review.
53. Hirakawa S, Kojima T, Obata K, et al. Marked induction of matrix metalloproteinase-10 by respiratory syncytial virus infection in human nasal epithelial cells. *J Med Virol*. 2013;85(12):2141–2150.
54. Zhang G, Miyake M, Lawton A, Goodison S, Rosser CJ. Matrix metalloproteinase-10 promotes tumor progression through regulation of angiogenic and apoptotic pathways in cervical tumors. *BMC Cancer*. 2014;14:310.
55. Lim CS, Qiao X, Reslan OM, et al. Prolonged mechanical stretch is associated with upregulation of hypoxia-inducible factors and reduced contraction in rat inferior vena cava. *J Vasc Surg*. 2011;53(3):764–773.
56. Raffetto JD, Barros YV, Wells AK, Khalil RA. MMP-2 induced vein relaxation via inhibition of [Ca2+]e-dependent mechanisms of venous smooth muscle contraction. Role of RGD peptides. *J Surg Res*. 2010;159(2):755–764.
57. Chew DK, Conte MS, Khalil RA. Matrix metalloproteinase-specific inhibition of Ca2+ entry mechanisms of vascular contraction. *J Vasc Surg*. 2004;40(5):1001–1010.
58. Waitkus-Edwards KR, Martinez-Lemus LA, Wu X, et al. alpha(4)beta(1) Integrin activation of L-type calcium channels in vascular smooth muscle causes arteriole vasoconstriction. *Circ Res*. 2002;90(4):473–480.
59. Raffetto JD, Ross RL, Khalil RA. Matrix metalloproteinase 2-induced venous dilation via hyperpolarization and activation of K+ channels: relevance to varicose vein formation. *J Vasc Surg*. 2007;45(2):373–380.
60. Deng DX, Spin JM, Tsalenko A, et al. Molecular signatures determining coronary artery and saphenous vein smooth muscle cell phenotypes: distinct responses to stimuli. *Arterioscler Thromb Vasc Biol*. 2006;26(5):1058–1065.
61. Turner NA, Ho S, Warburton P, O'Regan DJ, Porter KE. Smooth muscle cells cultured from human saphenous vein exhibit increased proliferation, invasion, and mitogen-activated protein kinase activation in vitro compared with paired internal mammary artery cells. *J Vasc Surg*. 2007;45(5):1022–1028.
62. Feletou M, Vanhoutte PM. Endothelium-derived hyperpolarizing factor: where are we now? *Arterioscler Thromb Vasc Biol*. 2006;26(6):1215–1225.
63. McNamara CA, Sarembock IJ, Gimple LW, Fenton 2nd JW, Coughlin SR, Owens GK. Thrombin stimulates proliferation of cultured rat aortic smooth muscle cells by a proteolytically activated receptor. *J Clin Invest*. 1993;91(1):94–98.
64. Coughlin SR. Thrombin signalling and protease-activated receptors. *Nature*. 2000;407(6801):258–264.
65. Garcia JG, Patterson C, Bahler C, Aschner J, Hart CM, English D. Thrombin receptor activating peptides induce Ca2+ mobilization, barrier dysfunction, prostaglandin synthesis, and platelet-derived growth factor mRNA expression in cultured endothelium. *J Cell Physiol*. 1993;156(3):541–549.
66. Boire A, Covic L, Agarwal A, Jacques S, Sherifi S, Kuliopulos A. PAR1 is a matrix metalloprotease-1 receptor that promotes invasion and tumorigenesis of breast cancer cells. *Cell*. 2005;120(3):303–313.
67. Lee HY, You HJ, Won JY, et al. Forkhead factor, FOXO3a, induces apoptosis of endothelial cells through activation of matrix metalloproteinases. *Arterioscler Thromb Vasc Biol*. 2008;28(2):302–308.
68. Nagase H, Fushimi K. Elucidating the function of non catalytic domains of collagenases and aggrecanases. *Connect Tissue Res*. 2008;49(3):169–174.

69. Ayuk SM, Abrahamse H, Houreld NN. The role of matrix metalloproteinases in diabetic wound healing in relation to photobiomodulation. *J Diabetes Res*. 2016; 2016:2897656.
70. Nam SI, Yu GI, Kim HJ, et al. A polymorphism at -1607 2G in the matrix metalloproteinase-1 (MMP-1) increased risk of sudden deafness in Korean population but not at -519A/G in MMP-1. *Laryngoscope*. 2011;121(1):171–175.
71. Chen X, Thibeault SL. Role of tumor necrosis factor-alpha in wound repair in human vocal fold fibroblasts. *Laryngoscope*. 2010;120(9):1819–1825.
72. Rahmani-Neishaboor E, Hartwell R, Jalili R, Jackson J, Brown E, Ghahary A. Localized controlled release of stratifin reduces implantation-induced dermal fibrosis. *Acta Biomater*. 2012;8(10):3660–3668.
73. Poormasjedi-Meibod MS, Hartwell R, Kilani RT, Ghahary A. Anti-scarring properties of different tryptophan derivatives. *PLoS One*. 2014;9(3): e91955.
74. Li Y, Kilani RT, Rahmani-Neishaboor E, Jalili RB, Ghahary A. Kynurenine increases matrix metalloproteinase-1 and -3 expression in cultured dermal fibroblasts and improves scarring in vivo. *J Invest Dermatol*. 2014;134(3):643–650.
75. Wada Y, Shimada K, Sugimoto K, Kimura T, Ushiyama S. Novel p38 mitogen-activated protein kinase inhibitor R-130823 protects cartilage by down-regulating matrix metalloproteinase-1,-13 and prostaglandin E2 production in human chondrocytes. *Int Immunopharmacol*. 2006;6(2):144–155.
76. Shimizu Y, Kondo S, Shirai A, Furukawa M, Yoshizaki T. A single nucleotide polymorphism in the matrix metalloproteinase-1 and interleukin-8 gene promoter predicts poor prognosis in tongue cancer. *Auris Nasus Larynx*. 2008;35(3):381–389.
77. Murray GI, Duncan ME, O'Neil P, McKay JA, Melvin WT, Fothergill JE. Matrix metalloproteinase-1 is associated with poor prognosis in oesophageal cancer. *J Pathol*. 1998;185(3):256–261.
78. Mizrachi A, Koren R, Hadar T, Yaniv E, Morgenstern S, Shvero J. Expression of MMP-1 in invasive well-differentiated thyroid carcinoma. *Eur Arch Otorhinolaryngol*. 2011;268(1):131–135.
79. Hasty KA, Pourmotabbed TF, Goldberg GI, et al. Human neutrophil collagenase. A distinct gene product with homology to other matrix metalloproteinases. *J Biol Chem*. 1990;265(20):11421–11424.
80. Knauper V, Murphy G, Tschesche H. Activation of human neutrophil procollagenase by stromelysin 2. *Eur J Biochem*. 1996;235(1–2):187–191.
81. Nwomeh BC, Liang HX, Diegelmann RF, Cohen IK, Yager DR. Dynamics of the matrix metalloproteinases MMP-1 and MMP-8 in acute open human dermal wounds. *Wound Repair Regen*. 1998;6(2):127–134.
82. Gutierrez-Fernandez A, Inada M, Balbin M, et al. Increased inflammation delays wound healing in mice deficient in collagenase-2 (MMP-8). *FASEB J*. 2007;21(10):2580–2591.
83. Lee W, Aitken S, Sodek J, McCulloch CA. Evidence of a direct relationship between neutrophil collagenase activity and periodontal tissue destruction in vivo: role of active enzyme in human periodontitis. *J Periodontal Res*. 1995;30(1):23–33.
84. Sorsa T, Hernandez M, Leppilahti J, Munjal S, Netuschil L, Mantyla P. Detection of gingival crevicular fluid MMP-8 levels with different laboratory and chair-side methods. *Oral Dis*. 2010;16(1):39–45.
85. Dalvie D, Cosker T, Boyden T, Zhou S, Schroeder C, Potchoiba MJ. Metabolism distribution and excretion of a matrix metalloproteinase-13 inhibitor, 4-[4-(4-fluorophenoxy)-benzenesulfonylamino]tetrahydropyran-4-carboxylic acid hydroxyamide (CP-544439), in rats and dogs: assessment of the metabolic profile of CP-544439 in plasma and urine of humans. *Drug Metab Dispos*. 2008;36(9): 1869–1883.

86. Wang G, Zhang Y, Zhao X, Meng C, Ma L, Kong Y. MicroRNA-411 inhibited matrix metalloproteinase 13 expression in human chondrocytes. *Am J Transl Res.* 2015;7(10):2000–2006.
87. Yu H, Li Y, Ma L, et al. A low ratio of n-6/n-3 polyunsaturated fatty acids suppresses matrix metalloproteinase 13 expression and reduces adjuvant-induced arthritis in rats. *Nutr Res.* 2015;35(12):1113–1121.
88. Mashimo Y, Sakurai-Yageta M, Watanabe M, et al. Induction of the matrix metalloproteinase 13 gene in bronchial epithelial cells by interferon and identification of its novel functional polymorphism. *Inflammation.* 2016;39(3):949–962.
89. Chen JH, Tsai CH, Lin HY, et al. Interleukin-18 is a pivot regulatory factor on matrix metalloproteinase-13 expression and brain astrocytic migration. *Mol Neurobiol.* 2016;53(9):6218–6227.
90. Abe H, Kamimura K, Kobayashi Y, et al. Effective prevention of liver fibrosis by liver-targeted hydrodynamic gene delivery of matrix metalloproteinase-13 in a rat liver fibrosis model. *Mol Ther Nucleic Acids.* 2016;5: e276.
91. You Y, Shan Y, Chen J, et al. Matrix metalloproteinase 13-containing exosomes promote nasopharyngeal carcinoma metastasis. *Cancer Sci.* 2015;106(12):1669–1677.
92. Cossins J, Dudgeon TJ, Catlin G, Gearing AJ, Clements JM. Identification of MMP-18, a putative novel human matrix metalloproteinase. *Biochem Biophys Res Commun.* 1996;228(2):494–498.
93. Tomlinson ML, Garcia-Morales C, Abu-Elmagd M, Wheeler GN. Three matrix metalloproteinases are required in vivo for macrophage migration during embryonic development. *Mech Dev.* 2008;125(11–12):1059–1070.
94. Tonge D, Zhu N, Lynham S, et al. Axonal growth towards Xenopus skin in vitro is mediated by matrix metalloproteinase activity. *Eur J Neurosci.* 2013;37(4):519–531.
95. Steffensen B, Wallon UM, Overall CM. Extracellular matrix binding properties of recombinant fibronectin type II-like modules of human 72-kDa gelatinase/type IV collagenase. High affinity binding to native type I collagen but not native type IV collagen. *J Biol Chem.* 1995;270(19):11555–11566.
96. Shipley JM, Doyle GA, Fliszar CJ, et al. The structural basis for the elastolytic activity of the 92-kDa and 72-kDa gelatinases. Role of the fibronectin type II-like repeats. *J Biol Chem.* 1996;271(8):4335–4341.
97. Koyama H, Iwata H, Kuwabara Y, Iwase H, Kobayashi S, Fujii Y. Gelatinolytic activity of matrix metalloproteinase-2 and -9 in oesophageal carcinoma; a study using in situ zymography. *Eur J Cancer.* 2000;36(16):2164–2170.
98. Bjorklund M, Koivunen E. Gelatinase-mediated migration and invasion of cancer cells. *Biochim Biophys Acta.* 2005;1755(1):37–69.
99. Overall CM, Tam E, McQuibban GA, et al. Domain interactions in the gelatinase A. TIMP-2.MT1-MMP activation complex. The ectodomain of the 44-kDa form of membrane type-1 matrix metalloproteinase does not modulate gelatinase A activation. *J Biol Chem.* 2000;275(50):39497–39506.
100. Augoff K, Grabowski K, Rabczynski J, Kolondra A, Tabola R, Sikorski AF. Expression of decorin in esophageal cancer in relation to the expression of three isoforms of transforming growth factor-beta (TGF-beta1, -beta2, and -beta3) and matrix metalloproteinase-2 activity. *Cancer Invest.* 2009;27(4):443–452.
101. Talvensaari-Mattila A, Paakko P, Turpeenniemi-Hujanen T. Matrix metalloproteinase-2 (MMP-2) is associated with survival in breast carcinoma. *Br J Cancer.* 2003;89(7): 1270–1275.
102. Chetty C, Bhoopathi P, Joseph P, Chittivelu S, Rao JS, Lakka S. Adenovirus-mediated small interfering RNA against matrix metalloproteinase-2 suppresses tumor growth and lung metastasis in mice. *Mol Cancer Ther.* 2006;5(9):2289–2299.

103. Morozevich G, Kozlova N, Cheglakov I, Ushakova N, Berman A. Integrin alpha5beta1 controls invasion of human breast carcinoma cells by direct and indirect modulation of MMP-2 collagenase activity. *Cell Cycle*. 2009;8(14):2219–2225.
104. Gu G, Xia H, Hu Q, et al. PEG-co-PCL nanoparticles modified with MMP-2/9 activatable low molecular weight protamine for enhanced targeted glioblastoma therapy. *Biomaterials*. 2013;34(1):196–208.
105. Kesanakurti D, Chetty C, Dinh DH, Gujrati M, Rao JS. Role of MMP-2 in the regulation of IL-6/Stat3 survival signaling via interaction with alpha5beta1 integrin in glioma. *Oncogene*. 2013;32(3):327–340.
106. Nam SI, Kwon TK. Dexamethasone inhibits interleukin-1beta-induced matrix metalloproteinase-9 expression in cochlear cells. *Clin Exp Otorhinolaryngol*. 2014;7(3): 175–180.
107. Olszewska E, Matulka M, Mroczko B, et al. Diagnostic value of matrix metalloproteinase 9 and tissue inhibitor of matrix metalloproteinases 1 in cholesteatoma. *Histol Histopathol*. 2016;31(3):307–315.
108. Detwiller KY, Smith TL, Mace JC, Trune DR, Sautter NB. Steroid-independent upregulation of matrix metalloproteinase 9 in chronic rhinosinusitis patients with radiographic evidence of osteitis. *Int Forum Allergy Rhinol*. 2013;3(5):364–368.
109. Eguchi T, Kubota S, Kawata K, et al. Novel transcription-factor-like function of human matrix metalloproteinase 3 regulating the CTGF/CCN2 gene. *Mol Cell Biol*. 2008;28(7):2391–2413.
110. Si-Tayeb K, Monvoisin A, Mazzocco C, et al. Matrix metalloproteinase 3 is present in the cell nucleus and is involved in apoptosis. *Am J Pathol*. 2006;169(4): 1390–1401.
111. Ding L, Guo D, Homandberg GA, Buckwalter JA, Martin JA. A single blunt impact on cartilage promotes fibronectin fragmentation and upregulates cartilage degrading stromelysin-1/matrix metalloproteinase-3 in a bovine ex vivo model. *J Orthop Res*. 2014;32(6):811–818.
112. Ye S, Watts GF, Mandalia S, Humphries SE, Henney AM. Preliminary report: genetic variation in the human stromelysin promoter is associated with progression of coronary atherosclerosis. *Br Heart J*. 1995;73(3):209–215.
113. Jiang YN, Yan HQ, Huang XB, Wang YN, Li Q, Gao FG. Interleukin 6 trigged ataxia-telangiectasia mutated activation facilitates lung cancer metastasis via MMP-3/ MMP-13 up-regulation. *Oncotarget*. 2015;6(38):40719–40733.
114. Kolomecki K, Stepien H, Bartos M, Kuzdak K. Usefulness of VEGF, MMP-2, MMP-3 and TIMP-2 serum level evaluation in patients with adrenal tumours. *Endocr Regul*. 2001;35(1):9–16.
115. Choi DH, Kim JH, Seo JH, Lee J, Choi WS, Kim YS. Matrix metalloproteinase-3 causes dopaminergic neuronal death through Nox1-regenerated oxidative stress. *PLoS One*. 2014;9(12): e115954.
116. Sokai A, Handa T, Tanizawa K, et al. Matrix metalloproteinase-10: a novel biomarker for idiopathic pulmonary fibrosis. *Respir Res*. 2015;16:120.
117. Martinez-Aguilar E, Gomez-Rodriguez V, Orbe J, et al. Matrix metalloproteinase 10 is associated with disease severity and mortality in patients with peripheral arterial disease. *J Vasc Surg*. 2015;61(2):428–435.
118. Rohani MG, McMahan RS, Razumova MV, et al. MMP-10 regulates collagenolytic activity of alternatively activated resident macrophages. *J Invest Dermatol*. 2015;135(10):2377–2384.
119. Wang H, Zhang ZQ, Wang SZ, Lu JL, Wang XL, Zhang ZY. Association of matrix metalloproteinase-10 polymorphisms with susceptibility to pelvic organ prolapse. *J Obstet Gynaecol Res*. 2015;41(12):1972–1981.

120. Kadeh H, Saravani S, Heydari F, Keikha M, Rigi V. Expression of matrix metalloproteinase-10 at invasive front of squamous cell carcinoma and verrucous carcinoma in the oral cavity. *Asian Pac J Cancer Prev*. 2015;16(15):6609–6613.
121. Basset P, Bellocq JP, Wolf C, et al. A novel metalloproteinase gene specifically expressed in stromal cells of breast carcinomas. *Nature*. 1990;348(6303):699–704.
122. Fiorentino M, Fu L, Shi YB. Mutational analysis of the cleavage of the cancer-associated laminin receptor by stromelysin-3 reveals the contribution of flanking sequences to site recognition and cleavage efficiency. *Int J Mol Med*. 2009;23(3):389–397.
123. Lin CW, Yang SF, Chuang CY, Lin HP, Hsin CH. Association of matrix metalloproteinase-11 polymorphisms with susceptibility and clinicopathologic characteristics for oral squamous cell carcinoma. *Head Neck*. 2015;37(10):1425–1431.
124. Sharma R, Chattopadhyay TK, Mathur M, Ralhan R. Prognostic significance of stromelysin-3 and tissue inhibitor of matrix metalloproteinase-2 in esophageal cancer. *Oncology*. 2004;67(3–4):300–309.
125. Zhang X, Huang S, Guo J, et al. Insights into the distinct roles of MMP-11 in tumor biology and future therapeutics (Review). *Int J Oncol*. 2016;48(5):1783–1793.
126. Li Z, Ding S, Zhong Q, Li G, Zhang Y, Huang XC. Significance of MMP11 and P14 (ARF) expressions in clinical outcomes of patients with laryngeal cancer. *Int J Clin Exp Med*. 2015;8(9):15581–15590.
127. Parks WC, Wilson CL, Lopez-Boado YS. Matrix metalloproteinases as modulators of inflammation and innate immunity. *Nat Rev Immunol*. 2004;4(8):617–629.
128. Acioglu E, Yigit O, Alkan Z, Server EA, Uzun H, Gelisgen R. The role of matrix metalloproteinases in recurrent tonsillitis. *Int J Pediatr Otorhinolaryngol*. 2010;74(5): 535–539.
129. Jiang T, Xie P, Liu H. Circulating anti-matrix metalloproteinase-7 antibodies may be a potential biomarker for oral squamous cell carcinoma. *J Oral Maxillofac Surg*. 2015;74(3):650–657.
130. Uria JA, Lopez-Otin C. Matrilysin-2, a new matrix metalloproteinase expressed in human tumors and showing the minimal domain organization required for secretion, latency, and activity. *Cancer Res*. 2000;60(17):4745–4751.
131. de Coignac AB, Elson G, Delneste Y, et al. Cloning of MMP-26. A novel matrilysin-like proteinase. *Eur J Biochem*. 2000;267(11):3323–3329.
132. Marchenko GN, Ratnikov BI, Rozanov DV, Godzik A, Deryugina EI, Strongin AY. Characterization of matrix metalloproteinase-26, a novel metalloproteinase widely expressed in cancer cells of epithelial origin. *Biochem J*. 2001;356(pt 3):705–718.
133. Marchenko ND, Marchenko GN, Weinreb RN, et al. Beta-catenin regulates the gene of MMP-26, a novel metalloproteinase expressed both in carcinomas and normal epithelial cells. *Int J Biochem Cell Biol*. 2004;36(5):942–956.
134. Gutschalk CM, Yanamandra AK, Linde N, Meides A, Depner S, Mueller MM. GM-CSF enhances tumor invasion by elevated MMP-2, -9, and -26 expression. *Cancer Med*. 2013;2(2):117–129.
135. Bister V, Skoog T, Virolainen S, Kiviluoto T, Puolakkainen P, Saarialho-Kere U. Increased expression of matrix metalloproteinases-21 and -26 and TIMP-4 in pancreatic adenocarcinoma. *Mod Pathol*. 2007;20(11):1128–1140.
136. Zhao YG, Xiao AZ, Park HI, et al. Endometase/matrilysin-2 in human breast ductal carcinoma in situ and its inhibition by tissue inhibitors of metalloproteinases-2 and -4: a putative role in the initiation of breast cancer invasion. *Cancer Res*. 2004;64(2):590–598.
137. Sohail A, Sun Q, Zhao H, Bernardo MM, Cho JA, Fridman R. MT4-(MMP17) and MT6-MMP (MMP25), a unique set of membrane-anchored matrix metalloproteinases: properties and expression in cancer. *Cancer Metastasis Rev*. 2008; 27(2):289–302.

138. Knauper V, Will H, Lopez-Otin C, et al. Cellular mechanisms for human procollagenase-3 (MMP-13) activation. Evidence that MT1-MMP (MMP-14) and gelatinase a (MMP-2) are able to generate active enzyme. *J Biol Chem.* 1996;271(29): 17124–17131.
139. Itoh Y, Kajita M, Kinoh H, Mori H, Okada A, Seiki M. Membrane type 4 matrix metalloproteinase (MT4-MMP, MMP-17) is a glycosylphosphatidylinositol-anchored proteinase. *J Biol Chem.* 1999;274(48):34260–34266.
140. Radichev IA, Remacle AG, Shiryaev SA, et al. Biochemical characterization of the cellular glycosylphosphatidylinositol-linked membrane type-6 matrix metalloproteinase. *J Biol Chem.* 2010;285(21):16076–16086.
141. Kayano K, Shimada T, Shinomiya T, et al. Activation of pro-MMP-2 mediated by MT1-MMP in human salivary gland carcinomas: possible regulation of pro-MMP-2 activation by TIMP-2. *J Pathol.* 2004;202(4):403–411.
142. Imanishi Y, Fujii M, Tokumaru Y, et al. Clinical significance of expression of membrane type 1 matrix metalloproteinase and matrix metalloproteinase-2 in human head and neck squamous cell carcinoma. *Hum Pathol.* 2000;31(8):895–904.
143. Johnson JL, Jenkins NP, Huang WC, et al. Relationship of MMP-14 and TIMP-3 expression with macrophage activation and human atherosclerotic plaque vulnerability. *Mediators Inflamm.* 2014;2014:276457.
144. Wiesner C, El Azzouzi K, Linder S. A specific subset of RabGTPases controls cell surface exposure of MT1-MMP, extracellular matrix degradation and three-dimensional invasion of macrophages. *J Cell Sci.* 2013;126(pt 13):2820–2833.
145. Sato H, Seiki M. Membrane-type matrix metalloproteinases (MT-MMPs) in tumor metastasis. *J Biochem.* 1996;119(2):209–215.
146. Szabova L, Son MY, Shi J, et al. Membrane-type MMPs are indispensable for placental labyrinth formation and development. *Blood.* 2010;116(25):5752–5761.
147. Sena P, Mariani F, Marzona L, et al. Matrix metalloproteinases 15 and 19 are stromal regulators of colorectal cancer development from the early stages. *Int J Oncol.* 2012;41(1):260–266.
148. Zhang H, Liu M, Sun Y, Lu J. MMP-14 can serve as a prognostic marker in patients with supraglottic cancer. *Eur Arch Otorhinolaryngol.* 2009;266(9):1427–1434.
149. Sun YN, Li Y. Expression of mRNA for membrane-type 1, 2, and 3 matrix metalloproteinases in human laryngeal cancer. *Chin Med Sci J.* 2004;19(3):170–173.
150. Liu J, van Mil A, Aguor EN, et al. MiR-155 inhibits cell migration of human cardiomyocyte progenitor cells (hCMPCs) via targeting of MMP-16. *J Cell Mol Med.* 2012;16(10):2379–2386.
151. Tatti O, Gucciardo E, Pekkonen P, et al. MMP16 mediates a proteolytic switch to promote cell-cell adhesion, collagen alignment, and lymphatic invasion in melanoma. *Cancer Res.* 2015;75(10):2083–2094.
152. Chen MF, Zeng F, Qi L, et al. Transforming growth factorbeta1 induces epithelial mesenchymal transition and increased expression of matrix metalloproteinase16 via miR200b downregulation in bladder cancer cells. *Mol Med Rep.* 2014;10(3):1549–1554.
153. Liu C, Wu F, Liu Y, Meng C. Catalpol suppresses proliferation and facilitates apoptosis of MCF-7 breast cancer cells through upregulating microRNA-146a and downregulating matrix metalloproteinase-16 expression. *Mol Med Rep.* 2015;12(5): 7609–7614.
154. Hadchouel A, Decobert F, Franco-Montoya ML, et al. Matrix metalloproteinase gene polymorphisms and bronchopulmonary dysplasia: identification of MMP16 as a new player in lung development. *PLoS One.* 2008;3(9): e3188.
155. Puente XS, Pendas AM, Llano E, Velasco G, Lopez-Otin C. Molecular cloning of a novel membrane-type matrix metalloproteinase from a human breast carcinoma. *Cancer Res.* 1996;56(5):944–949.

156. Llano E, Pendas AM, Freije JP, et al. Identification and characterization of human MT5-MMP, a new membrane-bound activator of progelatinase a overexpressed in brain tumors. *Cancer Res.* 1999;59(11):2570–2576.
157. Pei D. Identification and characterization of the fifth membrane-type matrix metalloproteinase MT5-MMP. *J Biol Chem.* 1999;274(13):8925–8932.
158. Porlan E, Marti-Prado B, Morante-Redolat JM, et al. MT5-MMP regulates adult neural stem cell functional quiescence through the cleavage of N-cadherin. *Nat Cell Biol.* 2014;16(7):629–638.
159. Warren KM, Reeves TM, Phillips LL. MT5-MMP, ADAM-10, and N-cadherin act in concert to facilitate synapse reorganization after traumatic brain injury. *J Neurotrauma.* 2012;29(10):1922–1940.
160. Komori K, Nonaka T, Okada A, et al. Absence of mechanical allodynia and Abeta-fiber sprouting after sciatic nerve injury in mice lacking membrane-type 5 matrix metalloproteinase. *FEBS Lett.* 2004;557(1–3):125–128.
161. Folgueras AR, Valdes-Sanchez T, Llano E, et al. Metalloproteinase MT5-MMP is an essential modulator of neuro-immune interactions in thermal pain stimulation. *Proc Natl Acad Sci USA.* 2009;106(38):16451–16456.
162. Benson CS, Babu SD, Radhakrishna S, Selvamurugan N, Ravi SB. Expression of matrix metalloproteinases in human breast cancer tissues. *Dis Markers.* 2013;34(6): 395–405.
163. Zhao H, Sohail A, Sun Q, et al. Identification and role of the homodimerization interface of the glycosylphosphatidylinositol-anchored membrane type 6 matrix metalloproteinase (MMP25). *J Biol Chem.* 2008;283(50):35023–35032.
164. Fortin CF, Sohail A, Sun Q, McDonald PP, Fridman R, Fulop T. MT6-MMP is present in lipid rafts and faces inward in living human PMNs but translocates to the cell surface during neutrophil apoptosis. *Int Immunol.* 2010;22(8):637–649.
165. Velasco G, Cal S, Merlos-Suarez A, et al. Human MT6-matrix metalloproteinase: identification, progelatinase A activation, and expression in brain tumors. *Cancer Res.* 2000;60(4):877–882.
166. Sun Q, Weber CR, Sohail A, et al. MMP25 (MT6-MMP) is highly expressed in human colon cancer, promotes tumor growth, and exhibits unique biochemical properties. *J Biol Chem.* 2007;282(30):21998–22010.
167. Shipley JM, Wesselschmidt RL, Kobayashi DK, Ley TJ, Shapiro SD. Metalloelastase is required for macrophage-mediated proteolysis and matrix invasion in mice. *Proc Natl Acad Sci USA.* 1996;93(9):3942–3946.
168. Kerkela E, Bohling T, Herva R, Uria JA, Saarialho-Kere U. Human macrophage metalloelastase (MMP-12) expression is induced in chondrocytes during fetal development and malignant transformation. *Bone.* 2001;29(5):487–493.
169. Hou P, Troen T, Ovejero MC, et al. Matrix metalloproteinase-12 (MMP-12) in osteoclasts: new lesson on the involvement of MMPs in bone resorption. *Bone.* 2004;34(1):37–47.
170. Marchant DJ, Bellac CL, Moraes TJ, et al. A new transcriptional role for matrix metalloproteinase-12 in antiviral immunity. *Nat Med.* 2014;20(5):493–502.
171. Li W, Li J, Wu Y, et al. Identification of an orally efficacious matrix metalloprotease 12 inhibitor for potential treatment of asthma. *J Med Chem.* 2009;52(17):5408–5419.
172. Chelluboina B, Klopfenstein JD, Pinson DM, Wang DZ, Vemuganti R, Veeravalli KK. Matrix metalloproteinase-12 induces blood-brain barrier damage after focal cerebral ischemia. *Stroke.* 2015;46(12):3523–3531.
173. Kim JM, Kim HJ, Koo BS, Rha KS, Yoon YH. Expression of matrix metalloproteinase-12 is correlated with extracapsular spread of tumor from nodes with metastasis in head and neck squamous cell carcinoma. *Eur Arch Otorhinolaryngol.* 2013;270(3):1137–1142.

174. Stracke JO, Hutton M, Stewart M, et al. Biochemical characterization of the catalytic domain of human matrix metalloproteinase 19. Evidence for a role as a potent basement membrane degrading enzyme. *J Biol Chem*. 2000;275(20):14809–14816.
175. Stracke JO, Fosang AJ, Last K, et al. Matrix metalloproteinases 19 and 20 cleave aggrecan and cartilage oligomeric matrix protein (COMP). *FEBS Lett*. 2000;478(1–2): 52–56.
176. Sadowski T, Dietrich S, Muller M, et al. Matrix metalloproteinase-19 expression in normal and diseased skin: dysregulation by epidermal proliferation. *J Invest Dermatol*. 2003;121(5):989–996.
177. Sadowski T, Dietrich S, Koschinsky F, Sedlacek R. Matrix metalloproteinase 19 regulates insulin-like growth factor-mediated proliferation, migration, and adhesion in human keratinocytes through proteolysis of insulin-like growth factor binding protein-3. *Mol Biol Cell*. 2003;14(11):4569–4580.
178. Sadowski T, Dietrich S, Koschinsky F, et al. Matrix metalloproteinase 19 processes the laminin 5 gamma 2 chain and induces epithelial cell migration. *Cell Mol Life Sci*. 2005;62(7–8):870–880.
179. Brauer R, Beck IM, Roderfeld M, Roeb E, Sedlacek R. Matrix metalloproteinase-19 inhibits growth of endothelial cells by generating angiostatin-like fragments from plasminogen. *BMC Biochem*. 2011;12:38.
180. Jara P, Calyeca J, Romero Y, et al. Matrix metalloproteinase (MMP)-19-deficient fibroblasts display a profibrotic phenotype. *Am J Physiol Lung Cell Mol Physiol*. 2015;308(6):L511–L522.
181. Yu G, Kovkarova-Naumovski E, Jara P, et al. Matrix metalloproteinase-19 is a key regulator of lung fibrosis in mice and humans. *Am J Respir Crit Care Med*. 2012;186(8):752–762.
182. Jirouskova M, Zbodakova O, Gregor M, et al. Hepatoprotective effect of MMP-19 deficiency in a mouse model of chronic liver fibrosis. *PLoS One*. 2012;7(10): e46271.
183. Hazlewood RJ, Roos BR, Solivan-Timpe F, et al. Heterozygous triplication of upstream regulatory sequences leads to dysregulation of matrix metalloproteinase 19 in patients with cavitary optic disc anomaly. *Hum Mutat*. 2015;36(3):369–378.
184. Pendas AM, Folgueras AR, Llano E, et al. Diet-induced obesity and reduced skin cancer susceptibility in matrix metalloproteinase 19-deficient mice. *Mol Cell Biol*. 2004;24(12):5304–5313.
185. Lee DG, Lee SH, Kim JS, et al. Loss of NDRG2 promotes epithelial-mesenchymal transition of gallbladder carcinoma cells through MMP-19-mediated Slug expression. *J Hepatol*. 2015;63(6):1429–1439.
186. Yu G, Herazo-Maya JD, Nukui T, et al. Matrix metalloproteinase-19 promotes metastatic behavior in vitro and is associated with increased mortality in non-small cell lung cancer. *Am J Respir Crit Care Med*. 2014;190(7):780–790.
187. Chan KC, Ko JM, Lung HL, et al. Catalytic activity of matrix metalloproteinase-19 is essential for tumor suppressor and anti-angiogenic activities in nasopharyngeal carcinoma. *Int J Cancer*. 2011;129(8):1826–1837.
188. Ryu OH, Fincham AG, Hu CC, et al. Characterization of recombinant pig enamelysin activity and cleavage of recombinant pig and mouse amelogenins. *J Dent Res*. 1999;78(3):743–750.
189. Llano E, Pendas AM, Knauper V, et al. Identification and structural and functional characterization of human enamelysin (MMP-20). *Biochemistry*. 1997;36(49): 15101–15108.
190. Barron LA, Giardina JB, Granger JP, Khalil RA. High-salt diet enhances vascular reactivity in pregnant rats with normal and reduced uterine perfusion pressure. *Hypertension*. 2001;38(3 pt 2):730–735.

191. Caterina JJ, Skobe Z, Shi J, et al. Enamelysin (matrix metalloproteinase 20)-deficient mice display an amelogenesis imperfecta phenotype. *J Biol Chem.* 2002;277(51): 49598–49604.
192. Skoog T, Ahokas K, Orsmark C, Jeskanen L, Isaka K, Saarialho-Kere U. MMP-21 is expressed by macrophages and fibroblasts in vivo and in culture. *Exp Dermatol.* 2006;15(10):775–783.
193. Ahokas K, Lohi J, Illman SA, et al. Matrix metalloproteinase-21 is expressed epithelially during development and in cancer and is up-regulated by transforming growth factor-beta1 in keratinocytes. *Lab Invest.* 2003;83(12):1887–1899.
194. Pu Y, Wang L, Wu H, Feng Z, Wang Y, Guo C. High MMP-21 expression in metastatic lymph nodes predicts unfavorable overall survival for oral squamous cell carcinoma patients with lymphatic metastasis. *Oncol Rep.* 2014;31(6):2644–2650.
195. Zhao Z, Yan L, Li S, Sun H, Zhou Y, Li X. Increased MMP-21 expression in esophageal squamous cell carcinoma is associated with progression and prognosis. *Med Oncol.* 2014;31(8):91.
196. Wu T, Li Y, Liu X, et al. Identification of high-risk stage II and stage III colorectal cancer by analysis of MMP-21 expression. *J Surg Oncol.* 2011;104(7):787–791.
197. Suomela S, Koljonen V, Skoog T, Kukko H, Bohling T, Saarialho-Kere U. Expression of MMP-10, MMP-21, MMP-26, and MMP-28 in Merkel cell carcinoma. *Virchows Arch.* 2009;455(6):495–503.
198. Gururajan R, Grenet J, Lahti JM, Kidd VJ. Isolation and characterization of two novel metalloproteinase genes linked to the Cdc2L locus on human chromosome 1p36.3. *Genomics.* 1998;52(1):101–106.
199. Velasco G, Pendas AM, Fueyo A, Knauper V, Murphy G, Lopez-Otin C. Cloning and characterization of human MMP-23, a new matrix metalloproteinase predominantly expressed in reproductive tissues and lacking conserved domains in other family members. *J Biol Chem.* 1999;274(8):4570–4576.
200. Hegedus L, Cho H, Xie X, Eliceiri GL. Additional MDA-MB-231 breast cancer cell matrix metalloproteinases promote invasiveness. *J Cell Physiol.* 2008;216(2):480–485.
201. Cominelli A, Halbout M, N'Kuli F, et al. A unique C-terminal domain allows retention of matrix metalloproteinase-27 in the endoplasmic reticulum. *Traffic.* 2014;15(4):401–417.
202. Bar-Or A, Nuttall RK, Duddy M, et al. Analyses of all matrix metalloproteinase members in leukocytes emphasize monocytes as major inflammatory mediators in multiple sclerosis. *Brain.* 2003;126(pt 12):2738–2749.
203. Cominelli A, Gaide Chevronnay HP, Lemoine P, Courtoy PJ, Marbaix E, Henriet P. Matrix metalloproteinase-27 is expressed in CD163+/CD206+ M2 macrophages in the cycling human endometrium and in superficial endometriotic lesions. *Mol Hum Reprod.* 2014;20(8):767–775.
204. Kohrmann A, Kammerer U, Kapp M, Dietl J, Anacker J. Expression of matrix metalloproteinases (MMPs) in primary human breast cancer and breast cancer cell lines: new findings and review of the literature. *BMC Cancer.* 2009;9:188.
205. Lohi J, Wilson CL, Roby JD, Parks WC. Epilysin, a novel human matrix metalloproteinase (MMP-28) expressed in testis and keratinocytes and in response to injury. *J Biol Chem.* 2001;276(13):10134–10144.
206. Saarialho-Kere U, Kerkela E, Jahkola T, Suomela S, Keski-Oja J, Lohi J. Epilysin (MMP-28) expression is associated with cell proliferation during epithelial repair. *J Invest Dermatol.* 2002;119(1):14–21.
207. Marchenko GN, Strongin AY. MMP-28, a new human matrix metalloproteinase with an unusual cysteine-switch sequence is widely expressed in tumors. *Gene.* 2001;265(1–2):87–93.

208. Padmavati P, Savita S, Shivaprasad BM, Kripal K, Rithesh K. mRNA expression of MMP-28 (Epilysin) in gingival tissues of chronic and aggressive periodontitis patients: a reverse transcriptase PCR study. *Dis Markers*. 2013;35(2):113–118.
209. Li QL, Illman SA, Wang HM, Liu DL, Lohi J, Zhu C. Matrix metalloproteinase-28 transcript and protein are expressed in rhesus monkey placenta during early pregnancy. *Mol Hum Reprod*. 2003;9(4):205–211.
210. Ma Y, Halade GV, Zhang J, et al. Matrix metalloproteinase-28 deletion exacerbates cardiac dysfunction and rupture after myocardial infarction in mice by inhibiting M2 macrophage activation. *Circ Res*. 2013;112(4):675–688.
211. Bister VO, Salmela MT, Karjalainen-Lindsberg ML, et al. Differential expression of three matrix metalloproteinases, MMP-19, MMP-26, and MMP-28, in normal and inflamed intestine and colon cancer. *Dig Dis Sci*. 2004;49(4):653–661.
212. Kevorkian L, Young DA, Darrah C, et al. Expression profiling of metalloproteinases and their inhibitors in cartilage. *Arthritis Rheum*. 2004;50(1):131–141.
213. Momohara S, Okamoto H, Komiya K, et al. Matrix metalloproteinase 28/epilysin expression in cartilage from patients with rheumatoid arthritis and osteoarthritis: comment on the article by Kevorkian et al. *Arthritis Rheum*. 2004;50(12):4074–4075. author reply 4075.
214. Baker AH, Edwards DR, Murphy G. Metalloproteinase inhibitors: biological actions and therapeutic opportunities. *J Cell Sci*. 2002;115(pt 19):3719–3727.
215. Batra J, Robinson J, Soares AS, Fields AP, Radisky DC, Radisky ES. Matrix metalloproteinase-10 (MMP-10) interaction with tissue inhibitors of metalloproteinases TIMP-1 and TIMP-2: binding studies and crystal structure. *J Biol Chem*. 2012;287(19):15935–15946.
216. Meng Q, Malinovskii V, Huang W, et al. Residue 2 of TIMP-1 is a major determinant of affinity and specificity for matrix metalloproteinases but effects of substitutions do not correlate with those of the corresponding P1' residue of substrate. *J Biol Chem*. 1999;274(15):10184–10189.
217. Hu J, Van den Steen PE, Sang QX, Opdenakker G. Matrix metalloproteinase inhibitors as therapy for inflammatory and vascular diseases. *Nat Rev Drug Discov*. 2007;6(6):480–498.

CHAPTER TWO

Matrix Metalloproteinases in Myocardial Infarction and Heart Failure

Kristine Y. DeLeon-Pennell*,†, Cesar A. Meschiari*, Mira Jung*, Merry L. Lindsey*,†,1

*Mississippi Center for Heart Research, UMMC, Jackson, MS, United States
†Research Service, G.V. (Sonny) Montgomery Veterans Affairs Medical Center, Jackson, MS, United States
[1]Corresponding author: e-mail address: mllindsey@umc.edu

Contents

Abstract

Cardiovascular disease is the leading cause of death, accounting for 600,000 deaths each year in the United States. In addition, heart failure accounts for 37% of health care spending. Matrix metalloproteinases (MMPs) increase after myocardial infarction (MI) and correlate with left ventricular dysfunction in heart failure patients. MMPs regulate the remodeling process by facilitating extracellular matrix turnover and inflammatory signaling. Due to the critical role MMPs play during cardiac remodeling, there is a need to

Progress in Molecular Biology and Translational Science, Volume 147
ISSN 1877-1173
http://dx.doi.org/10.1016/bs.pmbts.2017.02.001

better understand the pathophysiological mechanism of MMPs, including the biological function of the downstream products of MMP proteolysis. Future studies developing new therapeutic targets that inhibit specific MMP actions to limit the development of heart failure post-MI are warranted. This chapter focuses on the role of MMPs post-MI, the efficiency of MMPs as biomarkers for MI or heart failure, and the future of MMPs and their cleavage products as targets for prevention of post-MI heart failure.

1. INTRODUCTION

Despite significant advancements in risk prediction, cardiovascular disease (CVD) remains a leading cause of death.[1] Adverse cardiac remodeling that involves excessive extracellular matrix (ECM) turnover contributes to high morbidity and mortality in patients with myocardial infarction (MI).[2] Elevated matrix metalloproteinase (MMP) levels strongly correlate with left ventricular dysfunction in CVD patients.

MMPs are a family of 25 proteolytic enzymes that regulate ECM turnover and inflammatory signaling. Only about half of the known MMPs have been measured in the post-MI left ventricle (LV), which leaves a significant knowledge gap in the post-MI MMP literature.[3] Following MI, MMPs facilitate ECM degradation and recruit inflammatory cells for removal of necrotic cardiomyocytes. The upregulation of proinflammatory cytokines initially results in robust MMP activation; however, long-term stimulation increases tissue inhibitor of metalloproteinase (TIMP) levels. This ultimately leads to a decrease in the MMP/TIMP ratio and results in ongoing long-term remodeling.[4,5]

The fate of the myocardium post-MI depends on the balance between several competing events that occur during the wound-healing response to form the ECM scar (Fig. 1). Development of heart failure post-MI can be induced by exaggerated cardiac remodeling leading to impaired cardiac physiology. In response to myocyte injury induced by ischemia and infarction, a series of events occur in three distinct, but temporally overlapping, phases of wound healing: inflammation, proliferation, and maturation.[6] Each phase contributes to the temporal changes in MMP levels in the post-MI infarct. Multiple cell types in the post-MI myocardium express MMPs, including neutrophils, macrophages, endothelial cells, myocytes, and fibroblasts, making MMPs key regulators in the cardiac remodeling progression.[7–10] This chapter focuses on the role of MMPs during the post-MI development of heart failure and discusses the future of MMP inhibitors (MMPi) to prevent the development of heart failure.

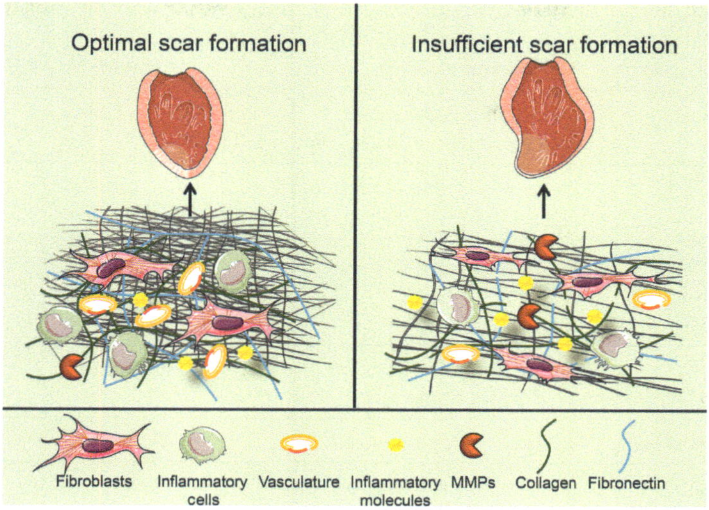

Fig. 1 Myocardial wound healing is dependent on the balance between extracellular matrix (ECM) breakdown and synthesis. Matrix metalloproteinases (MMPs) are critical during this process as key regulators of inflammation, fibrosis, angiogenesis, and collagen degradation. Optimal scar formation (*left*) requires (1) appropriate inflammation; (2) fibroblast differentiation, proliferation, and migration to the wound site; (3) suitable angiogenesis; and (4) proper synthesis, cross-linking, and alignment of collagen at the infarct site. Too much or too little of these events will result in insufficient scar formation (*right*) and can facilitate in the development of heart failure postmyocardial infarction.

2. MMPs AS BIOMARKERS FOR HEART FAILURE

MMPs have been widely studied as possible markers to predict the development of CVD, particularly in post-MI remodeling and heart failure. The use of proteomic techniques over the last 10 years has amplified the discover of candidate biomarkers, due to enhanced sample preparation protocols, improvements in database searching, capabilities of mass spectroscopy, and bioinformatics analytic tools. Combined, these improvements have made identification of biomarkers for heart failure post-MI more attainable. This is especially true when it comes to biomarkers associated with cardiac ECM.[11,12]

Discovering novel substrates and the biological functions of peptide fragments generated by MMPs is vital to fully comprehend MMP function post-MI. MMPs have a broad number of substrates that contribute to scar formation and wound-healing post-MI. Table 1 lists known MMP substrates and their biological function. MMPs also release ECM fragments called matricryptins or matrikines that are key during the

Table 1 Summary of MMP Substrates and Their Postmyocardial Infarction (MI) Functions

MMP Substrates	MI Functional Roles	Cleaved by MMP
Angiostatin	Angiogenesis inhibitor, cardiomyocytes death, ↑ heart failure	-2, -3, and -9
C-1158/59	Increased migration rate of fibroblast cells, ↑ wound healing	-2 and -9
C-terminal telopeptide of collagen I	Exaggerated myocardial fibrosis	-1, -2, -8, and -9
CD36	↓ Macrophage phagocytosis and neutrophil apoptosis	-9 and -12
Citrate synthase	↓ Mitochondrial function	-9
Endostatin	Suppresses proliferation and migration of endothelial cells	-2, -9, and -13
Fibronectin	Act as chemoattractant, ↑ inflammation, migration of monocytes	-2, -7, -9, -12, and -13
Galectin-3	↑ Collagen deposition, ↓ LVEF	-9
Hyaluronan	↑ Inflammation, ↓ neutrophil apoptosis, induce cardiac dysfunction	-9 and -12
Laminin	Inhibit migration of macrophages into the inflammatory region	-2
Osteopontin	↑ Migration rate of cardiac fibroblast, ↑ wound healing	-2, -3, -7, -9, and -12
Periostin	↑ Myocardial fibrosis, ↑ heart failure	-2, -9, and -14
SPARC	Antiangiogenic effect, maturation of ECM	-2, -3, -7, -9, and -13
Tenascin-C	Unknown	-3, -4, -7, and -9

See manuscript text for references.
SPARC, secreted protein acidic and rich in cysteine.

development of heart failure post-MI.[13–17] Multiple studies have suggested that matricryptins could be potential therapeutic targets for heart failure patients.[14,18,19] We summarize our current knowledge on the involvement of MMPs in post-MI remodeling.

2.1 MMP-1

Understanding the role of MMP-1 in the post-MI LV has been hindered due to humans only having one isoform of MMP-1 and mice having two: MMP-1a (59% homology with human MMP-1) and MMP-1b (57% homology with human MMP-1).[20] MMP-1 is mainly expressed by leukocytes, fibroblasts, and endothelial cells.[21] In serum of post-MI patients who undergo reperfusion, MMP-1 increases 4 days after admission reaching a peak concentration around day 14. By day 28 MMP-1 levels decrease by 50% compared to day 14.[22] In addition, serum MMP-1 levels negatively correlate with the LV end-systolic volume index and positively correlate with LV ejection fraction.[22]

MMP-1 preferentially degrades collagens I and III; compared to the 25 known MMPs, MMP-1 has the highest affinity for fibrillar collagen. MMP-1 initiates the degradation of collagen fibers within the LV by cleaving collagen into 3/4 and 1/4 fragments. These fragments then become unfolded and degraded by MMP-2, -9, and -3.[23]

2.2 MMP-2

MMP-2 is expressed by cardiomyocytes, endothelial cells, vascular smooth muscle cells, macrophages, and fibroblasts.[24–26] Due to its high constitutive activity, MMP-2 is considered a MMP housekeeping gene that helps regulate normal tissue turnover.[27] Post-MI, MMP-2 levels increase both in plasma and within the infarct due to stimulation of the cardiomyocyte and cardiac fibroblast.[28–30] In patients diagnosed with heart failure there is a fourfold increase in MMP-2 expression compared to controls.[31] In rats, MMP-2 mRNA and protein levels elevate within 24 h post-MI and peak around day 14 post-MI.[32] Similar to rats, MMP-2 activity in mice rapidly increases within 4 days post-MI, peaks at day 7, and remains elevated until day 14.[33]

Matsumura *et al.* demonstrated the MMP-2-generated fragments of laminin inhibited migration of macrophages into the inflammatory region and resulted in delayed wound healing after MI.[34] MMP-2, in addition to MMP-3, -7, and -9, processes vitronectin into multiple fragments; however, the biological function of these fragments in the post-MI environment has not been evaluated.[35] Recently, Zhao *et al.* showed that periostin increases collagen fibrogenesis in the human failing heart and was associated with elevated MMP-2 levels.[31] MMP-2 also generates the matricryptin C-1158/59 from collagen.[14] In vitro stimulation with the downstream peptide from

C-1158/59 increases fibroblast migration rate and angiogenesis to improve wound healing.

There are multiple MMP-2 polymorphisms, of which only five have been associated with MI. Elevations in MMP-2 levels due to the −1575 A/G gene polymorphism increase risk for MI by fourfold in Hispanic males, indicating MMP-2 may act as a strong biomarker for MI incidence in this population.[36] Similar to the −1575 A/G polymorphism, a MMP-2 single-nucleotide polymorphism, −1306 T/C, displayed a twofold increase in promoter activity resulting in increased MMP-2 expression and enzymatic activity. In Hispanics, this polymorphism also associates with increased risk for MI and coronary triple-vessel disease.[37] In France, the −1306 T/C showed no association with heart failure-related deaths.[38]

A study in African- and Caucasian-Brazilian patients diagnosed with heart failure of any etiology and reduced ejection fraction (<45%) implicates the −1575 A/G, −1059 A/G, and −790 T/G MMP-2 polymorphisms with a 2.5-fold increase in heart failure risk compared to nondiseased controls.[39] Interestingly, all three MMP-2 polymorphisms associate with heart failure-related deaths only in Caucasians.[39] Similarly, the −790 T/G MMP-2 polymorphism, in addition to the −735 C/T polymorphism, also associates with chronic heart failure in patients from the Czech Republic.[40] How these polymorphisms affect heart failure-related deaths is still unclear.

2.3 MMP-3

MMP-3 is secreted by cardiac fibroblasts and macrophages.[28] In post-MI patients, circulating MMP-3 concentrations increase steadily between admission and discharge and are higher at 3 months compared to 48 h after MI.[41] MMP-3 levels at 72–96 h post-MI associate with left ventricular dysfunction, adverse left ventricular remodeling, and prognosis of heart failure.[41] Although MMP-3 correlates with MI severity, MMP-3 mechanisms of action are not clear.

MMP-3 breaks down multiple ECM components, including collagen, fibronectin, laminins, proteoglycans, and vitronectin.[42] MMP-3 also activates a number of MMPs including MMP-1, -7, and -9. As such, MMP-3 is considered an upstream MMP activator. MMP-3 proteolytic action on pro-MMP-1 is critical for the generation of fully active MMP-1.[43] Secreted protein acidic and rich in cysteine (SPARC) can also be cleaved by MMP-3 generating three biologically active peptides (Z-1, Z-2, and Z-3).[44] Fragment Z-1 increases angiogenesis and vascular growth, whereas fragments Z-2 and Z-3 inhibit cell proliferation. MMP-3 also generates tenascin-C fragments.[45] The significance of these fragments in cardiac pathology is not yet clear.

Due to their ability to regulate MMP-3 activity, MMP-3 polymorphisms have been implicated as regulators of MI prevalence and heart failure outcomes.[46,47] For example, the −1171 6A allele has lower promoter activity compared to the 5A allele and is found significantly less frequent in MI patients than in control subjects.[46,48] This polymorphism impacts on cardiac survival in heart failure patients with ischemic and nonischemic cardiomyopathy differ. MMP-3 6A allele is an independent predictor of cardiac mortality in patients with nonischemic heart failure. In contrast, there is no evidence for any effect of the MMP-3 genotype on cardiac events in patients with ischemic cardiomyopathy.[49]

2.4 MMP-7

MMP-7 is expressed in endothelial cells, cardiomyocytes, and macrophages.[9,50,51] In animal models of MI, MMP-7 increases threefold in both remote and infarct regions at 7 days post-MI.[9] MMP-7 activity is linked to increased risk for major adverse cardiac events, including decreased post-MI survival and increased hospitalization for congestive heart failure.[9,52–54] Elevated serum MMP-7 levels are associated with LV structural remodeling in 144 patients with LV hypertrophy.[53]

MMP-7 has an extensive portfolio of substrates, including collagen IV, connexin-43, fibronectin, laminin, peroxiredoxins, tenascin-C, and tumor necrosis factor-α.[9,54–56] MMP-7 can also cleave other MMPs, including MMP-1, -2, and -9, leading to their activation and implicating MMP-7 as both a direct and an indirect regulator for LV remodeling.[57] Of these substrates, MMP-7 has major effects on connexin-43 and serves as a predominant mechanism for post-MI arrhythmias.[9]

2.5 MMP-8

MMP-8 is expressed by neutrophils and macrophages.[58,59] MMP-8 is a major player during the inflammatory response. Serum levels of MMP-8 are a significant predictor of LV remodeling, cardiac rupture, and development of heart failure after MI.[58,60] MMP-8 increases sixfold within 6 h post-MI and peaks at 12 h due to infiltration of neutrophils.[59,61] At day 3 post-MI, MMP-8 spikes again most likely due to macrophage expression during the later stages of remodeling.[62]

MMP-8 coordinates leukocyte trafficking through cleavage of collagen and chemokine-binding proteins.[63–65] MMP-8 processes bioactive molecules, such as LPS-induced CXC chemokine (LIX).[66] Cleavage of the N-terminus of LIX by MMP-8 enhances neutrophil chemotaxis in response to lipopolysaccharide stimulation. MMP-8 cleavage of interleukin (IL)-8 and CXCL5, the human orthologues of LIX, also increases neutrophil chemotaxis.[66] How MMP-8 substrates affect the wound-healing response post-MI has not been studied.

2.6 MMP-9

MMP-9 is secreted by a wide number of cell types, including cardiomyocytes, endothelial cells, neutrophils, macrophages, and fibroblasts.[67] Circulating MMP-9 increases at day 1 and remains elevated until day 7 post-MI in mice.[68] Blankenberg and colleagues were the first to implicate MMP-9 as a novel prognostic biomarker for the development of LV dysfunction and late survival in patients with CVD.[69,70] MMP-9 correlates with IL-6, C-reactive protein, and fibrinogen concentrations in the plasma, indicating MMP-9 can predict cardiovascular outcome independent of an association with inflammatory markers.[69] Squire *et al.* demonstrated that increased MMP-9 correlates with larger LV volumes and greater LV dysfunction following MI.[71]

MMP-9 regulates tissue remodeling by directly degrading ECM and activating cytokines and chemokines.[67] Using an ECM-targeted proteomic approach, Zamilpa *et al.* identified multiple in vivo MMP-9 substrates in the post-MI setting including fibronectin, a known in vitro MMP-9 substrate.[72] Cleavage of MMP-9 substrates is both detrimental and beneficial for wound-healing post-MI. For example, MMP-9-mediated degradation of CD36 post-MI decreases macrophage phagocytosis and prolongs neutrophil inflammation leading to an enlarged LV post-MI.[73] In contrast, MMP-9 cleavage of osteopontin generates two biologically active peptides that increased the migration rate of cardiac fibroblasts resulting in enhanced infarct wound healing.[74] For this reason, targeting substrates downstream of MMP-9 may serve as a feasible alternative for predicting and preventing LV dysfunction post-MI.

There is clinical evidence that MMP genetic polymorphisms can contribute to MMP protein levels and thus influence cardiovascular outcomes.[75] The MMP-9 −1562 C/T polymorphism associates with increased MI incidence.[76] Associations vary across ethnic populations. For example, in healthy white subjects, MMP-9 genetic polymorphisms did not associate with plasma MMP-9 levels, whereas a positive association was discovered in healthy African American subjects.[77–79]

In respect to heart failure, the MMP-9 −1562 C/T polymorphisms have had conflicting results.[49,76] In a study performed in patients in Brazil, this MMP-9 polymorphism did not associate with heart failure susceptibility or heart failure-related survival.[49] In a separate study completed in France, increases in the T allele of the MMP-9 polymorphism did not associate with lower ejection fraction or higher end-diastolic diameter, but are an independent predictor of cardiac mortality.[38] Both studies had similar patient characteristics excluding patients who had an ejection fraction higher than 45%. This suggests that the differences observed were most likely due to the differences in genetic ancestrality, and that the −1562 C/T polymorphism may not be a strong biomarker across multiple populations. Before this conclusion can be made, the alternative explanation that the two studies measured different end points needs to be ruled out.

2.7 MMP-12

Historically macrophages have been regarded as the main cell type to express MMP-12 and hence the earlier term macrophage metalloelastase.[80–82] In recent literature, neutrophils have been identified as an early source of MMP-12 post-MI.[81] MMP-12 is also expressed by endothelial cells, fibroblasts, and vascular smooth muscle cells.[83,84] Post-MI, MMP-12 protein levels increase in the infarct at day 1 and remain elevated through day 7 in mice.[81] MMP-12 inhibition in mice led to impaired cardiac function post-MI compared to saline controls, revealing a protective role for MMP-12 in the post-MI setting.[81]

MMP-12 has broad substrate specificity, including type IV collagen fibronectin, heparan sulfate, laminin, and vitronectin.[85–87] Cleavage of these ECM proteins plays a vital role during cardiac remodeling. For example, MMP-12 cleavage of type IV collagen disrupts the basement membrane and enables fibroblast and macrophages to access to the injured site.[87] MMP-12 can process pro-TNFα into mature TNFα, indicating that MMP-12 has the potential to amplify to TNFα-driven inflammation.[85] MMP-12 also cleaves CD36 into one major fragment, but whether this peptide has biological activity is not known.[73]

While little is known about MMP-12 mechanisms in the post-MI LV remodeling process, recent evidence indicates that the MMP-12

−82 A/G polymorphism increases MI risk, as the MMP-12 −82 AG and GG genotypes were associated with a 3.7-fold increase in risk of having two or three occluded vessels.[36] This polymorphism is located at the activator protein-1 transcription factor-binding site for MMP-12. In vitro studies have indicated that the G allele of this polymorphism results in lower MMP-12 promoter activity and thus lower transcriptional activity.[36]

2.8 MMP-14

MMP-14 is expressed in cardiomyocytes, macrophages, and fibroblasts.[27] MMP-14 increases 20-fold at 3 days post-MI and peaks at 16 weeks post-MI in noninfarcted LV regions, indicating critical roles in both early and late remodeling events and in both infarct and remote regions.[8,24,32,88] Elevated MMP-14 in both plasma and the LV infarct post-MI correlates with extensive LV remodeling including significant cardiac fibrosis, reduced LV function, and lower survival.[62,88]

Post-MI, MMP-14 degrades collagen, fibronectin, and gelatin leading to a loss of ECM structure and support.[89] In addition to ECM structural proteins, MMP-14 can process profibrotic signaling molecules, such as transforming growth factor-β and periostin, leading to increased fibrillar collagen synthesis and accumulation.[90] MMP-14 can also activate MMP-2 and -13.[91] Subsequent activation of these MMPs would result in continued ECM degradation and instability thereby contributing to adverse LV remodeling, dilation, and possibly heart failure.[89]

2.9 MMP-28

MMP-28 is the newest identified member of the MMP family. MMP-28 is expressed by cardiomyocytes, neutrophils, and macrophages.[92] In contrast to what is generally observed for MMPs post-MI, total MMP-28 actually decreases post-MI due to the loss of myocytes.[93] While myocyte-derived MMP-29 decreases, macrophage-derived MMP-29 increases in the post-MI LV.[93] In vitro, MMP-28 has been shown to proteolytically processes casein, Nogo-A (a myelin component), and neural cell adhesion molecule-1; however, little is known about the role of MMP-28 post-MI.[94] Deletion of MMP-28 exaggerates LV dysfunction and cardiac rupture post-MI by reducing the inflammatory and fibrotic response and tilting the balance away from adequate wound healing and high-quality scar formation.[93] Inhibition of MMP-28, therefore, would most likely not be a successful target for improving patient outcomes post-MI.

3. CLINICAL USE OF MMP INHIBITORS POST-MI

MMPs can process a variety of substrates, and the same substrate may be processed by a variety of MMPs; combined, this complexity yields a post-MI in vivo environment that is still not fully understood. For this reason, the clinical use of MMPi in post-MI patients is still under investigation (Fig. 2).[95]

3.1 Direct Nonselective Inhibition

Broad-spectrum MMPi have been used in clinical trials as an attempt to prevent heart failure post-MI. To date, more than 25 MMPi have been investigated in post-MI clinical trials. However, the majority of MMPi developed to date have not proved efficacious. Side effects due to MMPi treatment include joint pain, stiffness, edema, skin discoloration, and reduced patient mobility.[96] However, this musculoskeletal syndrome is reversible on cessation of the drug intake.[97] Insufficient knowledge about complex biological role of MMPs and lack of specificity are main reasons for clinical failure.[98,99]

Animals treated with CP-471,474 (a broad-spectrum MMPi) had attenuated end-diastolic and end-systolic dimensions and increased the number of vessels post-MI.[100] Similarly, pigs, mice, and rabbits treated with

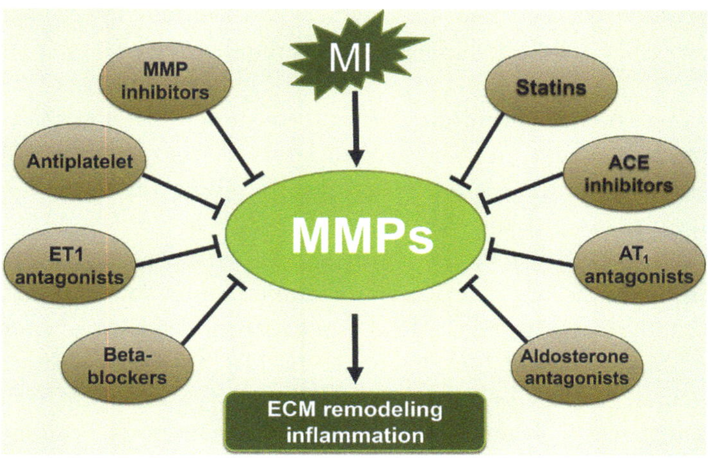

Fig. 2 Therapeutics given to postmyocardial infarction (MI) patients either directly or indirectly inhibit matrix metalloproteinase (MMP) activity leading to extracellular matrix (ECM) remodeling of the left ventricle (LV).

PD166793 (another broad-spectrum MMPi) at a concentration that can inhibit MMP-2, -3, -9, and -13, but not MMP-1, had lower LV end-diastolic dimension and increased TIMP-1 concentration in infarct region post-MI.[100–102]

Pigs treated 3 days pre- or post-MI with an orally available broad-spectrum MMPi (PGE-530742) at a concentration that inhibited MMP-2, -3, -8, -9, and -13, but not MMP-1 and -7, decreased post-MI end-systolic volume, and attenuated fibrillar collagen content in the infarct zone.[103] Pre-MI treatment, however, increased collagen content in the border zone and decreased collagen content in the remote zone in comparison to post-MI treatment, demonstrating that timing plays an important role on MMPi effects on ECM remodeling. In a similar study, post-MI mice were given 2*R*-2-[5-[4-[ethyl-methylamino]phenyl] thiophene-2-sulfonylamino]-3-methylbutyric acid (TISAM) orally, at a dose that inhibited MMP-2 and -9, but not MMP-1, -3, or -7. The treatment improved survival rate, decreased cardiac rupture rate, and decreased macrophage infiltration, but increased necrotic area at day 7 post-MI.[34]

PG-116800 is an oral MMPi of the hydroxamic acid class with high affinity for MMP-2, -3, -8, -9, -13, and -14 and low affinity for MMP-1 and -7. In a phase II double-blind, multicenter RCT PREMIER (Prevention of Myocardial Infarction Early Remodeling) trial, post-MI patients did not improve in echocardiography parameters or in clinical outcomes after 90 days treatment with PG-116800.[97] This may be due in part by the clinical dose used, which was fourfold lower than what was shown effective for achieving a LV antiremodeling effect in preclinical studies.[103]

Doxycycline is an antibiotic used at subantimicrobial doses as a broad-spectrum MMPi. Its use is approved by FDA for periodontal disease treatment.[104] Rats treated with doxycycline showed attenuated mRNA transcription and protein expressions of MMP-8, MMP-13, TIMP-1, TIMP-2, and type I collagen content in the remote zone post-MI.[105] In another study, rats treated with doxycycline showed attenuated MMP-2 activity, increased TIMP-1 expression, and improved endothelial dysfunction post-MI.[106] In the phase II TIPTOP (Early Short-Term Doxycycline Therapy in Patients with Acute Myocardial Infarction and Left Ventricular Dysfunction to Prevent the Ominous Progression to Adverse Remodeling) trial, treatment with doxycycline (100 mg twice daily) reduced end-diastolic volumes index, infarct size, and infarct severity in comparison to patients who received only the standard treatment.[107]

3.2 Direct Selective Inhibition

First-generation MMPi targeted a broad range of MMPs and therefore lacked specificity and selectivity, leading to less effective use and inconclusive results.[108] The Fields laboratory developed a potent and highly selective MMP-9 inhibitor (MMP-9i).[109,110] A recent study in mice showed that early inhibition of MMP-9 had no effect on infarct size or survival. By day 7 post-MI, MMP-9i enhanced infarct wall thinning and worsened cardiac dysfunction opposing studies utilizing MMP-9-deleted mice.[73,108] This was due in part by MMP-9i treatment increasing expression of *Mmp8*, *Mmp12*, and *Mmp14*, and decreasing collagen deposition. MMP-9i treatment also increased neutrophil numbers at day 1 post-MI and macrophage infiltration at day 7 post-MI.

A selective MMP-12 inhibitor (RXP 470.1; MMP-12i) was also found to promote adverse cardiac function post-MI.[81] This is in contrast to an atherosclerotic model, where MMP-12 inhibition was found to reduce atherosclerosis progression in apo-E-null mice by inhibiting MMP-12 activity in macrophages.[98] Post-MI, MMP-12i repressed neutrophil functions leading to impaired cardiac wound healing. This suggests that MMP-12i in an acute inflammation model may be harmful, while MMP-12i in a chronic inflammation model may be favorable.

These studies reveal that using a selective MMPi will require complete assessment and complex data interpretation. Studies utilizing MMPi diverge greatly from those seen with global or cell-specific MMP deletion.[108,111] In addition, the overall effect of MMPi differs among CVD pathologies and timing of intervention. This is due in part by the fact that MMPs have both detrimental and beneficial mechanisms that are dependent on injury stimuli and on substrate availability.[81]

3.3 Indirect MMP Inhibition

While the use of direct MMPi is still under examination, MMP inhibition is being achieved through indirect mechanisms. The majority of current medications used for MI and heart failure (e.g., angiotensin-converting enzyme (ACE) inhibitors, angiotensin receptor antagonists, beta-blockers, and statins) act as indirect inhibitors for MMPs.

ACE inhibitors are well documented to improve post-MI outcomes.[112] The catalytic domain of ACE is similar to that of MMPs, thus ACE inhibitors simultaneously have an inhibitory effect on MMPs.[113] For example, captopril and lisinopril inhibit MMP-2 activity at concentrations greater than 4 and 1 mmol/L, whereas MMP-9 was inhibited by 87 nmol/L of captopril.[114,115] ACE inhibitors bind to S1 and S1′ subsite of MMP-9, which forms a deep hydrophobic pocket similar to the hydrophobic moieties in the ACE active site.[113] Lisinopril is stabilized to the MMP-9 active site by specific hydrogen bonds and hydrophobic interactions, and its hydrophobic group showed greater affinity with the S1 site in comparison to the S1′ site.[113] In addition, imidapril is stabilized to the MMP-9 active site with less molecular distortion, which may explain the greater MMP-9 inhibitory activity of imidapril compared to lisinopril.[116] In fact, in vivo experiments showed that hamsters treated with lisinopril or imidapril had decreased MMP-9 activity in the post-MI LV, with MMP-9 activity being lower in the group that received imidapril compared to lisinopril.[116] In a separate study, post-MI patients treated with perindopril showed decreased levels of MMP-1 in plasma and attenuated LV dysfunction.[117]

Similar to ACE inhibitors, angiotensin II receptor antagonists inhibit MMPs levels and improve ECM remodeling post-MI.[118] Rats treated with losartan showed attenuated mRNA transcription and protein expression of MMP-8, MMP-13, TIMP-1, TIMP-2, and type I collagen content in the remote zone post-MI.[118] Treatment with valsartan (a selective angiotensin II type 1 receptor antagonist, AT_1) but not PD123319 (a selective angiotensin II type 2 receptor antagonist) decreased levels of MMP-2, -3, and -9 post-MI in rats, indicating that angiotensin II receptor blockage effects on ECM remodeling are due to inhibition of the AT_1 receptor.[119] Deletion of the AT_1 receptor in mice increased survival rate and decreased LV remodeling post-MI.[120] Moreover, patients treated with trandolapril and valsartan showed attenuated MMP-9 levels in plasma and suppressed LV remodeling post-MI.[121] While further investigation is needed to assess the mechanism whereby ACE inhibitors and angiotensin II receptor antagonists decrease MMPs and ameliorate LV dysfunction post-MI, the therapeutical inhibition of MMPs by inhibition of the angiotensin pathway is encouraging.

The National Institutes for Health and Clinical Excellence (NICE) recommends aldosterone antagonist therapy should be given within 3–14 days post-MI (preferably after treatment with ACE inhibitor has been initiated) to treat and prevent LV systolic dysfunction and heart failure in post-MI patients.[122] Aldosterone increases MMP-2 and -9 activity in cultured adult rat ventricular myocytes through modulating mitogen/extracellular signal-regulated kinase and extracellular signal-regulated kinase 1/2 phosphorylation, and increasing ROS production.[123] Pretreatment with spironolactone (aldosterone receptor antagonist) abolished the aldosterone-induced increase in MMP activity and decreased collagen deposition post-MI.[123,124] Long-term treatment (24 weeks) with spironolactone in heart failure patients improved LV dysfunction and attenuated plasma MMP-9, TIMP-1, and type I collagen carboxyterminal telopeptide concentrations.[125]

Others vasoactive peptides, such as endothelin-1, can alter MMP expression and ECM remodeling post-MI. The endothelin receptor subtype A (ET_A) is predominate on myocytes and its activation induces myocyte hypertrophy.[126,127] Thus, endothelin increases collagen expression and collagenase activity in cardiac fibroblasts.[128] Rats treated with sitaxsentan (ET_A receptor antagonist) attenuated MMP-1, -2, and -9, increased TIMP-1, and improved post-MI cardiac dilation.[129]

The American Heart Association and American College of Cardiology recommend initiating beta-adrenergic antagonists in all post-MI patients and continuing therapy indefinitely.[130] Dogs treated with atenolol, at supratherapeutic doses, showed decreased MMP activity and improved LV stiffness in an experimental pacing-induced heart failure model.[131] Rats treated with metoprolol showed decreased MMP-2 mRNA levels and decreased oxidative stress markers post-MI.[132] Similar results were observed in post-MI pigs treated with carvedilol or metoprolol, both of which decreased MMP-2 activity, MCP-1 expression, and macrophage infiltration.[133] Patients with heart failure treated with standard therapy plus carvedilol showed reduced MMP-9 activity in plasma.[134]

Antiplatelet, anticoagulant, and thrombolytic therapies are administrated after cardiovascular events to improve patient outcome.[135] Post-MI patients treated with tissue plasminogen activator showed increased plasma MMP-9 levels.[136] Direct thrombin inhibitors and heparin are commonly used in post-MI therapy.[137,138] Heparin treatment increases MMP-1 and -2 in cultured fibroblasts, and MMP-1, -2, -3, and -9 in mesangial cells.[139,140] The effect of heparin therapy on MMPs in vivo has not been determined.

The use of nonsteroidal antiinflammatory drugs (NSAIDs) is highly associated with risk of MI; however, selective cyclooxygenase-2 inhibitors reduce C-reactive protein levels and improve endothelial function.[141] Aspirin is an NSAID administrated as an antiplatelet and is highly recommended for patients after first MI or with heart failure.[142] MMP-2 gene expression was attenuated with NSAID treatment by inhibiting specific protein (SP)-1 transcription factor from binding to the MMP-2 promoter site.[143] Up to date, no studies have examined the effects of NSAIDs on MMPs post-MI.

In vitro experiments in human endothelial cells indicate that aspirin attenuates MMP-1 but not MMP-2 or -9.[144,145]

Statins (hydroxymethylglutaryl coenzyme A reductase inhibitors) exert a variety of pleiotropic effects, such as inhibiting isoprenoid intermediates production, which decrease the activation of Rho-family small GTP-binding proteins.[146] The inhibitory effect of statins on MMP-9 expression is dependent on nitrite-mediated mechanisms.[147] Interestingly, macrophages treated with statins showed increased MMP-12 expression, which is consistent with the concept that MMP-12 may be beneficial post-MI.[80] Human cardiac myofibroblasts treated with simvastatin showed reduced TNFα-induced invasion by MMP-9-dependent mechanisms.[148] Patients treated with pravastatin showed decreased MMP-2, MMP-9, C-reactive protein, and CD40L levels post-MI.[149–151] Combined there is ample evidence to support MMPs as off targets of a variety of CVD drugs.

3.4 General Considerations

Multiple clinical studies have demonstrated the role of direct and indirect MMPi during the development of end-stage heart failure and LV remodeling. However, clinic trials using nonspecific MMPi have been inconclusive.[97] In addition, animal studies with direct MMPi have given mixed results. To develop a successful therapeutic treatment, additional studies are needed to understand the entire composite of MMP roles. Identifying the downstream effects of MMPs may provide answers and serve as novel and more selective treatment strategies for post-MI patients.[73]

4. FUTURE DIRECTIONS

While the cardiac MMP literature field has exploded over the last 15 years, future research is needed. There is a need to better understand the biological function of MMPs in cardiac maintenance and tissue repair after injury. Out of the 25 MMPs identified to date, about half of the known MMPs have been characterized post-MI with MMP-2 and -9 being the main focus. In addition, the role of the MMPs that have been studied is complicated by interactions between MMP family members. For example, inhibition of a specific MMP can result in an increase of other MMPs due to compensatory effects. MMPs also compete with each other for the same substrate. Competition assays evaluating the interaction with other MMPs are necessary in order to fully characterize the role of MMPs in the post-MI environment.

Furthermore, each MMP has a broad range of substrates that include chemokines, cytokines, adhesion molecules, and growth factors as well as ECM components such as collagen and fibronectin.[72] Post-MI, small fragments or peptides generated by MMP proteolysis are increased both in the plasma and in the infarct area. Evaluating the efficacy of using MMP cleavage products for diagnostic and therapeutic purposes in heart failure is warranted.

Temporal and spatial MMP patterns during LV remodeling need to be determined to gain a more extensive understanding of MMPs. For example, MMP-9 gene deletion has proven beneficial post-MI in mice, yet, macrophage-specific transgenic overexpression of MMP-9 has also shown to be beneficial by attenuating the post-MI inflammatory response.[111] This suggests that timing and concentration of MMPs may dictate divergent mechanisms of response.[108,111] Combined, the elucidation of these MMP mechanisms will help us to more completely understand how MMPs coordinate the post-MI response.

5. CONCLUSION

In conclusion, understanding the role of MMPs during post-MI remodeling remains an important issue.[152] A better understanding of pathophysiological processes, including the biological function of the downstream products of MMPs, may lead to new strategies for the post-MI patient, particularly therapies limiting heart failure progression.

ACKNOWLEDGMENTS

This work was supported by the National Institute of Health [HL075360, HL129823, and GM114833 to M.L.L.] and from the Biomedical Laboratory Research and Development Service of the Veterans Affairs Office of Research and Development Award [5I01BX000505 to M.L.L.], and by HL051971 and GM104357.

REFERENCES

1. Roger VL, Go AS, Lloyd-Jones DM, et al. Heart disease and stroke statistics—2012 update: a report from the American heart association. *Circulation*. 2012;125:e2–e220.
2. Cohn JN, Ferrari R, Sharpe N. Cardiac remodeling—concepts and clinical implications: a consensus paper from an international forum on cardiac remodeling. Behalf of an international forum on cardiac remodeling. *J Am Coll Cardiol*. 2000;35:569–582.
3. Lindsey ML, Iyer RP, Jung M, DeLeon-Pennell KY, Ma Y. Matrix metalloproteinases as input and output signals for post-myocardial infarction remodeling. *J Mol Cell Cardiol*. 2016;91:134–140.
4. Wilson EM, Gunasinghe HR, Coker ML, et al. Plasma matrix metalloproteinase and inhibitor profiles in patients with heart failure. *J Card Fail*. 2002;8:390–398.

5. Spinale FG, Coker ML, Krombach SR, et al. Matrix metalloproteinase inhibition during the development of congestive heart failure: effects on left ventricular dimensions and function. *Circ Res*. 1999;85:364–376.
6. Ertl G, Frantz S. Healing after myocardial infarction. *Cardiovasc Res*. 2005;66:22–32.
7. Lindsey M, Wedin K, Brown MD, et al. Matrix-dependent mechanism of neutrophil-mediated release and activation of matrix metalloproteinase 9 in myocardial ischemia/reperfusion. *Circulation*. 2001;103:2181–2187.
8. Coker ML, Jolly JR, Joffs C, et al. Matrix metalloproteinase expression and activity in isolated myocytes after neurohormonal stimulation. *Am J Physiol Heart Circ Physiol*. 2001;281:H543–H551.
9. Lindsey ML, Escobar GP, Mukherjee R, et al. Matrix metalloproteinase-7 affects connexin-43 levels, electrical conduction, and survival after myocardial infarction. *Circulation*. 2006;113:2919–2928.
10. Heymans S, Luttun A, Nuyens D, et al. Inhibition of plasminogen activators or matrix metalloproteinases prevents cardiac rupture but impairs therapeutic angiogenesis and causes cardiac failure. *Nat Med*. 1999;5:1135–1142.
11. Lindsey ML, Hall ME, Harmancey R, Ma Y. Adapting extracellular matrix proteomics for clinical studies on cardiac remodeling post-myocardial infarction. *Clin Proteomics*. 2016;13:19.
12. Barallobre-Barreiro J, Didangelos A, Schoendube FA, et al. Proteomics analysis of cardiac extracellular matrix remodeling in a porcine model of ischemia/reperfusion injury. *Circulation*. 2012;125:789–802.
13. Davis GE, Bayless KJ, Davis MJ, Meininger GA. Regulation of tissue injury responses by the exposure of matricryptic sites within extracellular matrix molecules. *Am J Pathol*. 2000;156:1489–1498.
14. Lindsey ML, Iyer RP, Zamilpa R, et al. A novel collagen matricryptin reduces left ventricular dilation post-myocardial infarction by promoting scar formation and angiogenesis. *J Am Coll Cardiol*. 2015;66:1364–1374.
15. Ma Y, de Castro Bras LE, Toba H, et al. Myofibroblasts and the extracellular matrix network in post-myocardial infarction cardiac remodeling. *Pflugers Arch*. 2014;466:1113–1127.
16. Yabluchanskiy A, Chilton RJ, Lindsey ML. Left ventricular remodeling: one small step for the extracellular matrix will translate to a giant leap for the myocardium. *Congest Heart Fail*. 2013;19:E5–E8.
17. Ma Y, Halade GV, Lindsey ML. Extracellular matrix and fibroblast communication following myocardial infarction. *J Cardiovasc Transl Res*. 2012;5:848–857.
18. Diez J, Panizo A, Gil MJ, Monreal I, Hernandez M, Pardo Mindan J. Serum markers of collagen type I metabolism in spontaneously hypertensive rats: relation to myocardial fibrosis. *Circulation*. 1996;93:1026–1032.
19. McGavigan AD, Maxwell PR, Dunn FG. Serological evidence of altered collagen homeostasis reflects early ventricular remodeling following acute myocardial infarction. *Int J Cardiol*. 2006;111:267–274.
20. Balbin M, Fueyo A, Knauper V, et al. Identification and enzymatic characterization of two diverging murine counterparts of human interstitial collagenase (MMP-1) expressed at sites of embryo implantation. *J Biol Chem*. 2001;276:10253–10262.
21. Spinale FG, Coker ML, Bond BR, Zellner JL. Myocardial matrix degradation and metalloproteinase activation in the failing heart: a potential therapeutic target. *Cardiovasc Res*. 2000;46:225–238.
22. Hirohata S, Kusachi S, Murakami M, et al. Time dependent alterations of serum matrix metalloproteinase-1 and metalloproteinase-1 tissue inhibitor after successful reperfusion of acute myocardial infarction. *Heart*. 1997;78:278–284.

23. Visse R, Nagase H. Matrix metalloproteinases and tissue inhibitors of metalloproteinases: structure, function, and biochemistry. *Circ Res*. 2003;92:827–839.
24. Lindsey ML, Goshorn DK, Squires CE, et al. Age-dependent changes in myocardial matrix metalloproteinase/tissue inhibitor of metalloproteinase profiles and fibroblast function. *Cardiovasc Res*. 2005;66:410–419.
25. Porter KE, Turner NA. Cardiac fibroblasts: at the heart of myocardial remodeling. *Pharmacol Ther*. 2009;123:255–278.
26. Chow AK, Cena J, Schulz R. Acute actions and novel targets of matrix metalloproteinases in the heart and vasculature. *Br J Pharmacol*. 2007;152:189–205.
27. Vanhoutte D, Schellings M, Pinto Y, Heymans S. Relevance of matrix metalloproteinases and their inhibitors after myocardial infarction: a temporal and spatial window. *Cardiovasc Res*. 2006;69:604–613.
28. Lindner D, Zietsch C, Becher PM, et al. Differential expression of matrix metalloproteases in human fibroblasts with different origins. *Biochem Res Int*. 2012;2012:875742.
29. Baghirova S, Hughes BG, Poirier M, Kondo MY, Schulz R. Nuclear matrix metalloproteinase-2 in the cardiomyocyte and the ischemic-reperfused heart. *J Mol Cell Cardiol*. 2016;94:153–161.
30. Lovett DH, Mahimkar R, Raffai RL, et al. A novel intracellular isoform of matrix metalloproteinase-2 induced by oxidative stress activates innate immunity. *PLoS One*. 2012;7:e34177.
31. Zhao S, Wu H, Xia W, et al. Periostin expression is upregulated and associated with myocardial fibrosis in human failing hearts. *J Cardiol*. 2014;63:373–378.
32. Peterson JT, Li H, Dillon L, Bryant JW. Evolution of matrix metalloprotease and tissue inhibitor expression during heart failure progression in the infarcted rat. *Cardiovasc Res*. 2000;46:307–315.
33. Tao ZY, Cavasin MA, Yang F, Liu YH, Yang XP. Temporal changes in matrix metalloproteinase expression and inflammatory response associated with cardiac rupture after myocardial infarction in mice. *Life Sci*. 2004;74:1561–1572.
34. Matsumura SI, Iwanaga S, Mochizuki S, Okamoto H, Ogawa S, Okada Y. Targeted deletion or pharmacological inhibition of MMP-2 prevents cardiac rupture after myocardial infarction in mice. *J Clin Invest*. 2005;115:599–609.
35. Imai K, Shikata H, Okada Y. Degradation of vitronectin by matrix metalloproteinases-1, -2, -3, -7 and -9. *FEBS Lett*. 1995;369:249–251.
36. Perez-Hernandez N, Vargas-Alarcon G, Martinez-Rodriguez N, et al. The matrix metalloproteinase 2-1575 gene polymorphism is associated with the risk of developing myocardial infarction in Mexican patients. *J Atheroscler Thromb*. 2012;19:718–727.
37. Delgado-Enciso I, Gonzalez-Hernandez NA, Baltazar-Rodriguez LM, et al. Association of matrix metalloproteinase-2 gene promoter polymorphism with myocardial infarction susceptibility in a Mexican population. *J Genet*. 2009;88:249–252.
38. Mizon-Gerard F, de Groote P, Lamblin N, et al. Prognostic impact of matrix metalloproteinase gene polymorphisms in patients with heart failure according to the aetiology of left ventricular systolic dysfunction. *Eur Heart J*. 2004;25:688–693.
39. Beber AR, Polina ER, Biolo A, et al. Matrix metalloproteinase-2 polymorphisms in chronic heart failure: relationship with susceptibility and long-term survival. *PLoS One*. 2016;11:e0161666.
40. Vasku A, Goldbergova M, Holla LI, et al. Two MMP-2 promoter polymorphisms (−790 T/G and −735 C/T) in chronic heart failure. *Clin Chem Lab Med*. 2003;41:1299–1303.
41. Kelly D, Khan S, Cockerill G, et al. Circulating stromelysin-1 (MMP-3): a novel predictor of LV dysfunction, remodelling and all-cause mortality after acute myocardial infarction. *Eur J Heart Fail*. 2008;10:133–139.

42. Visse R, Nagase H. Matrix metalloproteinases and tissue inhibitors of metalloproteinases: structure, function, and biochemistry. *Circ Res*. 2003;92:827–839.
43. Suzuki K, Enghild JJ, Morodomi T, Salvesen G, Nagase H. Mechanisms of activation of tissue procollagenase by matrix metalloproteinase 3 (stromelysin). *Biochemistry*. 1990;29:10261–10270.
44. Sage EH, Reed M, Funk SE, et al. Cleavage of the matricellular protein SPARC by matrix metalloproteinase 3 produces polypeptides that influence angiogenesis. *J Biol Chem*. 2003;278:37849–37857.
45. Midwood KS, Hussenet T, Langlois B, Orend G. Advances in tenascin-C biology. *Cell Mol Life Sci*. 2011;68:3175–3199.
46. Terashima M, Akita H, Kanazawa K, et al. Stromelysin promoter 5A/6A polymorphism is associated with acute myocardial infarction. *Circulation*. 1999;99:2717–2719.
47. Wang J, Xu D, Wu X, et al. Polymorphisms of matrix metalloproteinases in myocardial infarction: a meta-analysis. *Heart*. 2011;97:1542–1546.
48. Ye S, Eriksson P, Hamsten A, Kurkinen M, Humphries SE, Henney AM. Progression of coronary atherosclerosis is associated with a common genetic variant of the human stromelysin-1 promoter which results in reduced gene expression. *J Biol Chem*. 1996;271:13055–13060.
49. Velho FM, Cohen CR, Santos KG, et al. Polymorphisms of matrix metalloproteinases in systolic heart failure: role on disease susceptibility, phenotypic characteristics, and prognosis. *J Card Fail*. 2011;17:115–121.
50. Furman C, Copin C, Kandoussi M, et al. Rosuvastatin reduces MMP-7 secretion by human monocyte-derived macrophages: potential relevance to atherosclerotic plaque stability. *Atherosclerosis*. 2004;174:93–98.
51. Boixel C, Fontaine V, Rucker-Martin C, et al. Fibrosis of the left atria during progression of heart failure is associated with increased matrix metalloproteinases in the rat. *J Am Coll Cardiol*. 2003;42:336–344.
52. Sato A, Aonuma K, Imanaka-Yoshida K, et al. Serum tenascin-C might be a novel predictor of left ventricular remodeling and prognosis after acute myocardial infarction. *J Am Coll Cardiol*. 2006;47:2319–2325.
53. Zile MR, Desantis SM, Baicu CF, et al. Plasma biomarkers that reflect determinants of matrix composition identify the presence of left ventricular hypertrophy and diastolic heart failure. *Circ Heart Fail*. 2011;4:246–256.
54. Chiao YA, Zamilpa R, Lopez EF, et al. In vivo matrix metalloproteinase-7 substrates identified in the left ventricle post-myocardial infarction using proteomics. *J Proteome Res*. 2010;9:2649–2657.
55. Agnihotri R, Crawford HC, Haro H, Matrisian LM, Havrda MC, Liaw L. Osteopontin, a novel substrate for matrix metalloproteinase-3 (stromelysin-1) and matrix metalloproteinase-7 (matrilysin). *J Biol Chem*. 2001;276:28261–28267.
56. von Bredow DC, Nagle RB, Bowden GT, Cress AE. Degradation of fibronectin fibrils by matrilysin and characterization of the degradation products. *Exp Cell Res*. 1995;221:83–91.
57. von Bredow DC, Cress AE, Howard EW, Bowden GT, Nagle RB. Activation of gelatinase-tissue-inhibitors-of-metalloproteinase complexes by matrilysin. *Biochem J*. 1998;331(pt 3):965–972.
58. Fertin M, Lemesle G, Turkieh A, et al. Serum MMP-8: a novel indicator of left ventricular remodeling and cardiac outcome in patients after acute myocardial infarction. *PLoS One*. 2013;8:e71280.
59. Webb CS, Bonnema DD, Ahmed SH, et al. Specific temporal profile of matrix metalloproteinase release occurs in patients after myocardial infarction: relation to left ventricular remodeling. *Circulation*. 2006;114:1020–1027.

60. Iyer RP, de Castro Bras LE, Jin YF, Lindsey ML. Translating Koch's postulates to identify matrix metalloproteinase roles in postmyocardial infarction remodeling: cardiac metalloproteinase actions (CarMA) postulates. *Circ Res.* 2014;114:860–871.
61. Kandalam V, Basu R, Abraham T, Wang S, Soloway PD, Jaworski DM, Oudit GY, Kassiri Z. TIMP2 Deficiency Accelerates Adverse Post–Myocardial Infarction Remodeling Because of Enhanced MT1-MMP Activity Despite Lack of MMP2 Activation. *Circ Res.* 2010;106:796–808.
62. Wilson EM, Moainie SL, Baskin JM, et al. Region- and type-specific induction of matrix metalloproteinases in post-myocardial infarction remodeling. *Circulation.* 2003;107:2857–2863.
63. Marini S, Fasciglione GF, de Sanctis G, D'Alessio S, Politi V, Coletta M. Cleavage of bovine collagen I by neutrophil collagenase MMP-8. Effect of pH on the catalytic properties as compared to synthetic substrates. *J Biol Chem.* 2000;275:18657–18663.
64. Van Lint P, Wielockx B, Puimege L, Noel A, Lopez-Otin C, Libert C. Resistance of collagenase-2 (matrix metalloproteinase-8)-deficient mice to TNF-induced lethal hepatitis. *J Immunol.* 2005;175:7642–7649.
65. Van Lint P, Libert C. Chemokine and cytokine processing by matrix metalloproteinases and its effect on leukocyte migration and inflammation. *J Leukoc Biol.* 2007;82:1375–1381.
66. Tester AM, Cox JH, Connor AR, et al. LPS responsiveness and neutrophil chemotaxis in vivo require PMN MMP-8 activity. *PLoS One.* 2007;2:e312.
67. Yabluchanskiy A, Ma Y, Iyer RP, Hall ME, Lindsey ML. Matrix metalloproteinase-9: many shades of function in cardiovascular disease. *Phys Chem Chem Phys.* 2013;28:391–403.
68. DeLeon-Pennell KY, de Castro Bras LE, Iyer RP, et al. P. gingivalis lipopolysaccharide intensifies inflammation post-myocardial infarction through matrix metalloproteinase-9. *J Mol Cell Cardiol.* 2014;76:218–226.
69. Blankenberg S, Rupprecht HJ, Poirier O, et al. Plasma concentrations and genetic variation of matrix metalloproteinase 9 and prognosis of patients with cardiovascular disease. *Circulation.* 2003;107:1579–1585.
70. Kelly D, Cockerill G, Ng LL, et al. Plasma matrix metalloproteinase-9 and left ventricular remodelling after acute myocardial infarction in man: a prospective cohort study. *Eur Heart J.* 2007;28:711–718.
71. Squire IB, Evans J, Ng LL, Loftus IM, Thompson MM. Plasma MMP-9 and MMP-2 following acute myocardial infarction in man: correlation with echocardiographic and neurohumoral parameters of left ventricular dysfunction. *J Card Fail.* 2004;10:328–333.
72. Zamilpa R, Lopez EF, Chiao YA, et al. Proteomic analysis identifies in vivo candidate matrix metalloproteinase-9 substrates in the left ventricle post-myocardial infarction. *Proteomics.* 2010;10:2214–2223.
73. DeLeon-Pennell KY, Tian Y, Zhang B, et al. CD36 is a matrix metalloproteinase-9 substrate that stimulates neutrophil apoptosis and removal during cardiac remodeling. *Circ Cardiovasc Genet.* 2016;9:14–25.
74. Lindsey ML, Zouein FA, Tian Y, Padmanabhan Iyer R, de Castro Bras LE. Osteopontin is proteolytically processed by matrix metalloproteinase 9. *Can J Physiol Pharmacol.* 2015;93:879–886.
75. Spinale FG. Matrix metalloproteinase gene polymorphisms in heart failure: new pieces to the myocardial matrix puzzle. *Eur Heart J.* 2004;25:631–633.
76. Wang X, Shi LZ. Association of matrix metalloproteinase-9 C1562T polymorphism and coronary artery disease: a meta-analysis. *J Zhejiang Univ Sci B.* 2014;15:256–263.
77. Lacchini R, Metzger IF, Luizon M, Ishizawa M, Tanus-Santos JE. Interethnic differences in the distribution of matrix metalloproteinases genetic polymorphisms are consistent with interethnic differences in disease prevalence. *DNA Cell Biol.* 2010;29:649–655.

78. Metzger IF, Luizon MR, Lacchini R, Tanus-Santos JE. Genetic variants in matrix metalloproteinase-9 gene modify metalloproteinase-9 levels in black subjects. *DNA Cell Biol.* 2012;31:504–510.
79. Demacq C, Vasconcellos VB, Marcaccini AM, Gerlach RF, Silva Jr WA, Tanus-Santos JE. Functional polymorphisms in the promoter of the matrix metalloproteinase-9 (MMP-9) gene are not linked with significant plasma MMP-9 variations in healthy subjects. *Clin Chem Lab Med.* 2008;46:57–63.
80. Arikan MC, Shapiro SD, Mariani TJ. Induction of macrophage elastase (MMP-12) gene expression by statins. *J Cell Physiol.* 2005;204:139–145.
81. Iyer RP, Patterson NL, Zouein FA, et al. Early matrix metalloproteinase-12 inhibition worsens post-myocardial infarction cardiac dysfunction by delaying inflammation resolution. *Int J Cardiol.* 2015;185:198–208.
82. Molet S, Belleguic C, Lena H, et al. Increase in macrophage elastase (MMP-12) in lungs from patients with chronic obstructive pulmonary disease. *Inflamm Res.* 2005;54:31–36.
83. Serrati S, Cinelli M, Margheri F, et al. Systemic sclerosis fibroblasts inhibit in vitro angiogenesis by MMP-12-dependent cleavage of the endothelial cell urokinase receptor. *J Pathol.* 2006;210:240–248.
84. Wu L, Tanimoto A, Murata Y, et al. Matrix metalloproteinase-12 gene expression in human vascular smooth muscle cells. *Genes Cells.* 2003;8:225–234.
85. Chandler S, Cossins J, Lury J, Wells G. Macrophage metalloelastase degrades matrix and myelin proteins and processes a tumour necrosis factor-alpha fusion protein. *Biochem Biophys Res Commun.* 1996;228:421–429.
86. Gronski Jr TJ, Martin RL, Kobayashi DK, et al. Hydrolysis of a broad spectrum of extracellular matrix proteins by human macrophage elastase. *J Biol Chem.* 1997;272:12189–12194.
87. Shipley JM, Wesselschmidt RL, Kobayashi DK, Ley TJ, Shapiro SD. Metalloelastase is required for macrophage-mediated proteolysis and matrix invasion in mice. *Proc Natl Acad Sci USA.* 1996;93:3942–3946.
88. Lindsey ML, Zamilpa R. Temporal and spatial expression of matrix metalloproteinases and tissue inhibitors of metalloproteinases following myocardial infarction. *Cardiovasc Ther.* 2012;30:31–41.
89. Spinale FG, Janicki JS, Zile MR. Membrane-associated matrix proteolysis and heart failure. *Circ Res.* 2013;112:195–208.
90. Mu D, Cambier S, Fjellbirkeland L, et al. The integrin alpha(v)beta8 mediates epithelial homeostasis through MT1-MMP-dependent activation of TGF-beta1. *J Cell Biol.* 2002;157:493–507.
91. Knauper V, Will H, Lopez-Otin C, et al. Cellular mechanisms for human procollagenase-3 (MMP-13) activation. Evidence that MT1-MMP (MMP-14) and gelatinase a (MMP-2) are able to generate active enzyme. *J Biol Chem.* 1996;271:17124–17131.
92. Ma Y, Chiao YA, Zhang J, Manicone AM, Jin YF, Lindsey ML. Matrix metalloproteinase-28 deletion amplifies inflammatory and extracellular matrix responses to cardiac aging. *Microsc Microanal.* 2012;18:81–90.
93. Ma Y, Halade GV, Zhang J, et al. Matrix metalloproteinase-28 deletion exacerbates cardiac dysfunction and rupture after myocardial infarction in mice by inhibiting M2 macrophage activation. *Circ Res.* 2013;112:675–688.
94. Werner SR, Mescher AL, Neff AW, et al. Neural MMP-28 expression precedes myelination during development and peripheral nerve repair. *Dev Dyn.* 2007;236:2852–2864.
95. Iyer RP, Jung M, Lindsey ML. Using the laws of thermodynamics to understand how matrix metalloproteinases coordinate the myocardial response to injury. *Metalloproteinases Med.* 2015;2:75–82.

96. Kaludercic N. Inhibiting metalloproteases with PD 166793 in heart failure: impact on cardiac remodeling and beyond. *Cardiovasc Ther.* 2008;26:24–37.
97. Hudson MP, Armstrong PW, Ruzyllo W, et al. Effects of selective matrix metalloproteinase inhibitor (PG-116800) to prevent ventricular remodeling after myocardial infarction: results of the PREMIER (Prevention of Myocardial Infarction Early Remodeling) trial. *J Am Coll Cardiol.* 2006;48:15–20.
98. Johnson JL, Devel L, Czarny B, et al. A selective matrix metalloproteinase-12 inhibitor retards atherosclerotic plaque development in apolipoprotein E-knockout mice. *Arterioscler Thromb Vasc Biol.* 2011;31:528–535.
99. Robichaud TK, Steffensen B, Fields GB. Exosite interactions impact matrix metalloproteinase collagen specificities. *J Biol Chem.* 2011;286:37535–37542.
100. Lindsey ML, Gannon J, Aikawa M, et al. Selective matrix metalloproteinase inhibition reduces left ventricular remodeling but does not inhibit angiogenesis after myocardial infarction. *Circulation.* 2002;105:753–758.
101. Mukherjee R, Brinsa TA, Dowdy KB, et al. Myocardial infarct expansion and matrix metalloproteinase inhibition. *Circulation.* 2003;107:618–625.
102. Rohde LE, Ducharme A, Arroyo LH, et al. Matrix metalloproteinase inhibition attenuates early left ventricular enlargement after experimental myocardial infarction in mice. *Circulation.* 1999;99:3063–3070.
103. Yarbrough WM, Mukherjee R, Escobar GP, et al. Selective targeting and timing of matrix metalloproteinase inhibition in post-myocardial infarction remodeling. *Circulation.* 2003;108:1753–1759.
104. Peterson JT. Matrix metalloproteinase inhibitor development and the remodeling of drug discovery. *Heart Fail Rev.* 2004;9:63–79.
105. Zhang P, Yang YJ, Chen X, et al. Effects of doxycycline on the expression of matrix metalloproteinase and tissue inhibitor of MMP in myocardium after acute myocardial infarction: an animal experiment with rats. *Zhonghua Yi Xue Za Zhi.* 2004;84: 1288–1293.
106. Camp TM, Tyagi SC, Aru GM, Hayden MR, Mehta JL, Tyagi SC. Doxycycline ameliorates ischemic and border-zone remodeling and endothelial dysfunction after myocardial infarction in rats. *J Heart Lung Transplant.* 2004;23:729–736.
107. Cerisano G, Buonamici P, Valenti R, et al. Early short-term doxycycline therapy in patients with acute myocardial infarction and left ventricular dysfunction to prevent the ominous progression to adverse remodelling: the tiptop trial. *Eur Heart J.* 2014;35:184–191.
108. Iyer RP, de Castro Bras LE, Patterson NL, et al. Early matrix metalloproteinase-9 inhibition post-myocardial infarction worsens cardiac dysfunction by delaying inflammation resolution. *J Mol Cell Cardiol.* 2016;100:109–117.
109. Lauer-Fields JL, Sritharan T, Stack MS, Nagase H, Fields GB. Selective hydrolysis of triple-helical substrates by matrix metalloproteinase-2 and -9. *J Biol Chem.* 2003;278: 18140–18145.
110. Lauer-Fields JL, Whitehead JK, Li S, Hammer RP, Brew K, Fields GB. Selective modulation of matrix metalloproteinase 9 (MMP-9) functions via exosite inhibition. *J Biol Chem.* 2008;283:20087–20095.
111. Zamilpa R, Ibarra J, de Castro Bras LE, et al. Transgenic overexpression of matrix metalloproteinase-9 in macrophages attenuates the inflammatory response and improves left ventricular function post-myocardial infarction. *J Mol Cell Cardiol.* 2012;53:599–608.
112. Pfeffer MA. ACE inhibitors in acute myocardial infarction: patient selection and timing. *Circulation.* 1998;97:2192–2194.
113. Yamamoto D, Takai S, Miyazaki M. Prediction of interaction mode between a typical ACE inhibitor and MMP-9 active site. *Biochem Biophys Res Commun.* 2007;354: 981–984.

114. Kuntze LB, Antonio RC, Izidoro-Toledo TC, Meschiari CA, Tanus-Santos JE, Gerlach RF. Captopril and lisinopril only inhibit matrix metalloproteinase-2 (MMP-2) activity at millimolar concentrations. *Basic Clin Pharmacol Toxicol.* 2014;114:233–239.
115. Yamamoto D, Takai S, Miyazaki M. Inhibitory profiles of captopril on matrix metalloproteinase-9 activity. *Eur J Pharmacol.* 2008;588:277–279.
116. Yamamoto D, Takai S, Jin D, Inagaki S, Tanaka K, Miyazaki M. Molecular mechanism of imidapril for cardiovascular protection via inhibition of MMP-9. *J Mol Cell Cardiol.* 2007;43:670–676.
117. Papadopoulos DP, Economou EV, Makris TK, et al. Effect of angiotensin-converting enzyme inhibitor on collagenolytic enzyme activity in patients with acute myocardial infarction. *Drugs Exp Clin Res.* 2004;30:55–65.
118. Zhang P, Yang YJ, Chen X, et al. Comparison of doxycycline, losartan, and their combination on the expression of matrix metalloproteinase, tissue inhibitor of matrix metalloproteinase, and collagen remodeling in the noninfarcted myocardium after acute myocardial infarction in rats. *Zhongguo Yi Xue Ke Xue Yuan Xue Bao.* 2005;27:53–61.
119. Yang D, Ma S, Li D, Tang B, Yang Y. Angiotensin II receptor blockade improves matrix metalloproteinases/tissue inhibitor of matrix metalloproteinase-1 balance and restores fibronectin expression in rat infarcted myocardium. *Biochem Biophys Res Commun.* 2009;388:606–611.
120. Harada K, Sugaya T, Murakami K, Yazaki Y, Komuro I. Angiotensin II type 1A receptor knockout mice display less left ventricular remodeling and improved survival after myocardial infarction. *Circulation.* 1999;100:2093–2099.
121. Miyazaki S, Kasai T, Miyauchi K, et al. Changes of matrix metalloproteinase-9 level is associated with left ventricular remodeling following acute myocardial infarction among patients treated with trandolapril, valsartan or both. *Circ J.* 2010;74:1158–1164.
122. Jones K, Saxon L, Cunningham W, Adams P, Guideline Development Group. Secondary prevention for patients after a myocardial infarction: summary of updated nice guidance. *BMJ.* 2013;347:f6544.
123. Rude MK, Duhaney T-AS, Kuster GM, et al. Aldosterone stimulates matrix metalloproteinases and reactive oxygen species in adult rat ventricular cardiomyocytes. *Hypertension.* 2005;46:555–561.
124. Mill JG, Milanez Mda C, de Resende MM, Gomes Mda G, Leite CM. Spironolactone prevents cardiac collagen proliferation after myocardial infarction in rats. *Clin Exp Pharmacol Physiol.* 2003;30:739–744.
125. Li MJ, Huang CX, Okello E, Yanhong T, Mohamed S. Treatment with spironolactone for 24 weeks decreases the level of matrix metalloproteinases and improves cardiac function in patients with chronic heart failure of ischemic etiology. *Can J Cardiol.* 2009;25:523–526.
126. Modesti PA, Vanni S, Paniccia R, et al. Characterization of endothelin-1 receptor subtypes in isolated human cardiomyocytes. *J Cardiovasc Pharmacol.* 1999;34:333–339.
127. Shubeita HE, McDonough PM, Harris AN, et al. Endothelin induction of inositol phospholipid hydrolysis, sarcomere assembly, and cardiac gene expression in ventricular myocytes. A paracrine mechanism for myocardial cell hypertrophy. *J Biol Chem.* 1990;265:20555–20562.
128. Guarda E, Katwa LC, Myers PR, Tyagi SC, Weber KT. Effects of endothelins on collagen turnover in cardiac fibroblasts. *Cardiovasc Res.* 1993;27:2130–2134.
129. Podesser BK, Siwik DA, Eberli FR, et al. ET(A)-receptor blockade prevents matrix metalloproteinase activation late postmyocardial infarction in the rat. *Am J Physiol Heart Circ Physiol.* 2001;280:H984–H991.

130. Gheorghiade M, Goldstein S. Beta-blockers in the post-myocardial infarction patient. *Circulation*. 2002;106:394–398.
131. Senzaki H, Paolocci N, Gluzband YA, et al. β-Blockade prevents sustained metalloproteinase activation and diastolic stiffening induced by angiotensin II combined with evolving cardiac dysfunction. *Circ Res*. 2000;86:807–815.
132. Deten A, Volz HC, Holzl A, Briest W, Zimmer HG. Effect of propranolol on cardiac cytokine expression after myocardial infarction in rats. *Mol Cell Biochem*. 2003;251(1–2): 127–137.
133. Cimmino G, Ibanez B, Giannarelli C, et al. Carvedilol administration in acute myocardial infarction results in stronger inhibition of early markers of left ventricular remodeling than metoprolol. *Int J Cardiol*. 2011;153:256–261.
134. Song G, Hennessy M, Zhao YL, et al. Adrenoceptor blockade alters plasma gelatinase activity in patients with heart failure and MMP-9 promoter activity in a human cell line (ECV304). *Pharmacol Res*. 2006;54:57–64.
135. McDaniel MC. Anticoagulation after anterior myocardial infarction: primum non nocere, or first do no harm. *JACC Cardiovasc Interv*. 2015;8:163–165.
136. Tziakas DN, Chalikias GK, Hatzinikolaou EI, et al. Alteplase treatment affects circulating matrix metalloproteinase concentrations in patients with ST segment elevation acute myocardial infarction. *Thromb Res*. 2006;118:221–227.
137. Nutescu EA, Shapiro NL, Chevalier A. New anticoagulant agents: direct thrombin inhibitors. *Cardiol Clin*. 2008;26:169–187. v-vi.
138. Nutescu EA, Shapiro NL, Chevalier A, Amin AN. A pharmacologic overview of current and emerging anticoagulants. *Cleve Clin J Med*. 2005;72(suppl 1):S2–S6.
139. Tyagi SC, Kumar S, Katwa L. Differential regulation of extracellular matrix metalloproteinase and tissue inhibitor by heparin and cholesterol in fibroblast cells. *J Mol Cell Cardiol*. 1997;29:391–404.
140. Zhou Q, Yuichiro Y, Masatoshi T, Terukatsu A. Effect of heparin on high glucose induced proliferation and expression of matrix metalloproteinases in normal human mesangial cells. *J Cent South Univ Technol*. 2005;12:359–364.
141. Bhatt DL. NSAIDS and the risk of myocardial infarction: do they help or harm? *Eur Heart J*. 2006;27:1635–1636.
142. Awtry EH, Loscalzo J. Aspirin. *Circulation*. 2000;101:1206–1218.
143. Pan M-R, Hung W-C. Nonsteroidal anti-inflammatory drugs inhibit matrix metalloproteinase-2 via suppression of the ERK/Sp1-mediated transcription. *J Biol Chem*. 2002;277:32775–32780.
144. Mehta JL, Chen J, Yu F, Li DY. Aspirin inhibits ox-LDL-mediated LOX-1 expression and metalloproteinase-1 in human coronary endothelial cells. *Cardiovasc Res*. 2004;64:243–249.
145. Kiran MS, Sameer Kumar VB, Viji RI, Sudhakaran PR. Temporal relationship between MMP production and angiogenic process in HUVECs. *Cell Biol Int*. 2006;30:704–713.
146. Koh KK. Effects of statins on vascular wall: vasomotor function, inflammation, and plaque stability. *Cardiovasc Res*. 2000;47:648–657.
147. Meschiari CA, Pinheiro LC, Guimaraes DA, Gerlach RF, Tanus-Santos JE. Sodium nitrite attenuates MMP-9 production by endothelial cells and may explain similar effects of atorvastatin. *Naunyn Schmiedebergs Arch Pharmacol*. 2016;389:223–231.
148. Turner NA, Aley PK, Hall KT, et al. Simvastatin inhibits TNFalpha-induced invasion of human cardiac myofibroblasts via both MMP-9-dependent and -independent mechanisms. *J Mol Cell Cardiol*. 2007;43:168–176.
149. Li J, Zhao SP, Peng DQ, Xu ZM, Zhou HN. Early effect of pravastatin on serum soluble CD40L, matrix metalloproteinase-9, and C-reactive protein in patients with acute myocardial infarction. *Clin Chem*. 2004;50:1696–1699.

150. Nakaya R, Uzui H, Shimizu H, et al. Pravastatin suppresses the increase in matrix metalloproteinase-2 levels after acute myocardial infarction. *Int J Cardiol*. 2005;105: 67–73.
151. Yasuda S, Miyazaki S, Kinoshita H, et al. Enhanced cardiac production of matrix metalloproteinase-2 and -9 and its attenuation associated with pravastatin treatment in patients with acute myocardial infarction. *Clin Sci (Lond)*. 2007;112:43–49.
152. Deleon KY, de Castro Bras LE, Lange RA, Lindsey ML. Extracellular matrix proteomics in cardiac ischemia/reperfusion: the search is on. *Circulation*. 2012;125:746–748.

CHAPTER THREE

The Balance Between Metalloproteinases and TIMPs: Critical Regulator of Microvascular Endothelial Cell Function in Health and Disease

Marcello G. Masciantonio*[,†], Christopher K.S. Lee*[,†], Valerie Arpino*[,†], Sanjay Mehta*, Sean E. Gill*[,†,1]

*Centre for Critical Illness Research, Lawson Health Research Institute, London, ON, Canada
†Schulich School of Medicine and Dentistry, Western University, London, ON, Canada
[1]Corresponding author: e-mail address: sgill8@uwo.ca

Contents

Progress in Molecular Biology and Translational Science, Volume 147
ISSN 1877-1173
http://dx.doi.org/10.1016/bs.pmbts.2017.01.001

Abstract

Endothelial cells (EC), especially the microvascular EC (MVEC), have critical functions in health and disease. For example, healthy MVEC provide a barrier between the fluid and protein found within the blood, and the surrounding tissue. Following tissue injury or infection, the microvascular barrier is often disrupted due to activation and dysfunction of the MVEC. Multiple mechanisms promote MVEC activation and dysfunction, including stimulation by cytokines, mechanical interaction with activated leukocytes, and exposure to harmful leukocyte-derived molecules, which collectively result in a loss of MVEC barrier function. However, MVEC activation is also critical to facilitate recruitment of inflammatory cells, such as neutrophils (PMNs) and monocytes, into the injured or infected tissue. Metalloproteinases, including the matrix metalloproteinases (MMPs) and the closely related, a disintegrin and metalloproteinases (ADAMs), have been implicated in regulating both MVEC barrier function, through cleavage of adherens and tight junctions proteins between adjacent MVEC and through degradation of the extracellular matrix, as well as PMN–MVEC interaction, through shedding of cell surface PMN receptors. Moreover, the tissue inhibitors of metalloproteinases (TIMPs), which collectively inhibit most MMPs and ADAMs, are critical regulators of MVEC activation and dysfunction through their ability to inhibit metalloproteinases and thereby promote MVEC stability. However, TIMPs have been also found to modulate MVEC function through metalloproteinase-independent mechanisms, such as regulation of vascular endothelial growth factor signaling. This chapter is focused on examining the role of the metalloproteinases and TIMPs in regulation of MVEC function in both health and disease.

1. INTRODUCTION

Endothelial cells (EC), especially those found within the microvasculature, referred to as the microvascular EC (MVEC), have critical functions in both health and disease. One of the primary functions of healthy MVEC is to provide a barrier between the fluid and protein found within the blood, and the surrounding tissue.[1] Following tissue injury or infection, the microvascular barrier is often disrupted due to activation and dysfunction of the MVEC.[1] Multiple mechanisms promote MVEC activation and dysfunction, including stimulation by cytokines, mechanical interaction with activated leukocytes, and exposure to harmful leukocyte-derived molecules, such as proteases and oxidants.[2–7] These factors result in MVEC abnormalities, including disruption of inter-MVEC junctions, cytoskeleton-driven retraction, and apoptotic death, which leads to a loss of barrier function.[2–7] However, while activation of the MVEC is associated with loss of barrier function, this process is also critical to facilitate recruitment of inflammatory cells, such as neutrophils (PMNs) and monocytes, into the injured or infected tissue.[2–7]

Recent work has demonstrated that metalloproteinases, specifically the matrix metalloproteinases (MMPs) and a disintegrin and metalloproteinases (ADAMs), as well as their specific inhibitors, the tissue inhibitors of metalloproteinases (TIMPs), are important regulators of MVEC function in health and disease (Fig. 1).

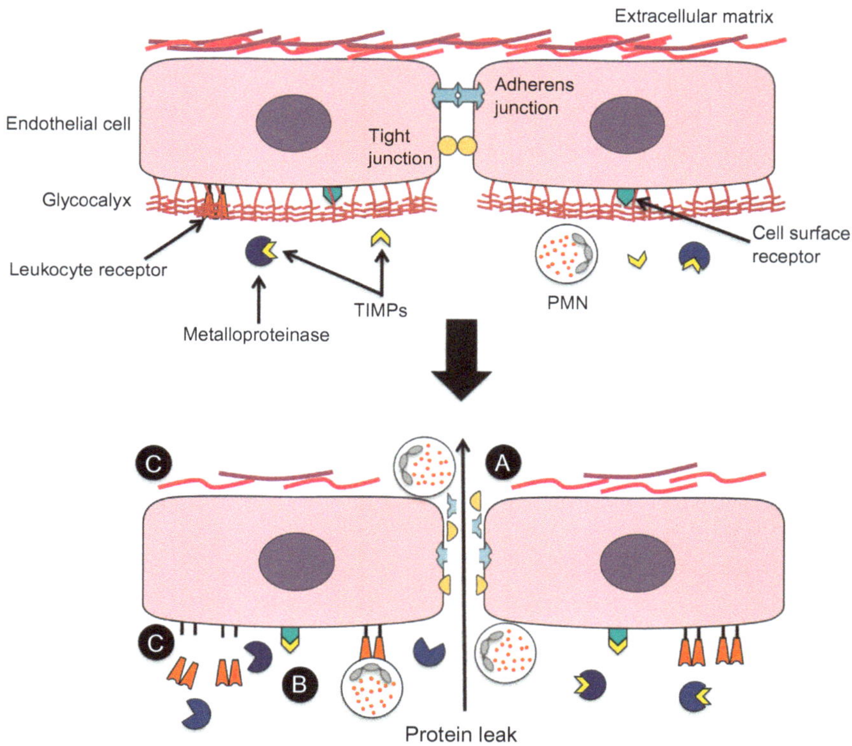

Fig. 1 Disruption in the balance between metalloproteinases and tissue inhibitors of metalloproteinases (TIMPs) leading to increased metalloproteinase activity promotes microvascular endothelial cell (MVEC) activation and dysfunction following tissue injury or infection. (A) Cleavage of inter-MVEC junctional proteins (i.e., vascular endothelial [VE]-cadherin, junctional adhesion molecule [JAM]-A) reduces MVEC barrier function and increases permeability for larger macromolecules. (B) Shedding of leukocyte receptors reduces neutrophil (PMN) adhesion to endothelial cells and allows for PMN extravasation across the MVEC into the surrounding tissue. (C) Cleavage of the extracellular matrix/glycocalyx increases MVEC permeability and promotes PMN–MVEC interaction. Importantly, TIMPs inhibit metalloproteinase activity, and thereby are critical to maintaining MVEC barrier function and restricting PMN–MVEC interaction.

This chapter will focus on examining the roles of metalloproteinases and TIMPs in MVEC function, specifically assessing the role of the balance between metalloproteinases and TIMPs in regulation of MVEC activation and dysfunction.

2. METALLOPROTEINASES

Metalloproteinases are an incredibly diverse group of enzymes comprised of more than 50 distinct families.[8] For the purpose of this chapter, our focus will be on the metzincin family of metalloproteinases, which are endopeptidases utilizing a Zn^{2+} or Ca^{2+} in their active site. Specifically, our focus will be the MMPs and the closely related ADAMs, which share several common structural domains with MMPs.[8] These domains include a signal peptide, the propeptide domain, and the catalytic domain. The propeptide domain contains a cysteine residue that ligates the active site Zn^{2+} or Ca^{2+} and maintains the enzyme in an inactive conformation until this interaction is disrupted.[8]

2.1 Matrix Metalloproteinases

MMPs were initially described for their role in processing (cleavage and degradation) the extracellular matrix (ECM); however, it is now understood that MMPs process a wide variety of extracellular proteins, including cytokines, cell surface receptors, and cell–cell junctional proteins.[9–11]

2.1.1 Structure/Regulation

MMPs are generally categorized by their domain structure.[12,13] In addition to the domains common to all MMPs and other metalloproteinases (i.e., signal peptide, prodomain, and catalytic domain), specific subsets of MMPs contain unique domains. Many MMPs contain a hemopexin-like C-terminal region, which is connected to the catalytic region by a short flexible hinge and can mediate protein–protein interactions (i.e., binding to TIMPs). Additionally, the membrane type (MT)-MMPs contain a furin-recognition domain allowing for activation by the proprotein convertase furin, as well as a C-terminal transmembrane domain.[10,13,14]

MMPs are regulated in four ways—gene expression, compartmentalization, pro-MMP activation, and enzyme inactivation.[13] MMP expression can be mediated by many factors, such as growth factors (e.g., transforming growth factor [TGF] β and vascular endothelial growth factor [VEGF]), cytokines (e.g., interferon [IFN] γ, interleukin [IL] 1β, and IL6), cell–ECM signaling through integrins, and pathogen-associated molecular patterns (PAMPs).[15] Thus, MMP expression is generally higher in tissues that are injured, inflamed, or undergoing repair or remodeling, relative to lower expression in healthy tissues.[13]

Studies have shown that MMPs demonstrate considerable overlap in their substrate selectivity in vitro. Thus, regulation of MMP activity is critical and likely occurs at multiple levels. Compartmentalization (e.g., the pericellular accumulation of MMPs in specific microenvironments through cell–enzyme interactions) is thought to be important in regulating extracellular substrate degradation.[13] Additionally, while some MMPs are secreted in their active form, many are secreted as zymogens, or inactive precursors, and remain in this inactive conformation due to the interaction between the propeptide and catalytic domains.[10] Disruption of this interaction, which occurs through multiple mechanisms including proteolytic removal or oxidation of the prodomain, allows for MMP activity.[10,13,15,16] Finally, MMP activity is directly inhibited by specific inhibitors, the TIMPs.[15,17]

2.1.2 Function

The primary function of MMPs was classically defined as degradation of the ECM through turnover of connective tissue proteins including collagens and elastins, as well as basement membrane components such as laminin and fibronectin.[18] Importantly, it is now recognized that the ECM is a complex and dynamic structure that functions as much more than a simple scaffold for cells. The ECM directly provides cues to the cells through cell–ECM interactions mediated by integrins.[19] Additionally, the ECM can bind growth factors, such as fibroblast growth factor and TGFβ, thus regulating their activity, including facilitating localization and presentation of these growth factors to the surrounding cells.[10,20] As such, remodeling of the ECM by MMPs likely regulates cell–ECM signaling, the release of cryptic ECM fragments, and the release of sequestered growth factors.[10,20]

While some MMPs, such as MMP14, appear to have a role in regulating ECM remodeling, it is now known that MMPs are also involved in the control of several other biological functions including, but not limited to, direct cleavage and activation of growth factors, such as TGFβ, chemokine processing, regulation of cellular apoptosis, and cleavage of cell surface-associated or transmembrane proteins.[13,18,21] Together, these functions strongly support a role for MMPs in regulation of inflammation and tissue injury/repair through control of a variety of cellular functions, such as cell migration, activation, differentiation, and apoptosis.

2.2 A Disintegrin and Metalloproteinases

ADAMs differ from MMPs, which are primarily secreted from the cell as "soluble" proteinases, as they are typically found anchored to the cellular membrane.[8] Further, ADAMs function as transmembrane proteases or "sheddases" that act to cleave the extracellular domain of several cell surface proteins and receptors, a process referred to as ectodomain shedding.[22]

2.2.1 Structure/Regulation

ADAMs contain a propeptide and catalytic domain, similar to MMPs. In addition, ADAMs have a disintegrin-binding region, which can interact with integrins and mediate cell–ECM interactions; a cysteine-rich domain; an epidermal growth factor (EGF)-like region; a transmembrane region (similar to MT-MMPs), which anchors the ADAM to the cell membrane; and a cytoplasmic tail, which for some ADAMs has been found to mediate intracellular signaling.[8]

Similar to MMPs, ADAMs are regulated by gene expression, pro-ADAM activation, and enzyme inactivation. For most ADAMs, the prodomain is removed intracellularly by the proprotein convertase furin.[8] The prodomain of ADAM8 and -28, however, is thought to be removed by autocatalysis due to mutations in the binding site of the catalytic domain.[23] ADAM activity, as well as localization, has also been found to be induced by a variety of stimuli, including G protein-coupled receptors and protein kinase C (PKC) signaling, likely through phosphorylation of the cytoplasmic domain.[8,24]

ADAMs are inhibited by TIMPs; however, this inhibition is mediated primarily by TIMP3.[25] For example, TIMP3 is thought to be the physiological inhibitor of both ADAM10 and ADAM17.[15] Further, TIMP2 and TIMP4 are capable of inhibiting some ADAMs, but no evidence exists of a role for TIMP1 in ADAM inhibition.[25] It has been suggested that TIMPs may bind ADAMs via multiple interactions, using both inhibitory N-terminus and the C-terminus, which has been found to be involved in protein–protein interactions.[26]

2.2.2 Function

Within the human genome, 25 ADAMs have been identified; however, 4 are considered to be pseudogenes.[8] Of the 21 remaining ADAMs, only 13 are functional proteases with the other 8 ADAMs directly involved in cell–cell and cell–ECM interactions. ADAMs effect many cellular processes by cleaving cell surface molecules, such as cytokines, cytokine receptors, and cell adhesion molecules.[8,27] The ability to shed or cleave proteins from the cell surface strongly supports the central role played by ADAMs in the control of inflammation. Further, ADAMs, such as ADAM17 (also known as tumor necrosis factor α [TNFα] converting enzyme or TACE), can promote both the initiation of inflammation by cleaving proinflammatory cytokines (i.e., TNFα) from the cell surface and the resolution of inflammation by cleaving cytokine receptors (i.e., TNFα receptor [TNFR1]) or leukocyte receptors (i.e., intracellular adhesion molecule [ICAM]1).[8,28] In addition to regulation of inflammation, the ability to shed proteins associated with cell–cell junctions supports a role for ADAMs in regulation of tissue injury/repair. Specifically, shedding proteins, such as the cadherins, can lead to tissue damage (i.e., through a loss of barrier function and the onset of edema) but is also required for cell migration during the initiation of wound repair.

3. TISSUE INHIBITOR OF METALLOPROTEINASES

The TIMP family is comprised of four members, TIMP1–4.[17,29] TIMPs inhibit active metalloproteinases in a 1:1 stoichiometric inhibitor-to-enzyme ratio.[17,29] Specifically, the N-terminal domain of TIMP molecules interacts with the active site of metalloproteinases in a fashion similar to metalloproteinase substrates, resulting in inhibition of catalytic activity.[17,29] While TIMP1, -2, and -4 are secreted and thus function in a soluble manner, TIMP3 is known to be bound to sulfated glycosaminoglycans (GAGs) located in the ECM.[17,29,30]

3.1 Structure/Regulation

Each TIMP contains two distinct domains, an N- and C-terminal domain, connected through six conserved disulfide bonds, which control TIMP function and localization.[25,31] The N-terminal domain is critical for the inhibition of metalloproteinase activity as well as the affinity of TIMPs for specific metalloproteinases.[32] For example, the N-terminal domain of TIMP3 has been found to interact with the C-terminal domains of ADAMTS4 and ADAMTS5 promoting interaction between the enzyme and inhibitor.[33] The C-terminal domain has been found to mediate protein–protein interactions, such as binding of TIMP1 and TIMP2 to pro-MMP9 and pro-MMP2, respectively.[17,34]

Regulation of TIMPs occurs through control of gene expression as well as through protein endocytosis and subsequent recycling or degradation. TIMPs are expressed constitutively in many mammalian tissues; however, TIMP expression can also be induced in most tissues during development as well as during injury and tissue repair.[35] TIMP expression is regulated at the transcriptional level by cytokines, growth factors, and as recently demonstrated, by microRNAs.[36]

The specific role for TIMPs (i.e., beneficial or detrimental) in healthy or diseased tissue is dependent on the specific TIMP and tissue being examined. For example, TIMP1 is constitutively expressed at a low level in many tissues, but in the setting of inflammation and tissue injury/disease, including lung injury or breast cancer, TIMP1 expression is generally increased over healthy tissue.[37] Additionally, TIMP3 is also expressed in many tissues; however, TIMP3 expression in the setting of inflammation and tissue injury/disease is dependent on the tissue and even model of injury. For example, TIMP3 expression has been found to be increased in kidney inflammatory diseases.[38] However, TIMP3 expression has been found to be decreased following bleomycin-induced lung injury (a model of pulmonary fibrosis).[39]

3.2 Function

As suggested by their name, TIMPs are the primary inhibitors of metalloproteinases, including both the MMPs and ADAMs. However, TIMPs, especially TIMP1 but also TIMP2 and -3, have been found to have functions that appear to be independent of their ability to inhibit metalloproteinases.

3.2.1 Metalloproteinase-Dependent TIMP Functions

In general, all four TIMPs have been shown to collectively inhibit all MMPs in vitro. There are, however, some differences between TIMPs. For example, TIMP1 does not appear to be able to regulate MT-MMPs. In addition, ADAMs appear to be primarily inhibited by TIMP3. Some TIMPs have also been found to form a complex between an active MMP and latent pro-MMP leading to activation of the latent MMP. This has been demonstrated extensively with TIMP2, which binds active MMP14 and latent MMP2, and facilitates MMP2 activation.[40]

The majority of our understanding of metalloproteinase and TIMP interactions arise from in vitro studies utilizing purified proteins. Hence, much of these data reveal the potential for metalloproteinase–TIMP interaction but do not provide information as to physiological relevance. There have been a number of in vivo studies using mice lacking specific TIMPs that have sought to identify physiological interactions. For example, TIMP1 has been found to interact with and inhibit MMP7 (also referred to as matrilysin) as MMP7 activity was increased in the airway epithelium of mice lacking TIMP1 following naphthalene exposure. Further, coimmunoprecipitation demonstrated that TIMP1 directly binds to MMP7.[41]

Use of different models of injury and inflammation with mice lacking TIMP3 has found that TIMP3 is the physiological inhibitor of ADAM17. Specifically, inflammation is generally increased in mice genetically lacking TIMP3 following multiple models of injury (i.e., partial hepatectomy, bleomycin-induced lung injury, etc.) and in most cases, this increased inflammation is associated with increased TNFα shedding resulting from augmented ADAM17 activity. TIMP3 has also been associated with multiple other metalloproteinases in vivo, primarily through the observation of increased metalloproteinase activity (e.g., MMP9) or increased substrate degradation (e.g., aggrecan cleavage potentially by ADAMTS5) in mice lacking TIMP3.[31,42]

Rescue of the basal and pathological phenotypes in mice lacking specific TIMPs with small molecule metalloproteinase inhibitors also provides further support for the metalloproteinase-dependent roles for TIMPs in vivo.[39,42–45] Thus, through regulation of metalloproteinase activity, TIMPs are able to finely tune processing of a wide variety of extracellular substrates ranging from degradation of the ECM to cleavage of cytokines and cytokine receptors from the cell surface, which suggests TIMPs are critical for the regulation of many cellular functions from cell proliferation and differentiation during development to influx of leukocytes into injured tissues.

3.2.2 Metalloproteinase-Independent TIMP Functions

TIMPs have been found to exert diverse biological functions independent of their ability to inhibit metalloproteinases.[35] These functions include control of cell growth, migration, differentiation, and apoptosis. TIMP1 and TIMP2 appear to have the most established roles in metalloproteinase-independent cell signaling mechanisms, although recent studies have also identified metalloproteinase-independent functions for TIMP3.[35] In many cases, these metalloproteinase-independent functions have been mediated by TIMP interaction with a cell surface receptor to either activate the receptor and initiate cell signaling (i.e., TIMP1 binds CD63 and β1 integrin to promote epithelial cell proliferation and inhibit apoptosis) or to block ligand–receptor interaction (i.e., TIMP3 binds VEGFR1 and inhibits VEGF–VEGFR1 interaction and subsequently, angiogenesis).[34,46] While not the focus of this chapter, these mechanisms must be taken into account in any consideration of the biological actions of the metalloproteinase–TIMP balance. Importantly, the combination of both metalloproteinase-dependent and -independent functions for the TIMPs illustrate the critical role these proteins have in regulating cellular function during health and disease.

4. MICROVASCULAR ENDOTHELIAL CELLS

4.1 Structure

The inner lining of all blood vessels consists of a monolayer of EC, which serves as a selectively permeable barrier between the circulation and surrounding tissue.[4,5,47–49] The microvasculature includes the arterioles, capillaries, and postcapillary venules.[49] In addition to the endothelial monolayer, blood vessels are comprised of multiple layers of smooth muscles cells, connective tissue, and pericytes.[49] However, capillaries, which are the smallest of all blood vessels, consist only of the single endothelial monolayer.[49]

The EC are a multifunctional cell that, in addition to maintaining a selectively permeable barrier, are involved in angiogenesis, the maintenance of a nonthrombogenic barrier, and modulation of blood flow and vascular resistance. EC also regulate host immune responses via release of pro- and antiinflammatory mediators, such as cytokines and chemokines, as well as leukocyte adhesion to the endothelial surface via upregulation of leukocyte adhesion molecules.[47,49,50]

4.1.1 Interactions of Adjacent EC

Interaction of adjacent EC through adherens and tight intercellular junctions is the key mechanism regulating microvascular endothelial permeability.[4–7,48,51,52] These transmembrane junctional structures are comprised of multiple cell surface proteins, including vascular endothelial (VE)-cadherin within adherens junctions, and the claudins and occludins within tight junctions.[4–7,48,52] In most vascular beds, adherens junctions, which facilitate the formation of cell–cell contacts and ensure their proper maturation, account for the majority (approximately 80%) of the intercellular junctions present between EC, while tight junctions make up the remaining 20% of intercellular junctions and act to regulate the passage of small ions and solutes via the paracellular route.[7] However, certain vascular beds, which must maintain tight control of permeability to ions and solutes, such as brain capillaries, are enriched with tight junctions.[6] In addition to mediating paracellular interactions, the adhesive molecules present in adherens and tight junctions interact with an intracellular network of cytoskeletal proteins in order to further regulate EC morphology and integrity.[4,5,7,48,53]

VE-cadherin is the primary transmembrane structural adhesive protein present in endothelial adherens junctions, as it plays a principal role in the assembly of adherens junctions and maintenance of a restrictive endothelial barrier.[4–7,48,52] Interendothelial tight junctions are formed by the homotypic interaction between occludins, junctional adhesion molecules (JAMs), and claudin-5.[4,6,7,48] Further, within both adherens and tight junctions, the scaffold proteins—catenins (α, β, and γ) in adherens junctions and ZO1–3 in tight junctions—connect the intercellular junctional membrane proteins to the intracellular actin cytoskeleton and this interaction maintains junctional stability and proper barrier function.[4,6,7,48]

4.1.2 The Matrix of the Microvasculature

Within the microvasculature, the luminal surface of the endothelial monolayer is lined with a mesh-like matrix known as the glycocalyx.[4,54] The glycocalyx is made up of glycoproteins and proteoglycans, including syndecans and glypicans, in addition to GAGs.[4,49,54] The glycocalyx modulates several functions of the microvasculature, including barrier function and circulating leukocyte–MVEC interactions such as trans-MVEC migration.[4,54] However, the ability of the glycocalyx to facilitate maintenance of a semipermeable barrier to fluid and proteins may not occur in all microvascular beds (i.e., pulmonary microvasculature) as degradation of the glycocalyx was not associated with increased albumin flux or edema within the lung.[55]

The ECM is also known to be a structural component required for the maintenance of endothelial barrier integrity. EC bound to the ECM are generally quiescent, as activation of integrins on the basolateral surface of EC following binding to the ECM generates signals that stabilize adjacent cell–cell and cell–ECM adhesions, which are vital to the formation and preservation of an endothelial barrier.[56] Thus, both the glycocalyx and the ECM are critical to establishing and maintaining endothelial integrity and barrier function.

4.2 Function

EC are multifunctional, however, the primary role is to act as a selective barrier between the contents of the vasculature and the surrounding tissue. In order to control the substances passing through the semipermeable layer, EC have selective transport systems to target specific substrates. For example, larger substrates are thought to pass through the endothelial barrier membrane via a transcellular route mediated by caveolae.[57]

4.2.1 MVEC Barrier Function

Inter-MVEC adherens and tight intercellular junctions are required for the establishment and maintenance of MVEC barrier function.[4–7,48,51,52] However, dynamic control of endothelial permeability, which occurs through both paracellular and transcellular pathways, does allow for the regulated exchange of plasma proteins, solutes, and liquid between the vasculature and surrounding tissues.[4,5,47–49] The paracellular pathway consists of transport through the intercellular space between adjacent EC, whereas the transcellular pathway, also known as transcytosis, consists of active receptor-mediated transport through the cell itself.[4,5,48,49] Under basal conditions, only small molecules (e.g., ions, glucose) cross the endothelial barrier through the paracellular pathway.[4,49] Larger macromolecules are actively transported through membrane receptor/vesicle-mediated transcytosis.[4,48,49,58]

4.2.2 Other MVEC Functions

In addition to its role in barrier function, EC have a role in controlling blood viscosity and coagulation through secretion or release of large molecules, such as heparan sulfate and tissue plasminogen activator, as well as smaller molecules, such as prostacyclin and nitric oxide (NO).[59] Prostacyclin is a potent vasodilator that prevents coagulation by activating adenylyl cyclase and thus, increasing 3′,5′-cyclic adenosine monophosphate expression. It also prevents release of growth factors that cause thickening of the vascular wall. EC do not store prostacyclin, but rather produce it following mechanical (pulsatile pressure) or chemical (bradykinin, thrombin, IL1 signals). Prostacyclin synthesis decreases with age and vascular diseases such as hypertension.[59] NO regulates blood pressure and coagulation, and EC release NO following mechanical stimulation, mainly by pulsatile flow and shear stress.[59] The endothelial layer is also involved in angiogenesis, maintenance of nonthrombogenic barriers, production of ECM components, and the regulation of vascular resistance and tone, and host immune responses.[47]

5. MVEC DYSFUNCTION

Dysfunction of the microvasculature under conditions of inflammation or infection is central to the resulting organ injury, dysfunction, and potentially failure. This microvascular dysfunction is primarily characterized by activation and dysfunction of MVEC, leading to a loss of microvascular barrier function as well as upregulation of cell surface leukocyte receptors and upregulation of prothrombotic pathways.[2–7] Importantly, this dysfunction results in increased edema and tissue inflammation due to enhanced adhesion and diapedesis of leukocytes as well as increased microvascular permeability, both of which can lead to organ dysfunction and failure.[2–7]

5.1 Causes of MVEC Dysfunction

EC dysfunction occurs under many different conditions, including but not limited to local tissue injury, infection, and sepsis, but is also a hallmark of vascular diseases such as atherosclerosis. Within the setting of injury and/or infection, MVEC dysfunction is initiated by the interaction of MVEC with circulating blood components, including PMNs and soluble inflammatory mediators (e.g., bacterial products like lipopolysaccharide (LPS), or cytokines like TNFα). In fact, MVEC dysfunction following injury or infection is highly PMN dependent.[60–64] There are several mechanisms of PMN-dependent cell/tissue injury. These include physical interaction (e.g., PMN–MVEC adhesion), release of reactive O_2 species (ROS), proteases/peptides (e.g., metalloproteinases), proinflammatory cytokines/chemokines, and cytotoxic neutrophil extracellular traps.[64,65]

5.2 Features of MVEC Dysfunction

The primary features of MVEC dysfunction are the loss of MVEC barrier function and the increased PMN–MVEC interaction. The loss of MVEC barrier function is mediated through several mechanisms, including but not limited to disassembly of inter-MVEC adherens and tight junctions due to cleavage or modification of membrane junctional proteins; gap formation due to actin cytoskeleton rearrangement and microtubule-dependent MVEC contraction; activation of trans-MVEC transport systems; and MVEC death.[7,51,66] Enhanced PMN–MVEC interaction is associated with increased PMN recruitment out of the vasculature into tissues and is thought to be dependent on a number of factors, including but not limited to increased release of proinflammatory cytokines and chemokines, upregulation of MVEC cell surface leukocyte receptors, and shedding of the glycocalyx.[67,68]

5.2.1 Mechanisms of MVEC Barrier Dysfunction

MVEC activation by inflammatory cytokines and leukocytes leads to increased MVEC permeability, and this increased permeability is primarily due to loss of inter-MVEC junctions and MVEC retraction.[2–7] As

VE-cadherin is critical to the formation of MVEC adherens junctions, modifications of its structure along with its associated intracellular catenins, including p120, β- and α-catenin, significantly attenuates barrier integrity.[4–7,48,52,69] For example, enhanced barrier permeability is correlated with tyrosine phosphorylation of the VE-cadherin/catenin complex.[4,5,48,70]

In addition to p120 catenin, phosphorylation of β- and α-catenin results in the reduction of their affinity for the VE-cadherin cytoplasmic tail leading to modified VE-cadherin interaction with the actin cytoskeleton.[4,5,7,48] Reorganization of the endothelial actin cytoskeleton into contractile stress fibers following MVEC activation also facilitates cellular retraction, and this disrupts intercellular junctional protein adhesion resulting in the formation of intercellular gaps.[4,48,71–75] Further, this actin-mediated endothelial retraction is caused by the direct phosphorylation of the regulatory myosin light chain by myosin light chain kinase (MLCK), whose activation is dependent on the binding of Ca^{2+} to calmodulin (CaM).[48] For example, mice lacking a specific MLCK, MLCK210, were less susceptible to septic lung injury with reduced albumin leak and microvascular hyperpermeability.[76]

Activation of MVEC by proinflammatory cytokines (e.g., TNFα) has also been shown to induce loss of barrier function by promoting cleavage VE-cadherin and the generation of a soluble extracellular fragment.[4,48,77,78] This VE-cadherin cleavage appears to be at least partially dependent on tyrosine phosphorylation of the VE-cadherin/catenin complex by several kinases (i.e., PKC and p38 mitogen-activated protein kinase).[4,48,77,78] Finally, thrombin has been found to bind protease-activated receptor (PAR)1 on the MVEC surface and promote barrier dysfunction by increasing cytosolic Ca^{2+} concentrations leading to PKCα-dependent VE-cadherin phosphorylation and subsequent internalization.[4,48,79] Collectively, these studies illustrate the complex intercellular signaling pathways that are involved in the control of MVEC barrier function and that are often disrupted under pathological conditions, such as tissue injury or infection.

5.2.2 Regulation of Leukocyte–MVEC Interactions

The initial inflammatory response and secretion of proinflammatory cytokines, including TNFα, IL1β, and IFNγ, are driven primarily by local inflammatory cells (e.g., macrophages) but are also propagated by the EC themselves. This response, which leads to MVEC activation, occurs in response to infection through recognition and response to PAMPs (bacterial cell wall components such as LPS) as well as to tissue injury through recognition of damage-associated molecular patterns (e.g., ECM fragments, such as hyaluronan proteolytic products).[66,80,81]

Under basal conditions, platelet endothelial cell adhesion molecule (PECAM)1 and ICAM1 are expressed on the MVEC surface.[80–82] Under conditions of inflammation (e.g., TNFα stimulation), PECAM1 expression decreases while ICAM1 is significantly upregulated.[83,84] Chemotactic stimuli released from the site of injury/infection, such as KC in mice or IL8 in humans, actively recruit PMN to the sites of injury/infection where they adhere to the endothelium in order to migrate into the surrounding tissue.[47,80–82] PMN adhesion to MVEC is a highly controlled process dependent on sequential activation of multiple selectins and integrins. Leukocyte (L)-selectin on the PMN cell surface binds glycoproteins on the MVEC surface. Endothelial (E)-selectin and platelet (P)-selectin are present on MVEC and bind specific carbohydrate motifs on the PMN cell surface.[80–82] These interactions allow the PMNs to make further contact with inflammatory mediators produced by the MVEC, while upregulating β1 and β2 integrins, which bind MVEC ICAM1 and vascular cell adhesion molecule 1 (VCAM1), respectively.[47,66,80–82,85]

The glycocalyx also has a significant role in PMN–MVEC interactions under inflammatory conditions, as it regulates access to MVEC surface adhesion molecules as well as the release of PMN-specific chemokines.[4,54] Moreover, inflammatory conditions, such as sepsis, appear to be associated with degradation of the glycocalyx, as indicated by the increased plasma concentrations of fragmented GAGs.[54,86] This glycocalyx degradation is associated with the activation of endothelial heparanase, a glucuronidase responsible for the hydrolysis and resulting degradation of heparan sulfate; however, there is also limited evidence for the involvement of MMPs in proteolytic processing of the glycocalyx.[54,87] Importantly, glycocalyx degradation allows for the greater exposure of endothelial surface adhesion molecules, such as ICAM1 and VCAM1, and thus facilitates the subsequent adhesion and extravasation of circulating PMNs.[54,87]

6. METALLOPROTEINASES IN ENDOTHELIAL DYSFUNCTION

Multiple metalloproteinases, including both MMPs and ADAMs, are expressed by MVEC under healthy conditions. Moreover, EC expression and release of these metalloproteinases has been found to change during MVEC activation and dysfunction in response to infection, inflammation, and injury. Although current data are limited and focused on only a few metalloproteinases, it is clear that metalloproteinases play an important role in regulation of MVEC activation and dysfunction, including both MVEC barrier dysfunction and recruitment of leukocytes (Fig. 1 and Table 1).[27,68]

6.1 EC Metalloproteinase Expression

Multiple MMPs are expressed by MVEC, some of which are localized to vesicles likely for rapid secretion following activation. While data on metalloproteinase-regulated MVEC dysfunction are limited, changes in metalloproteinase expression in response to proinflammatory cytokines has been characterized in other types of EC. For example, angiotensin II (ANG II) treatment of HUVEC stimulates TNFα production that leads to enhanced MMP2 production and release, and cotreatment with anti-TNFα antibodies blocks this increased MMP2 production.[88]

Table 1 Metalloproteinases Regulate EC Activation and Dysfunction, Including EC Barrier Dysfunction and Recruitment of Leukocytes

Biological Effect	MMP Activity	ADAM Activity
Increase vascular permeability	+	+
ECM degradation		
• Fibronectin, laminin, type IV and V collagens	+	+
Cleave inter-MVEC junctional proteins		
• Adherens junctions (VE-cadherin)	MMP7	ADAM10, -12
• Tight junctions (occludin, ZO1, JAMs)	MMP2, -9	ADAM17
• Tyrosine phosphorylation-dependent disassembly	+	
• F-actin stress fiber formation	MMP9	
Cleavage and release of VEGF	MMP9	
Mediate PMN–EC interaction	+	+
Shedding of EC surface receptors		
• Cleavage of glycocalyx (sulfated proteoglycans)	MMP2, -7, -9	
• Shedding leukocyte receptors (ICAM, VCAM)	MMP14	ADAM10, -17
Reduce inflammation	+	+

ADAM, a disintegrin and metalloproteinase; *ECM*, extracellular matrix; *EC*, endothelial cell; *ICAM*, intercellular adhesion molecule; *JAM*, juncitonal adhesion molecule; *MMP*, matrix metalloproteinase; *VCAM*, vascular cell adhesion molecule; *VEGF*, vascular endothelial growth factor; *ZO1*, zona occludens 1.

Interestingly, stimulation of brain EC with TNFα and IL1β, selectively upregulates MMP9 production, but does not alter MMP2 expression.[89] This suggests vascular bed specificity in MMP expression in response to cytokines. ADAM expression is also mediated by cytokines. Specifically, ADAM17 expression by murine brain EC is strongly upregulated by many proinflammatory cytokines (i.e., TNFα, IL1β, IFNγ) and growth factors (EGF, VEGF).[90]

EC, including MVEC, also express and release higher levels of metalloproteinases following infection. For example, LPS, a component of the outer membrane of gram-negative bacteria, elicits strong increases in the expression and activity of ADAM10 and ADAM17 in human pulmonary MVEC.[27] Further, Japanese encephalitis virus, a single-stranded RNA virus, increases expression of MMP9 in human brain MVEC.[91] Additionally, infection of HUVEC with *Chlamydophila pneumoniae*, a bacteria associated with lung infections such as pneumonia, leads to increased MMP9 production while infection of HUVEC with Dengue virus leads to overproduction of MMP2 and to a lesser extent of MMP9.[91–93] It is important to note, however, that at least part of the EC metalloproteinase response to infection may be mediated through autocrine signaling by cytokines expressed by EC in response to the infection.

6.2 Function of Metalloproteinases in EC Dysfunction

A summary of the role of metalloproteinases in EC activation and dysfunction can be found in Table 1. Multiple MMPs and ADAMs have been shown to cleave inter-MVEC junctional proteins, and thereby mediate microvascular dysfunction through increased vascular permeability.[27,79,94–96] For instance, MMP7, ADAM10, and ADAM12 are all capable of cleaving VE-cadherin, the integral cell surface transmembrane protein of endothelial adherens junctions.[79,95] Further, increased MMP2 and -9 expression following HUVEC infection by Dengue virus is correlated with a loss of expression of VE-cadherin cell–cell adhesion, suggesting that MMP2 and -9 may also be capable of cleaving VE-cadherin.[93]

MMP2 and -9 along with ADAM17 are also capable of cleaving tight junction-associated proteins such as occludin, ZO1, and JAMs.[27,94,96] Additionally, tyrosine phosphorylation-dependent disassembly of crucial adhesive junctional proteins may also be MMP dependent, as permeability induced by tyrosine phosphatase inhibitors have been shown to be blocked by MMP inhibitors.[78,97] Moreover, knockdown of ADAM10 following small interfering RNA treatment of either EC or activated leukocytes stabilizes the microvascular barrier as well as reduces leukocyte transendothelial migration.[79] Collectively, these studies suggest increased metalloproteinase expression and activity promotes MVEC dysfunction through the cleavage of critical inter-MVEC junctional proteins.

MMPs and ADAMs with thrombospondin motifs (ADAMTSs) have also been found to cleave the ECM surrounding the vasculature, which can also modify EC function through disrupted endothelial–ECM interactions, ultimately leading to increased vascular permeability.[98–103] For example, proteolytic degradation of ECM constituents, including fibronectin, laminin, and type IV and V collagens, by MMPs, including MMP2 and -9, results in increased permeability across EC monolayers in vitro as well as increased pulmonary edema in vivo.[102,103] The cleavage and degradation of other ECM protein constituents, including hyaluronan and aggrecan, have also been shown to augment endothelial permeability.[103,104] Additionally, MMP9 has been shown to be responsible for the cleavage and release of VEGF bound to the ECM, which results in increased microvascular permeability through internalization of VE-cadherin.[21] Thus, these studies suggest increased metalloproteinase activity drives MVEC dysfunction through the proteolytic processing of integral MVEC structural determinants (i.e., the ECM), resulting in MVEC barrier dysfunction and enhanced microvascular leak.

Supportive evidence for the role of metalloproteinases in microvascular and MVEC dysfunction includes data from several studies using synthetic metalloproteinase activity inhibitors, such as GM6001. For example, stimulating HUVEC with TNFα increases metalloproteinase expression and activity, and treatment of these HUVEC with GM6001 reduces the TNFα-induced increase in MVEC permeability through inhibition of VE-cadherin cleavage.[78,79] Additionally, in models of brain edema due to dysfunction of the blood–brain barrier, treatment of murine cerebrovascular EC with GM6001 leads to decreased microvascular leak due to the stabilization of tight junction adhesive proteins and decreased EC F-actin stress fiber formation through the inhibition of MMP9.[105,106] Collectively, these studies highlight the potential role of metalloproteinase inhibition in the control of MVEC barrier dysfunction.

In addition to controlling MVEC barrier dysfunction, metalloproteinases are also involved in mediating PMN–MVEC interaction through multiple potential mechanisms (Fig. 1). One mechanism through which they have been proposed to have a role is through cleavage of the glycocalyx; however, evidence for this is limited and primarily indirect. MMPs localized to vesicles for rapid secretion from EC, including MMP2, -7, and -9, have been shown to have the ability to cleave sulfated proteoglycans, such as syndecans. Previous studies, however, have demonstrated that the glycocalyx is degraded, at least in part, by heparanase.[48,68] Interestingly, degradation of the glycocalyx still occurs in mice lacking heparanase, and this degradation is associated with increased MMP activity.[68] Thus, this suggests that MMPs and heparanase may both regulate shedding of the glycocalyx.[68] Collectively, this potential degradation of the glycocalyx could lead to increased PMN–MVEC interaction; however, this remains to be confirmed.

Metalloproteinases, including both MMPs and ADAMs, can also mediate PMN–MVEC interaction through shedding of leukocyte receptors, such as ICAM and VCAM, from the MVEC surface. As an example, MMP14, as well as ADAM10 and -17 have been found to cleave ICAM1.[107–109] Further, the interaction between ICAM1 and MMP14 appears to be mediated by interaction between the cytoplasmic domains.[109] Additionally, ADAM17 has been shown to cleave VCAM1 from the surface of heart and aorta EC following stimulation with phorbol 12-myristate 13-acetate (PMA).[110] Cleavage of ICAM1 and VCAM1 by MMPs and ADAMs reduces MVEC surface expression of ICAM1 and VCAM1 leading to decreased PMN–MVEC interaction.[110] These studies suggest that MMPs and ADAMs have the ability to downregulate PMN–MVEC interactions, reduce inflammation, and could also be involved in the resolution of inflammation. Thus, the role of metalloproteinases, including both MMPs and ADAMs, in MVEC dysfunction is complex, including promoting the loss of MVEC barrier function and increased leukocyte recruitment through degradation of the glycocalyx, while also decreasing direct leukocyte–MVEC interaction. Further, the specific roles likely depend on the metalloproteinase in question, the specific vascular bed, and the type of inflammation or injury.

7. TIMPs IN ENDOTHELIAL DYSFUNCTION

7.1 EC TIMP Expression

TIMP expression varies in the different vascular beds throughout the body, including both micro- and macrovascular beds; however, all TIMP family members are expressed within at least one vascular bed supporting the importance of TIMPs to EC function. For example, TIMP1, -3, and -4 are expressed by brain MVEC, and TIMP1, -2, and -3 are expressed by

EC isolated from the aorta, iliac artery, and coronary artery.[111–113] Additionally, TIMP3 is expressed by pulmonary MVEC, whereas TIMP2 and -4 do not appear to be.[45,114]

Besides EC, TIMPs are also often expressed by multiple other cells associated with the vasculature. For example, TIMP3 is expressed by pericytes, and TIMP2 and -4 are expressed by vascular smooth muscle cells.[101,115,116] Importantly, TIMPs expressed by these adjacent cell populations are localized within the microenvironment surrounding the MVEC, and in many instances, have been found to be involved in regulation of MVEC function.[101,115–117]

Within the setting of tissue injury and infection, the expression of the various TIMPs appears to be differentially regulated (i.e., expression of different TIMPs does not change in the same direction).[45,46,88,101,111,112,118,119] For example, treatment of brain MVEC with proinflammatory cytokines, specifically IL1β and TNFα, led to an increase in *Timp1* and a decrease in *Timp3* mRNA expression.[111] Additionally, stimulation of brain MVEC with homocysteine, which is associated with endothelial dysfunction, increased TIMP1 expression but decreased TIMP4 expression.[112] Interestingly, the TIMP expression in response to various stimuli also appears to be cell dependent as stimulation of a transformed brain EC line (hCMEC/D3) with TNFα resulted in increased TIMP3 expression.[120] There is, however, some evidence suggesting that similar expression patterns for TIMPs may also be observed across different vascular beds. For example, while it has yet to be linked to MVEC-specific expression, TIMP1 levels in serum are increased in severe sepsis.[121,122] Further, brain MVEC treated with homocysteine have increased TIMP1 protein expression.[112] Conversely, similar to the reduction in *Timp3* expression in brain MVEC under inflammatory conditions, TIMP3 expression (mRNA and protein) was significantly decreased in pulmonary MVEC following treatment with IFNγ, IL1β, and TNFα.[45]

7.2 TIMP Function in EC Dysfunction

A summary of the role of TIMPs in EC activation and dysfunction can be found in Table 2. A number of studies have characterized TIMP expression within multiple EC types under various conditions; however, the direct evidence of a role for EC-derived TIMPs in endothelial dysfunction is more limited. Indirect evidence from mice globally deficient for individual TIMPs as well as from studies of recombinant TIMPs suggests that TIMPs are integral mediators of endothelial dysfunction following tissue injury or in the

Table 2 TIMPs Regulate MVEC Activation and Dysfunction via Metalloproteinase-Dependent and -Independent Mechanisms

Biological Effect	TIMP1	TIMP2	TIMP3	TIMP4
MMP inhibition	+	+	+	+
ADAM inhibition	ADAM10	ADAM12	+	Limited
EC functions				
Reduction in ECM degradation	+	+	+	+
Preserve blood–brain barrier integrity	+	+	+	
Protect ischemic injury	+	+		
Restoration of tight junctions	+			
Prevent vascular leak/increase in permeability	+		+	
Inhibit angiogenesis		+[a]	+[a]	
Inhibit VEGF signaling		+[a]	+[a]	
Inhibit endothelial cell proliferation		+[a]	+[a]	
Inhibit endothelial cell migration		+[a]	+[a]	
Establish a selectively permeable barrier			+	
Regulate PMN–EC interaction			+	

[a]Currently thought to be metalloproteinase independent.
ADAM, a disintegrin and metalloproteinase; *EC*, endothelial cell; *ECM*, extracellular matrix; *MMP*, matrix metalloproteinase; *TIMP*, tissue inhibitor of metalloproteinases; *VEGF*, vascular endothelial growth factor.

setting of inflammation. For example, TIMP1–3 all appear to support the blood–brain barrier following injury, although the mechanism may differ between TIMPs. Mice lacking TIMP1 have increased ischemic injury and blood–brain barrier disruption compared to WT mice following focal cerebral ischemia and this is associated with increased MMP9 secretion and activity.[123] Further, increased expression of TIMP1 in brain EC treated with angiotensin (1–7) following hypoxia-induced injury was responsible for restoration of tight junctions and reduced EC permeability.[124]

Mice lacking TIMP2 also exhibit blood–brain barrier disruption compared to WT mice following focal cerebral ischemia; however, no changes were observed in MMP activity.[123] Additionally, injection of recombinant TIMP3 led to decreased vascular leak across the blood–brain barrier in a model of traumatic brain injury, and this effect was not mimicked by use of a synthetic metalloproteinase inhibitor.[111,125] Together, these studies suggest that TIMP2 and -3 may promote brain EC barrier function through metalloproteinase-independent mechanisms.

TIMP3 also appears to mediate EC barrier function through metalloproteinase-dependent mechanisms. *Timp3*$^{-/-}$ mice have augmented leak of Evans blue (EB)-labeled albumin into their snout and kidneys at baseline, which is thought to be due to the absence of TIMP3 in pericytes resulting in enhanced microvascular permeability due to increased global metalloproteinase activity.[101] *Timp3*$^{-/-}$ mice also have increased pulmonary EB-albumin leak at baseline, which is likely due to loss of PMVEC-derived TIMP3 as PMVEC isolated from *Timp3*$^{-/-}$ mice also exhibit impaired barrier function in vitro compared to PMVEC from WT mice.[45] Further, treatment of *Timp3*$^{-/-}$ PMVEC with global metalloproteinase inhibitor GM6001 appears to restore basal PMVEC barrier function.[45] Together, these studies suggest that TIMP3 may be an important homeostatic regulator of EC function in healthy tissue, such as establishing a selectively permeable barrier, and that this role for TIMP3 is at least partly dependent on the ability of TIMP3 to inhibit metalloproteinases.[45,103]

While the classic function for TIMPs is inhibition of metalloproteinase activity, both TIMP2 and -3 have also been found to inhibit VEGF signaling through metalloproteinase-independent mechanisms.[46,119,126] For TIMP3, this inhibition occurs through TIMP3 binding to the VEGF receptor and blocking VEGF–VEGFR2 interaction, ultimately inhibiting angiogenesis.[46,119] During angiogenesis, VEGF acts a potent mediator of increased endothelial permeability through phosphorylation and internalization of VE-cadherin and subsequent disassembly of cell–cell adhesive contacts.[4,46,119,127] Importantly, both TIMP2 and -3 have been found to inhibit angiogenesis, including inhibition of both EC proliferation and migration, through inhibition of VEGF signaling.[46,119,128]

In addition to regulation of MVEC barrier function, TIMPs have also been associated with regulation of PMN–MVEC interaction. Specifically, TIMP3 has been found to inhibit shedding of VCAM1 from the surface of aortic EC following treatment with IL1β, TNFα, or the phorbol ester PMA.[129] Aortic EC from *Timp3*$^{-/-}$ mice, which had increased VCAM1 shedding compared to aortic EC from WT mice, suggested that the inhibition of VCAM1 shedding by TIMP3 was likely through inhibition of ADAM17 as knockdown of ADAM17 protected against increased VCAM1 shedding.[129] Thus, TIMPs appear to be critical regulators of MVEC dysfunction following tissue injury/infection, including protection against loss of MVEC barrier function through both metalloproteinase-dependent and -independent mechanisms, as well as the regulation of leukocyte recruitment by stabilization of leukocyte receptors on the MVEC surface.

8. CONCLUSION

8.1 Summary of Role of Metalloproteinase/TIMP Balance in MVEC Dysfunction

Metalloproteinases, including MMPs and ADAMs, as well as their endogenous inhibitors, the TIMPS, have diverse functions within the setting of inflammation and tissue injury/infection. Further, there is clear evidence that MVEC dysfunction is regulated by both metalloproteinases and TIMPs (Fig. 1). Traditionally, it was believed that a shift in the balance between metalloproteinases and TIMPs in favor of the metalloproteinases, either by increased expression of metalloproteinases or decreased expression of TIMPs, was associated with greater inflammation following injury. Within the context of MVEC dysfunction and the associated increase in vascular permeability, the evidence does support this traditional role for the metalloproteinase/TIMP balance. Specifically, MVEC activation and dysfunction are associated with increased metalloproteinase expression and activity combined with decreased TIMP expression, specifically TIMP3, leading to increased ECM and inter-MVEC junctional protein degradation and loss of MVEC barrier function. Mice lacking specific TIMPs, including TIMP1–3, which have the balance further shifted in favor of metalloproteinases, provide additional support for the traditional view of the role of the metalloproteinase/TIMP balance. Further, shifting the balance in favor of TIMPs through the use of synthetic metalloproteinase inhibitors rescues the increased permeability associated with MVEC activation and dysfunction.

There is, however, also evidence that suggests that this view may not always be the case for the individual TIMPs. Specifically, TIMP1 expression is generally increased in tissue injury and following MVEC activation and dysfunction. The specific function of the increased TIMP1, however, and how it affects MVEC dysfunction is not understood at this time. Thus, generally shifting the balance in favor of MMPs is deleterious, whereas shifting the balance in favor of TIMPs is protective when examining their role in MVEC barrier function.

The role of the metalloproteinase/TIMP balance in leukocyte recruitment is more uncertain. Metalloproteinases have been found to shed adhesion proteins (i.e., JAMs), cell surface chemokines (i.e., CX3CL1), cell surface leukocyte receptors (i.e., ICAM1), and potentially shed the glycocalyx (Fig. 1). Additionally, TIMPs, specifically TIMP3, have been found to directly regulate shedding of the leukocyte receptor, VCAM1, from the cell surface and loss of TIMP3 leads to increased VCAM1 release. Collectively, all of these studies suggest that shifting the balance in favor of metalloproteinases, either through increased metalloproteinase expression/activity or decreased TIMP expression would facilitate leukocyte transmigration and increased tissue inflammation. However, this effect on transmigration is likely very dependent on timing (i.e., early or late in the inflammatory response). Specifically, shedding of the above receptors following leukocyte–MVEC interaction would facilitate transmigration, but shedding of the receptors early in inflammation or during the resolution of inflammation blunts transmigration (i.e., blocks leukocyte–MVEC interaction).

Thus, the metalloproteinase/TIMP balance is, in fact, a critical mediator of MVEC activation and dysfunction. Shifting the balance in favor of metalloproteinases is associated with loss of MVEC barrier function and potentially increased leukocyte transmigration, while shifting the balance in favor of TIMPs stabilizes MVEC barrier function. It is also possible, however, that shifting the balance in favor of TIMPs may also serve to stabilize leukocyte receptors on the MVEC surface and thereby promote or prolong augmented tissue inflammation. This is currently unknown and hence, future studies must address the role of TIMPs in regulation of not just MVEC barrier function, but also in leukocyte–MVEC interaction.

8.2 Therapeutic Potential of Manipulating the Metalloproteinase/TIMP Balance

TIMPs and metalloproteinases are linked to a variety of diseases in humans, meaning they are appealing targets for the development of specific inhibitors or therapies. Historically, therapies have looked to reverse disease progression by creating small molecule inhibitors, or increasing the concentration of TIMPs by gene transfer or recombinant protein administration.[29] Unfortunately, synthetic MMP inhibitors have not had success in clinical trials due to a lack of efficacy and a variety of complications.[29] The biggest issue at present is that metalloproteinases in general share a very similar active site, have overlapping functions, and are important in a wide range of biological processes. One potential solution is to develop engineered TIMPs, with different mutations responsible for restricted inhibitory specificities. By increasing specificity, potential future treatment could better target a specific cell or metalloproteinase in order to address a key pathophysiologic event, permitting other metalloproteinases and/or TIMPs to continue to act homeostatically in other cells and tissues.

Additionally, there is controversy whether it is better to develop broad-spectrum or individual metalloproteinase-selective inhibitors. Even though broad-spectrum inhibitors may result in unwanted side effects, they may be more effective in treating certain diseases through the ability to target multiple metalloproteinases.[130] It is still believed that targeting a specific metalloproteinase or TIMP at early onset of a disease may be more effective, but it will be challenging to consistently identify patients at a specific time point of their disease course (e.g., early stage sepsis).

REFERENCES

1. Bhattacharya J, Matthay MA. Regulation and repair of the alveolar-capillary barrier in acute lung injury. *Annu Rev Physiol.* 2013;75:593–615.
2. Bernard G, Vincent J, Laterre PF, et al. Efficacy and safety of recombinant human activated protein C for severe sepsis. *N Engl J Med.* 2001;344(10):699–709.
3. De Backer D, Creteur J, Preiser J-C, Dubois M-J, Vincent J-L. Microvascular blood flow is altered in patients with sepsis. *Am J Respir Crit Care Med.* 2002;166 (1):98–104.
4. Mehta D, Malik AB. Signaling mechanisms regulating endothelial permeability. *Physiol Rev.* 2006;86:279–367.
5. Dejana E, Orsenigo F, Lampugnani MG. The role of adherens junctions and VE-cadherin in the control of vascular permeability. *J Cell Sci.* 2008;121:2115–2122.
6. Dejana E, Lampugnani MG, Martinez-Estrada O, Bazzoni G. The molecular organization of endothelial junctions and their functional role in vascular morphogenesis and permeability. *Int J Dev Biol.* 2000;44:743–748.
7. Dejana E, Tournier-Lasserve E, Weinstein BM. The control of vascular integrity by endothelial cell junctions: molecular basis and pathological implications. *Dev Cell.* 2009;16:209–221.
8. Edwards DR, Handsley MM, Pennington CJ. The ADAM metalloproteinases. *Mol Aspects Med.* 2008;29:258–289.
9. Löffek S, Schilling O, Franzke CW. Biological role of matrix metalloproteinases: a critical balance. *Eur Respir J.* 2011;38:191–208.
10. Visse R, Nagase H. Matrix metalloproteinases and tissue inhibitors of metalloproteinases: structure, function, and biochemistry. *Circ Res.* 2003;92(8):827–839.
11. Zitka O, Kukacka J, Krizkov S, et al. Matrix metalloproteinases. *Curr Med Chem.* 2010;17:3751–3768.
12. Overall CM, López-Otín C. Strategies for MMP inhibition in cancer: innovations for the post-trial era. *Nat Rev Cancer.* 2002;2:657–672.
13. Parks WC, Wilson CL, López-Boado YS. Matrix metalloproteinases as modulators of inflammation and innate immunity. *Nat Rev Immunol.* 2004;4(8):617–629.
14. Nagase H, Visse R, Murphy G. Structure and function of matrix metalloproteinases and TIMPs. *Cardiovasc Res.* 2006;69:562–573.
15. Khokha R, Murthy A, Weiss A. Metalloproteinases and their natural inhibitors in inflammation and immunity. *Nat Rev Immunol.* 2013;13:649–665.

16. Page-McCaw A, Ewald AJ, Werb Z. Matrix metalloproteinases and the regulation of tissue remodelling. *Nat Rev Mol Cell Biol.* 2007;8:221–233.
17. Brew K, Dinakarpandian D, Nagase H. Tissue inhibitors of metalloproteinases: evolution, structure and function. *Biochim Biophys Acta.* 2000;1477(1–2):267–283.
18. Gill SE, Kassim SY, Birkland TP, Parks WC. Mouse models of MMP and TIMP function. *Methods Mol Biol.* 2010;622:31–52.
19. Parsons JT, Martin KH, Slack JK, Taylor JM, Weed SA. Focal adhesion kinase: a regulator of focal adhesion dynamics and cell movement. *Oncogene.* 2000;19(49):5606–5613.
20. Bonnans C, Chou J, Werb Z. Remodelling the extracellular matrix in development and disease. *Nat Rev Mol Cell Biol.* 2014;15(12):786–801.
21. McCawley LJ, Matrisian LM. Matrix metalloproteinases: they're not just for matrix anymore! *Curr Opin Cell Biol.* 2001;13(5):534–540.
22. Gill SE, Parks WC. Metalloproteinases and their inhibitors: regulators of wound healing. *Int J Biochem Cell Biol.* 2008;40:1334–1347.
23. Schlomann U, Wildeboer D, Webster A, et al. The metalloprotease disintegrin ADAM8: processing by autocatalysis is required for proteolytic activity and cell adhesion. *J Biol Chem.* 2002;277(50):48210–48219.
24. Horiuchi K, Miyamoto T, Takaishi H, et al. Cell surface colony-stimulating factor 1 can be cleaved by TNF-α converting enzyme or endocytosed in a clathrin-dependent manner. *J Immunol.* 2007;179:6715–6724.
25. Murphy G. Tissue inhibitors of metalloproteinases. *Genome Biol.* 2011;12:233.
26. Wei S, Kashiwagi M, Kota S, Xie Z, Nagase H, Brew K. Reactive site mutations in tissue inhibitor of metalloproteinase-3 disrupt inhibition of matrix metalloproteinases but not tumor necrosis factor-α-converting enzyme. *J Biol Chem.* 2005;280(38):32877–32882.
27. Dreymueller D, Martin C, Kogel T, et al. Lung endothelial ADAM17 regulates the acute inflammatory response to lipopolysaccharide. *EMBO Mol Med.* 2012;4(5):412–423.
28. Murphy G. Regulation of the proteolytic disintegrin metalloproteinases, the "Sheddases" *Semin Cell Dev Biol.* 2009;20:138–145.
29. Baker A, Edwards D, Murphy G. Metalloproteinase inhibitors: biological actions and therapeutic opportunities. *J Cell Sci.* 2002;115:3719–3727.
30. Yu WH, Yu S, Meng Q, Brew K, Woessner JF. TIMP-3 binds to sulfated glycosaminoglycans of the extracellular matrix. *J Biol Chem.* 2000;275(40):31226–31232.
31. Brew K, Nagase H. The tissue inhibitors of metalloproteinases (TIMPs): an ancient family with structural and functional diversity. *Biochim Biophys Acta.* 2010;1803(1):55–71.
32. Nieuwesteeg M, Walsh L, Fox M, Damjanovski S. Domain specific overexpression of TIMP-2 and TIMP-3 reveals MMP-independent functions of TIMPs during *Xenopus laevis* development. *Biochem Cell Biol.* 2012;90:585–595.
33. Troeberg L, Nagase H. Analysis of TIMP expression and activity. *Arthritis Res.* 2007;135:251–267.
34. Jung K-K, Liu X-W, Chirco R, Fridman R, Kim H-RC. Identification of CD63 as a tissue inhibitor of metalloproteinase-1 interacting cell surface protein. *EMBO J.* 2006;25:3934–3942.
35. Stetler-Stevenson WG. Tissue inhibitors of metalloproteinases in cell signaling: metalloproteinase-independent biological activities. *Sci Signal.* 2008;1(27):re6.
36. Fu X, Zeng L, Liu Z, Ke X, Lei L, Li G. MicroRNA-206 regulates the secretion of inflammatory cytokines and MMP9 expression by targeting TIMP3 in Mycobacterium tuberculosis-infected THP-1 human macrophages. *Biochem Biophys Res Commun.* 2016;477:167–173.
37. Ulrich D, Hrynyschyn K, Pallua N. Matrix metalloproteinases and tissue inhibitors of metalloproteinases in sera and tissue of patients with Dupuytren's disease. *Plast Reconstr Surg.* 2003;112:1279–1286.

38. Kassiri Z, Oudit GY, Kandalam V, et al. Loss of TIMP3 enhances interstitial nephritis and fibrosis. *J Am Soc Nephrol.* 2009;20:1223–1235.
39. Gill SE, Huizar I, Bench EM, et al. Tissue inhibitor of metalloproteinases 3 regulates resolution of inflammation following acute lung injury. *Am J Pathol.* 2010;176(1):64–73.
40. English JL, Kassiri Z, Koskivirta I, et al. Individual TIMP deficiencies differentially impact pro-MMP-2 activation. *J Biol Chem.* 2006;281(15):10337–10346.
41. Chen P, McGuire JK, Hackman RC, et al. Tissue inhibitor of metalloproteinase-1 moderates airway re-epithelialization by regulating matrilysin activity. *Am J Pathol.* 2008;172(5):1256–1270.
42. Sahebjam S, Khokha R, Mort JS. Increased collagen and aggrecan degradation with age in the joints of Timp3$^{-/-}$ mice. *Arthritis Rheum.* 2007;56(3):905–909.
43. Gill SE, Pape MC, Khokha R, Watson AJ, Leco KJ. A null mutation for tissue inhibitor of metalloproteinases-3 (Timp-3) impairs murine bronchiole branching morphogenesis. *Dev Biol.* 2003;261(2):313–323.
44. Gill SE, Pape MC, Leco KJ. Tissue inhibitor of metalloproteinases 3 regulates extracellular matrix-cell signaling during bronchiole branching morphogenesis. *Dev Biol.* 2006;298:540–554.
45. Arpino V, Mehta S, Wang L, et al. Tissue inhibitor of metalloproteinases 3-dependent microvascular endothelial cell barrier function is disrupted under septic conditions. *Am J Physiol Heart Circ Physiol.* 2016;310:H1455–H1467.
46. Qi JH, Ebrahem Q, Moore N, et al. A novel function for tissue inhibitor of metalloproteinases-3 (TIMP3): inhibition of angiogenesis by blockage of VEGF binding to VEGF receptor-2. *Nat Med.* 2003;9(4):407–415.
47. Sumpio BE, Timothy Riley J, Dardik A. Cells in focus: endothelial cell. *Int J Biochem Cell Biol.* 2002;34:1508–1512.
48. Vandenbroucke E, Mehta D, Minshall R, Malik AB. Regulation of endothelial junctional permeability. *Ann N Y Acad Sci.* 2008;1123:134–145.
49. Ross MH, Pawlina W. *Histology: A Text and Atlas.* 6th ed. Baltimore, MD: Wolters Kluwer/Lippincott Williams & Wilkins Health; 2011.
50. Aird WC. The role of the endothelium in severe sepsis and multiple organ dysfunction syndrome. *Blood.* 2003;101(10):3765–3777.
51. Lee WL, Slutsky AS. Sepsis and endothelial permeability. *N Engl J Med.* 2010;363(7):689–691.
52. Corada M, Mariotti M, Thurston G, et al. Vascular endothelial-cadherin is an important determinant of microvascular integrity in vivo. *Proc Natl Acad Sci U S A.* 1999;96:9815–9820.
53. Petrache I, Birukova A, Ramirez SI, Garcia JGN, Verin AD. The role of the microtubules in tumor necrosis factor-α-induced endothelial cell permeability. *Am J Respir Cell Mol Biol.* 2003;28(5):574–581.
54. Yang Y, Schmidt EP. The endothelial glycocalyx: an important regulator of the pulmonary vascular barrier. *Tissue Barriers.* 2013;1(1):1–6.
55. Dull R, Cluff M, Kingston J. Lung heparan sulfates modulate K_{fc} during increased vascular pressure: evidence for glycocalyx-mediated mechanotransduction. *Am J Physiol Lung Cell Mol Physiol.* 2012;302(9):L816–L828.
56. Kalluri R. Basement membranes: structure, assembly and role in tumour angiogenesis. *Nat Rev Cancer.* 2003;3(6):422–433.
57. Galley HF, Webster NR. Physiology of the endothelium. *Br J Anaesth.* 2004;93(1):105–113.
58. Minshall RD, Sessa WC, Stan RV, Anderson RGW, Malik AB. Caveolin regulation of endothelial function. *Am J Physiol Lung Cell Mol Physiol.* 2003;285(6):L1179–L1183.
59. Vane JR, Anggard EE, Bottong RM. Regulatory functions of the vascular endothelium. *N Engl J Med.* 1990;323(1):27–36.

60. Mehta S. The effects of nitric oxide in acute lung injury. *Vascul Pharmacol.* 2005;43:390–403.
61. Wang L, Taneja R, Razavi HM, Law C, Gillis C, Mehta S. Specific role of neutrophil inducible nitric oxide synthase in murine sepsis-induced lung injury in vivo. *Shock.* 2012;37(5):539–547.
62. Gill SE, Taneja R, Rohan M, Wang L, Mehta S. Pulmonary microvascular albumin leak is associated with endothelial cell death in murine sepsis-induced lung injury in vivo. *PLoS One.* 2014;9(2):e88501.
63. Bone R. The sepsis syndrome: definition and general approach to management. *Clin Chest Med.* 1996;17(2):175–181.
64. Skibsted S, Jones A, Puskarich M, et al. Biomarkers of endothelial cell activation in early sepsis. *Shock.* 2013;39(5):427–432.
65. DiStasi MR, Ley K. Opening the flood-gates: how neutrophil-endothelial interactions regulate permeability. *Trends Immunol.* 2009;30(11):547–556.
66. Schouten M, Wiersinga WJ, Levi M, van der Poll T. Inflammation, endothelium, and coagulation in sepsis. *J Leukoc Biol.* 2008;83:536–545.
67. Scholz D, Devaux B, Hirche A, et al. Expression of adhesion molecules is specific and time-dependent in cytokine-stimulated endothelial cells in culture. *Cell Tissue Res.* 1996;284:415–423.
68. Lipowsky H. Protease activity and the role of the endothelial glycocalyx in inflammation. *Drug Discov Today Dis Model.* 2011;8(1):57–62.
69. Lum H, Malik AB. Regulation of vascular endothelial barrier function. *Am J Physiol.* 1994;267:L223–L241.
70. Esser S, Lampugnani MG, Corada M, Dejana E, Risau W. Vascular endothelial growth factor induces VE-cadherin tyrosine phosphorylation in endothelial cells. *J Cell Sci.* 1998;111:1853–1865.
71. Dudek S, Garcia JGN. Cytoskeletal regulation of pulmonary vascular permeability. *J Appl Physiol.* 2001;91:1487–1500.
72. Yuan SY. Protein kinase signaling in the modulation of microvascular permeability. *Vascul Pharmacol.* 2002;39:213–223.
73. Morel N, Petruzzo P, Hechtman H, Shepro D. Inflammatory agonists that increase microvascular permeability in vivo stimulate cultured pulmonary microvessel endothelial cell contraction. *Inflammation.* 1990;14(5):571–583.
74. McKenzie JA, Ridley AJ. Roles of Rho/ROCK and MLCK in TNF-a-induced changes in endothelial morphology and permeability. *J Cell Physiol.* 2007;213(1):221–228.
75. Petrache I, Verin AD, Crow MT, Birukova A, Liu F, Garcia JGN. Differential effect of MLC kinase in TNF-{α} -induced endothelial cell apoptosis and barrier dysfunction. *Am J Physiol Lung Cell Mol Physiol.* 2001;280(6):L1168–L1178.
76. Shen Q, Rigor RR, Pivetti CD, Wu MH, Yuan SY. Myosin light chain kinase in microvascular endothelial barrier function. *Cardiovasc Res.* 2010;87:272–280.
77. Angelini DJ, Hyun S-W, Grigoryev DN, et al. TNF-α increases tyrosine phosphorylation of vascular endothelial cadherin and opens the paracellular pathway through fyn activation in human lung endothelia. *Am J Physiol Lung Cell Mol Physiol.* 2006;291: L1232–L1245.
78. Sidibé A, Mannic T, Arboleas M, et al. Soluble VE-cadherin in rheumatoid arthritis patients correlates with disease activity: evidence for tumor necrosis factor α-induced VE-cadherin cleavage. *Arthritis Rheum.* 2012;64(1):77–87.
79. Schulz B, Pruessmeyer J, Maretzky T, et al. ADAM10 regulates endothelial permeability and T-Cell transmigration by proteolysis of vascular endothelial cadherin. *Circ Res.* 2008;102(10):1192–1201.
80. Brown KA, Brain SD, Pearson JD, Edgeworth JD, Lewis SM, Treacher DF. Neutrophils in development of multiple organ failure in sepsis. *Lancet.* 2006;368:157–169.
81. Cohen J. The immunopathogenesis of sepsis. *Nature.* 2002;420:885–891.

82. Hack CE, Zeerleder S. The endothelium in sepsis: source of and a target for inflammation. *Crit Care Med.* 2001;29(7):S21–S27.
83. Zhou Q, Einert M, Schmitt H, et al. MnTBAP increases BMPR-II expression in endothelial cells and attenuates vascular inflammation. *Vascul Pharmacol.* 2016;84:67–73.
84. Shen J, Ham RG, Karmiol S. Expression of adhesion molecules in cultured human pulmonary microvascular endothelial cells. *Microvasc Res.* 1995;50:360–372.
85. King EG, Bauzá GJ, Mella JR, Remick DG. Pathophysiologic mechanisms in septic shock. *Lab Invest.* 2013;94:4–12.
86. Nelson A, Berkestedt I, Schmidtchen A. Increased levels of glycosaminoglycans during septic shock: relation to mortality and the antibacterial actions of plasma. *Shock.* 2008;30(6):623–627.
87. Schmidt E, Yang Y, Janssen W. The pulmonary endothelial glycocalyx regulates neutrophil adhesion and lung injury during experimental sepsis. *Nat Med.* 2012;18(8):1217–1223.
88. Arenas IA, Xu Y, Lopez-Jaramillo P, Davidge ST. Angiotensin II-induced MMP-2 release from endothelial cells is mediated by TNF-α. *Am J Physiol Cell Physiol.* 2004;286:C779–C784.
89. Harkness K, Adamson P, Sussman J, Davies-Jones G, Greenwood J, Woodroofe M. Dexamethasone regulation of matrix metalloproteinase expression in CNS vascular endothelium. *Brain.* 2000;123:698–709.
90. Bzowska M, Jura N, Lassak A, Black RA, Bereta J. Tumour necrosis factor-α stimulates expression of TNF-α converting enzyme in endothelial cells. *Eur J Biochem.* 2004;271:2808–2820.
91. Shwetank, Date OS, Kim KS, Manjunath R. Infection of human endothelial cells by Japanese encephalitis virus: increased expression and release of soluble HLA-E. *PLoS One.* 2013;8(11):e79197.
92. Paolillo R, Iovene MR, Romano Carratelli C, Rizzo A. Induction of VEGF and MMP-9 expression by toll-like receptor 2/4 in human endothelial cells infected with *Chlamydia pneumoniae. Int J Immunopathol Pharmacol.* 2012;25(2):377–386.
93. Luplertlop N, Missé D. MMP cellular responses to dengue virus infection-induced vascular leakage. *Jpn J Infect Dis.* 2008;61:298–301.
94. Bauer AT, Bürgers HF, Rabie T, Marti HH. Matrix metalloproteinase-9 mediates hypoxia-induced vascular leakage in the brain via tight junction rearrangement. *J Cereb Blood Flow Metab.* 2010;30:837–848.
95. Ichikawa Y, Ishikawa T, Momiyama N, et al. Matrilysin (MMP-7) degrades VE-cadherin and accelerates accumulation of beta-catenin in the nucleus of human umbilical vein endothelial cells. *Oncol Rep.* 2006;15:311–315.
96. Liu J, Jin X, Liu KJ, Liu W. Matrix metalloproteinase-2-mediated occludin degradation and caveolin-1-mediated claudin-5 redistribution contribute to blood–brain barrier damage in early ischemic stroke stage. *J Neurosci.* 2012;32(9):3044–3057.
97. Alexander JS, Elrod JW. Extracellular matrix, junctional integrity and matrix metalloproteinase interactions in endothelial permeability regulation. *J Anat.* 2002;200:561–574.
98. Sounni NE, Paye A, Host L, Noël A. MT-MMPS as regulators of vessel stability associated with angiogenesis. *Front Pharmacol.* 2011;2(111):1–11.
99. Lemarchant S, Pruvost M, Montaner J, et al. ADAMTS proteoglycanases in the physiological and pathological central nervous system. *J Neuroinflammation.* 2013;10:133.
100. Jönsson-Rylander AC, Nilsson T, Fritsche-Danielson R, et al. Role of ADAMTS-1 in atherosclerosis: remodeling of carotid artery, immunohistochemistry, and proteolysis of versican. *Arterioscler Thromb Vasc Biol.* 2005;25:180–185.
101. Schrimpf C, Xin C, Campanholle G, et al. Pericyte TIMP3 and ADAMTS1 modulate vascular stability after kidney injury. *J Am Soc Nephrol.* 2012;23(5):868–883.

102. Partridge C, Jeffrey JJ, Malik AB. A 96-kDa gelatinase induced by TNF-α contributes to increased microvascular endothelial permeability. *Am J Physiol Lung Cell Mol Physiol.* 1993;265(5):L438–L447.
103. Passi A, Negrini D, Albertini R, De Luca G, Miserocchi G. Involvement of lung interstitial proteoglycans in development of hydraulic- and elastase-induced edema. *Am J Physiol.* 1998;275(3 pt 1):L631–L635.
104. Qiao R, Wang H, Yan W. Extracellular matrix hyaluronan is a determinant of the endothelial barrier. *Am J Physiol Cell Physiol.* 1995;269(1):C103–C109.
105. Chen F, Ohashi N, Li W, Eckman C, Nguyen JH. Disruptions of occludin and claudin-5 in brain endothelial cells in vitro and in brains of mice with acute liver failure. *Hepatology.* 2009;50(6):1914–1923.
106. Lominadze D, Roberts AM, Tyagi N, Moshal KS, Tyagi SC. Homocysteine causes cerebrovascular leakage in mice. *Am J Physiol Heart Circ Physiol.* 2006;290(3): H1206–H1213.
107. Ponnuchamy B, Khalil RA. Role of ADAMs in endothelial cell permeability: cadherin shedding and leukocyte rolling. *Circ Res.* 2008;102:1139–1142.
108. Rahman A, Fazal F. Hug tightly and say goodbye: role of endothelial ICAM-1 in leukocyte transmigration. *Antioxid Redox Signal.* 2009;11(4):823–839.
109. Sithu SD, English WR, Olson P, et al. Membrane-type 1-matrix metalloproteinase regulates intracellular adhesion molecule-1 (ICAM-1)-mediated monocyte transmigration. *J Biol Chem.* 2007;282(34):25010–25019.
110. Garton KJ, Gough PJ, Philalay J, et al. Stimulated shedding of vascular cell adhesion molecule 1 (VCAM- 1) is mediated by tumor necrosis factor-α-converting enzyme (ADAM 17). *J Biol Chem.* 2003;278:37459–37464.
111. Bugno M, Witek B, Bereta J, Bereta M, Edwards DR, Kordula T. Reprogramming of TIMP-1 and TIMP-3 expression profiles in brain microvascular endothelial cells and astrocytes in response to proinflammatory cytokines. *FEBS Lett.* 1999;448:9–14.
112. Shastry S, Tyagi SC. Homocysteine induces metalloproteinase and shedding of β-1 integrin in microvessel endothelial cells. *J Cell Biochem.* 2004;93:207–213.
113. Siemianowicz K, Likus W, Francuz T, Garczorz W. Effect of elastin-derived peptides on the production of tissue inhibitor of metalloproteinase-1, -2, and -3 and the ratios in various endothelial cell lines. *Exp Ther Med.* 2015;9:2245–2250.
114. Shen Q, Lee ES, Pitts RL, Wu MH, Yuan SY. Tissue inhibitor of metalloproteinase-2 regulates matrix metalloproteinase-2-mediated endothelial barrier dysfunction and breast cancer cell transmigration through lung microvascular endothelial cells. *Mol Cancer Res.* 2010;8(7):939–951.
115. Mascall KS, Small GR, Gibson G, Nixon GF. Sphingosine-1-phosphate-induced release of TIMP-2 from vascular smooth muscle cells inhibits angiogenesis. *J Cell Sci.* 2012;125(9):2267–2275.
116. Dollery CM, McEwan JR, Wang M, Sang QA, Liu YE, Shi YE. TIMP-4 is regulated by vascular injury in rats. *Ann N Y Acad Sci.* 1999;84:498–504.
117. Saunders WB, Bohnsack BL, Faske JB, et al. Coregulation of vascular tube stabilization by endothelial cell TIMP-2 and pericyte TIMP-3. *J Cell Biol.* 2006;175(1):179–191.
118. Menge T, Zhao Y, Zhao J, et al. Mesenchymal stem cells regulate blood brain barrier integrity through TIMP release after traumatic brain injury. *Sci Transl Med.* 2012;4(161):161ra150.
119. Ebrahem Q, Qi JH, Sugimoto M, et al. Increased neovascularization in mice lacking tissue inhibitor of metalloproteinases-3. *Invest Ophthalmol Vis Sci.* 2011;52(9):6117–6123.
120. Hurst LA, Bunning RAD, Couraud PO, et al. Expression of ADAM-17, TIMP-3 and fractalkine in the human adult brain endothelial cell line, hCMEC/D3, following pro-inflammatory cytokine treatment. *J Neuroimmunol.* 2009;210:108–112.

121. Lorente L, Martín MM, Solé-Violán J, et al. Association of sepsis-related mortality with early increase of TIMP-1/MMP-9 ratio. *PLoS One*. 2014;9(4):e94318.
122. Lorente L, Martín MM, Labarta L, et al. Matrix metalloproteinase-9, -10, and tissue inhibitor of matrix metalloproteinases-1 blood levels as biomarkers of severity and mortality in sepsis. *Crit Care*. 2009;13(5):R158.
123. Fujimoto M, Takagi Y, Aoki T, et al. Tissue inhibitor of metalloproteinases protect blood–brain barrier disruption in focal cerebral ischemia. *J Cereb Blood Flow Metab*. 2008;28:1674–1685.
124. Wu J, Zhao D, Wu S, Wang D. Ang-(1–7) exerts protective role in blood–brain barrier damage by the balance of TIMP-1/MMP-9. *Eur J Pharmacol*. 2015;748:30–36.
125. Menge T, Zhao Y, Zhao J, et al. Mesenchymal stem cells regulate blood–brain barrier integrity through TIMP3 release after traumatic brain injury. *Sci Transl Med*. 2012;4(161):161ra150:1–11.
126. Lee S-J, Tsang PS, Diaz TM, Wei B, Stetler-Stevenson WG. TIMP-2 modulates VEGFR-2 phosphorylation and enhances phosphodiesterase activity in endothelial cells. *Lab Invest*. 2010;90:374–382.
127. Darwish I, Liles WC. Emerging therapeutic strategies to prevent infection-related microvascular endothelial activation and dysfunction. *Virulence*. 2013;4(6):572–582.
128. Seo D-W, Li H, Guedez L, et al. TIMP-2 mediated inhibition of angiogenesis. *Cell*. 2003;114:171–180.
129. Singh RJR, Mason JC, Lidington EA, et al. Cytokine stimulated vascular cell adhesion molecule-1 (VCAM-1) ectodomain release is regulated by TIMP-3. *Cardiovasc Res*. 2005;67:39–49.
130. Vandenbroucke RE, Libert C. Is there new hope for therapeutic matrix metalloproteinase inhibition? *Nat Rev Drug Discov*. 2014;13:904–927.

CHAPTER FOUR

Matrix Metalloproteinases and Platelet Function

Paolo Gresele[1], Emanuela Falcinelli, Manuela Sebastiano, Stefania Momi
Section of Internal and Cardiovascular Medicine, University of Perugia, Perugia, Italy
[1]Corresponding author: e-mail address: paolo.gresele@unipg.it

Contents

Abstract

Platelets contain and release several matrix metalloproteinases (MMPs) and their tissue inhibitors of matrix metalloproteinases (TIMPs), including MMP-1, -2, -3, -9, and -14 and TIMP-1, -2, and -4. Although devoid of a nucleus, platelets also synthesize TIMP-2 upon activation. Platelet-released MMPs/TIMPs, as well as MMPs generated by other cells within the cardiovascular system, modulate platelet function in health and disease. In particular, a normal hemostatic platelet response to vessel wall injury may be transformed into pathologic thrombus formation by the release from platelets and/or by the local generation of some MMPs. Moreover, platelets may localize the production of leukocyte-derived MMPs to sites of vascular damage, contributing to atherosclerosis

Progress in Molecular Biology and Translational Science, Volume 147
ISSN 1877-1173
http://dx.doi.org/10.1016/bs.pmbts.2017.01.002

development and complications and to arterial aneurysm formation. Finally, the interaction between platelets and tumor cells is strongly influenced by MMPs/TIMPs.

All these mechanisms are emerging as important in atherothrombosis, inflammatory disease, and cancer growth and dissemination. Increasing knowledge of these mechanisms may open the way to novel therapeutic approaches.

1. PLATELETS: STRUCTURE AND FUNCTION

Anucleate platelets (2–3 μm diameter, 0.5 μm thickness, 6–10 fL volume) are generated by a nucleated parent cell, the megakaryocyte, in a process called thrombopoiesis that in humans yields ~100 billion platelets a day and a total mass of 1×10^{12} circulating platelets.[1,2] The platelet life span in blood is ~10 days in humans and ~5 days in mice.[3] Platelets are highly specialized hemostatic cells that act as first response to vascular injury and endothelial disruption.[1] Platelets are also immune cells and play an active role across the spectrum from acute inflammation to adaptive immunity.[4] Thus, platelets provide a central link between the inflammatory and the hemostatic arms of the host defense against injury, contributing to integrate vascular wound repair with the maintenance of tissue integrity.[5] Moreover platelets, although devoid of a nucleus, possess transcription factors, an active spliceosome, and an extensive repertoire of miRNAs and mRNAs that make them capable to synthesize proteins and to regulate protein synthesis by other cells.[6]

Platelets contribute to hemostasis by adhering with high efficiency to sites of vascular injury where subendothelial proteins, such as collagen and von Willebrand factor (VWF), are exposed.[7] Platelet adhesion is initially mediated by the glycoprotein (GP)lb-lX-V complex, interacting with subendothelial collagen-bound VWF, and subsequently by GPIa/IIa and GPVI, directly interacting with collagen fibrils and thus stabilizing the platelet–vessel wall interaction.[8] The ligation of GPVI by collagen fibrils as well as the local generation and the release from activated platelets of other agonists, such as thrombin, adenosine diphosphate (ADP), and thromboxane A_2 (TxA_2), acting on G protein-coupled and on immune receptor tyrosine-based activation motif receptors, trigger intracellular signaling that in turn starts activation-dependent effector functions.[9] Key functional responses triggered in platelets by activation are the shape change, spreading, degranulation, the exposure of a procoagulant surface, and integrin $\alpha_{IIb}\beta_3$ (GPIIb/IIIa) activation. GPIIb/IIIa is in fact normally present in a resting state and it requires inside-out signaling-triggered activation to become able to bind soluble fibrinogen, thus allowing the formation of platelet aggregates. Several mediators released by activated platelets, such as PGE_2, do not elicit aggregation but reinforce the platelet activation response induced by weak stimuli and recruit other platelets to the growing platelet plug (so-called primers of platelet activation).[10–12] Pathological thrombus formation occurs when the mechanisms designed to prevent unwarranted extension of platelet activation are impaired and/or when platelet primers accumulate.

Platelets contribute also to the coagulation cascade by expressing high-affinity receptors for coagulation proteases and cofactors, by providing surfaces for amplification of local thrombin generation, by releasing stored coagulation factors, and by regulating the activity of surface-bound coagulation enzymes.[13] Human platelets also synthesize and express tissue factor (TF),[14,15] in this way further contributing to clot propagation and stabilization.[16]

2. EXPRESSION OF MMPs BY PLATELETS AND MEGAKARYOCYTES

2.1 Protein Expression

Several MMPs have been identified in platelets, including MMP-1, MMP-2, MMP-3, MMP-9, MT1-MMP (MMP-14), and the endogenous inhibitors TIMP-1, TIMP-2, and TIMP-4 (Tables 1 and 2).[17]

Studies from nearly 40 years ago had shown that platelets harbor collagenases which can be released upon activation by ADP and epinephrine and are capable of degrading fibrillar collagen. The collagenase activity appeared in an early phase of platelet aggregation and was prevented by the calcium chelating agent EDTA.[18] The platelet collagenase was later identified as MMP-1.[19,20]

Resting platelets constitutively express significant amounts of MMP-1 (type I collagenase), primarily pro-MMP-1, which is released (and activated) upon thrombin stimulation. MMP-1 is much more abundant (10- to 20-fold) than MMP-2 and dramatically exceeds MMP-3. In addition to its collagenase activity, MMP-1 regulates outside-in signaling events in platelets leading to the phosphorylation of intracellular proteins with the consequent redistribution of β_3-integrins to areas of cellular contact, and primes platelets for aggregation.[19]

MMP-2 (gelatinase A) is abundant in humans and is constitutively expressed in cells of mesenchymal origin. Resting platelets contain small amounts of active MMP-2 (0.3 ± 0.1 ng/10^8 platelets). Upon platelet stimulation, e.g., with thrombin or collagen, pro-MMP-2 translocates from

Table 1 MMPs Present in Platelets

MMPs	Alternative Name	Substrates
MMP-1	Collagenase (Type I, interstitial)	Native collagens (Type II > I > II, VII, VIII, X, and XI), Denatured Collagens (Gelatin), Elastin, Fibronectin, Laminin-5, Aggrecan, Brevican Neurocan, BM-40, Decorin, Vitronectin, Entactin/Nidogen, Tenascin, Perlecan, Connective Tissue Growth Factor (CTGF), Link Protein, Myelin basic protein, Fibrin, Fibrinogen
MMP-2	Gelatinase A 72 kDa Gelatinase Type IV Collagenase	Native collagens (Type I, II, III, IV, V, VII, X, and XI), Gelatin, Elastin, Fibronectin, Entactin/Nidogen-1, Aggrecan, Decorin, Fibrillin, Fibulin 2, Laminin-5, Tenascin, SPARC, Vitronectin, Galectin-1, Galectin-3, Versican, BM-40, Brevican, Neurocan, CTGF, Chondroitin Sulfate Proteoglycan (CSPG)-4, Dystroglycan, Procollagen C-proteinase enhancer-1 (PCPE-1), Link Protein, Osteonectin, Myelin Basic Protein, Biglycan, Fibrin, Fibrinogen
MMP-3	Stromelysin-1; Proteoglycanase	Nontriple helical regions of native Collagens (Type III, IV, V, VII, IX, X, and XI), Gelatin, Collagen Telopeptides, Elastin, Fibronectin, Vitronectin, Laminin, Entactin/Nidogen-1, Tenascin, SPARC, Aggrecan, Decorin, Perlecan, Versican, Fibulin, Biglycan, Link Protein, Osteonectin, Myelin Basic Protein, Fibulin-2, Fibrin, Fibrinogen
MMP-9	Gelatinase B 92 kDa Gelatinase	Native collagens (Type I, IV, V, XI, and XIV), Gelatin, Elastin, Vitronectin, Laminin, Decorin, Fibrillin, Fibronectin, SPARC, Aggrecan, Link Protein, Galectin-1, Galectin-3, Versican, Decorin, Biglycan, Link Protein, Osteonectin, Myelin Basic Protein, Fibrin, Fibrinogen
MMP-14	MT1-MMP	Native collagens (Type I, II and III), Gelatin, Fibronectin, Tenascin, Vitronectin, Laminin, Entactin, Galectin-3, CTGF-L, Fibrillin, Aggrecan, Perlecan, Syndecan-1, Lumican, Myelin Basic Protein, Fibrinogen

Table 2 TIMPs Present in Platelets

TIMPs	Inhibition	Produced by
TIMP-1	MMP-14 -16, -19, -24 (weak) ADAM10 Pro-MMP-9	Fibroblasts Osteoblasts Endothelial cells Granulosa cells Dendritic cells Vascular smooth muscle cells Adipocytes Monocytes Platelets
TIMP-2	All MMPs ADAM12 Pro-MMP-2	
TIMP-4	Most MMPs ADAM17 and 28, ADAM33 (weak) Pro-MMP-2	

the cytosol to the platelet surface where it is activated, and a significant amount of active MMP-2 (17.3 ± 3.7 ng/10^8 platelets) is released to the extracellular space. Platelets also express MT1-MMP (MMP-14). MT1-MMP (MMP-14) has been found on the cell surface of various cell types and appears to form a trimolecular complex with pro-MMP-2 and TIMP-2.

Conflicting results exist concerning the presence of MMP-9 in platelets: some authors did not detect it,[21–24] while others did.[25,26] It is possible that MMP-9, if it exists in platelets, escaped the detection limits of some assays. It has also been suggested that MMP-9 is secreted during proplatelet formation from megakaryocytes, therefore, not being retained in mature platelets.[21] On the other hand the MMP-9 detected in platelet preparations may be an artifact deriving from residual leukocyte contamination.[27] It was recently observed that activated platelets bind plasma-derived MMP-9, suggesting that when MMP-9 is detected in platelets it is probably plasma derived.[28]

There are also discrepant findings concerning the presence of MMP-3 in platelets and megakaryocytes. Therefore, the absolute amounts of platelet-associated MMP-3 and MMP-9 vary widely in different publications and await clarification.

Resting platelets express the latent form of MT1-MMP (MMP-14) on their surface and this is activated during collagen-induced platelet aggregation.

2.2 Protein Regulation

MMP activity is regulated at three different levels: transcriptional, activation of the proenzyme, inhibition by aspecific and specific inhibitors. Transcriptional regulation differs depending on the cell implicated in MMP production. Genes encoding for MMPs are modulated by several stimuli, such as growth factors, cytokines, cell–cell or cell–matrix interactions, stress, and chemical agents. In the cardiovascular system, macrophages and smooth muscle cells are induced to express MMPs by some cytokines, such as IL-1β and TNF-α, by thrombin, high shear stress, and hypercholesterolemia.[29,30] Cecchetti *et al.* have shown that proplatelet-producing megakaryocytes differentially sort mRNAs for matrix metalloproteinases (MMPs) and tissue inhibitors of matrix metalloproteinases (TIMPs) into platelets. The mechanisms by which MMP and TIMP mRNAs are differentially transferred to platelets, however, remain unknown. There is also a paucity of studies examining whether changes in the megakaryocyte milieu alters the types and amounts of RNAs that are transferred to platelets.

MMPs are secreted as inactive zymogens that must be activated by cleavage of the N-terminal sequence of the propeptide domain, which allows the Zn^{+2}-binding site of the catalytic domain to become exposed. Proteolytic activation of MMPs is a stepwise process. The initial proteolytic attack occurs at an exposed loop region between the first and the second helices of the propeptide, with cleavage specificity of the bait region depending on the sequence typical of each MMP. Cleavage removes only a part of the propeptide, while complete removal of the propeptide is often realized in trans by the action of the MMP intermediate itself or by other active MMPs. Activation can be obtained in vitro by several agents, such as proteases (trypsin), detergents (SDS), APMA (*para*-aminophenylmercuric acetate), $HgCl_2$. Low pH and heat exposure can also lead to activation. One relevant pathway leading to pro-MMP activation in vivo is mediated by plasmin. Plasmin is generated from plasminogen by tissue plasminogen activator bound to fibrin and urokinase plasminogen activator bound to a specific cell surface receptor. Plasminogen is contained in platelet α granules and is released upon thrombin stimulation. Both plasminogen and urokinase plasminogen activator are membrane associated, thereby leading to localized pro-MMP activation.[31]

Finally, the activity of MMPs is regulated by two major types of endogenous inhibitors, α2-macroglobulin, a plasma protein that acts as a general proteinase inhibitor, and TIMPs, highly specific MMP inhibitors four of which have been identified so far (TIMP-1, -2, -3, -4).[32] TIMPs bind irreversibly to the catalytic domain of MMPs blocking it. This interaction is not selective for one specific MMP or family, with the exception of TIMP-1 which has stronger affinity for MT-MMPs as compared with the other MMPs. TIMP-1, -2, and -4 are secreted and circulate in soluble form, while TIMP-3 is associated with the extracellular matrix. TIMPs are secreted by different cells, including smooth muscle cells, macrophages, and platelets. Their activity is induced by PDGF and TGF-β and regulated by several cytokines. In particular, the expression of TIMP-1 and TIMP-2 was significantly elevated in PDGF-treated cells by the activation of the TGF-β1 pathway[33] and the mRNA levels of MMP-1, -2, and -3 and TIMP-1 in human retinal pigment epithelial cells were markedly increased by TNF-α.[34]

Immunohistochemical staining using monoclonal antibodies against TIMP-1 and TIMP-2 showed that megakaryocytes and platelets are positive for both TIMP-1 and TIMP-2, confirming that they are rich sources of TIMPs. Moreover, while serum levels of TIMP-1 and TIMP-2 were 101.1 ± 13.3 and 82.7 ± 26.3 ng/mL, respectively, in normal subjects, in patients with myeloproliferative disorders and an elevated platelet count they were 351.6 ± 200.9 and 148.9 ± 84 ng/mL, respectively. On the contrary, serum levels of TIMP-1 and TIMP-2 in patients with a low platelet count, such as in aplastic anemia or idiopathic thrombocytopenic purpura, were 57.2 ± 25.8 and 19.7 ± 7.68 ng/mL, respectively, showing that platelets contribute significantly to TIMP-1 and TIMP-2 accumulation in blood.[35]

The complete structure of TIMP-1 and of the TIMP-1–MMP-3 complex and that of the TIMP-2–MT1-MMP complex have been determined by X-ray crystallography.[36] In particular, in platelets, in response to stimulation TIMP-2 interacts with the catalytic site of MT1-MMP and the C-terminal domain of MMP-2 forming a trimolecular complex that controls the cleavage of pro-MMP-2.[37] When the three molecules come together, MT1-MMP activates MMP-2 utilizing TIMP-2 as a bridging molecule: TIMP2 interacts simultaneously with the catalytic site of MT1-MMP and with the C-terminal domain of pro-MMP2. Then, a second "free" MT1-MMP molecule cleaves and activates pro-MMP2, generating sequentially an intermediate 64-kDa form and then the active 62-kDa form.[38]

Interestingly, low concentrations of TIMP-2 induce MMP-2 activation, while higher concentrations inactivate it. The concentration-dependent divergent activities of TIMP-2 may allow to generate finely tuned levels of active MMP-2 during the dynamic process of platelet activation.[21,39] Kinetic studies have indicated that TIMP-3 is a better inhibitor of ADAM-17 and aggrecanases than of MMPs, whereas TIMP-4 is the endogenous inhibitor of MMP-9.[40]

The regulation of MMP levels and activity, either by their secretion or of that of their inhibitors or by the triggering of their synthesis by other cells, represents an important mechanism of platelet participation in disease. Any imbalance in this finely tuned system may have direct pathologic consequences.

2.3 Transcripts for MMPs

Platelets receive thousands of mRNAs from megakaryocytes,[41] still megakaryocytes transfer mRNAs to platelets in a selective fashion.[21] Megakaryocytes express mRNA for 10 MMPs and for 3 TIMPs, with variable abundance of the transcripts: some of these are found in platelets (MMP-1, -9, -11, -17, -19, -24, -25 and TIMPs-1, -2, and -3) while others are not (MMP-2, -14, -15), although there are divergent reports on the existence of mRNA for MMP-14 in platelets.[21,38]

Concerning the expression levels, MMP-1 and TIMP-1 transcripts are the most abundant in human platelets, followed by MMP-24 and TIMP-2.

Although mRNAs for MMP and TIMP family members are typically transferred to platelets with their corresponding protein, exceptions exist. One of these is represented by MMP-2, which is present both as mRNA and protein in megakaryocytes whereas platelets only express the protein.[21] The physiologic activator of MMP-2, MMP-14, has a similar expression pattern: platelets express the protein but not its mRNA. Moreover, megakaryocytes transfer TIMP-2 mRNA, but not its protein, to platelets that, however, synthesize de novo this protein in response to activating stimuli.[21] mRNA for MMP-3 and TIMP-4 has not been found either in megakaryocytes or in platelets.[21]

Unlike its corresponding protein, MMP-9 mRNA is present in platelets, albeit at levels far below those found in megakaryocytes.[21]

The mechanism by which MMP and TIMP mRNAs are differentially sorted to platelets, however, remains unknown. It is conceivable that megakaryocytes selectively transfer mRNAs to platelets through a dedicated mRNA transport machinery existing in other asymmetric mammalian cells, such as in neurons, where only mRNAs that have sequences interacting with localization factors are transported by cytoskeletal filaments.[42–44] In fact, megakaryocytes and platelets possess mRNA for 4 of the 5 mRNA-transport proteins described in mammalians, i.e., STAU1, STAU2, CASC3, and E1F4A3.[21]

It is conceivable that a dysregulated transfer of MMPs and TIMP transcripts from megakaryocytes to platelets may occur under pathologic conditions. This possibility is substantiated by studies demonstrating differences in the platelet transcriptome in the setting of inflammation, sepsis, malignancy, and vascular disease.[45–47]

Therefore, alterations in the pattern of expression of MMPs and TIMPs by platelets may contribute to plaque instability, favor thrombus formation, or boost tissue inflammation.

2.4 MMPs and TIMPs Localization

Conflicting data exist on the sites of storage of MMPs in platelets, because they have been reported to be found both in α-granules[48] and in the cytoplasm.[49] Although Sawicki *et al.* by transmission electron microscopy-immunogold analysis showed cytosolic localization of MMP-2, without apparent association with α or dense-granules, others using confocal microscopy have shown that platelets contain storage pools of MMP-1, MMP-2, MMP-3, and MMP-9, and that all these enzymes colocalize, although MMP-9 to a lesser degree, with α-granule markers such as VWF and P-selectin.[19,26,48]

More recently, although a granular distribution was confirmed, no significant overlap of MMP-1 and MMP-2 with the α-granule marker P-selectin was found, suggesting a localization distinct from α-granules.[48] At present, there are no explanations for these discrepant findings.

In resting platelets TIMP-3 seems to have an α-granule localization, while the other TIMPs are localized in submembranous structures separated from α-granules and are distributed independently of each other, showing a very low degree of colocalization. In the absence of α-granules, like in platelets from patients affected by the gray platelet syndrome (GPS), there is no evidence for a TIMP deficiency and these proteins continued to be found in distinct fluorescent patches.[48] Radomski *et al.* using immunogold electron microscopy showed that TIMP-4 colocalized with MMP-2 in resting platelets.[50]

Confocal microscopy demonstrated the presence of all four TIMPs in both mature MKs and in the CHRF-288-11 megakaryoblastic cell line, localized independently from VWF. The distribution of TIMP-2 in particular appeared to be peripheral.[48]

Platelet activation results in the translocation to the membrane of pro-MMP-1 and pro-MMP-2 where they colocalize with β3 integrin.[19,28,51] The translocation to the cell surface is the trigger for enzyme activation: proteolytic activation of pro-MMP-2 to active MMP-2 is mediated by MT1-MMP and TIMP-2 on the platelet surface.[38]

Translocation to the platelet surface of MMP-2 upon platelet activation by collagen and thrombin is downregulated by physiologic platelet inhibitors, such as nitric oxide and prostacyclin.[51]

Platelet activation by physiologic agonists causes also concentration-dependent release of MMP-1 and MMP-2 in the extracellular milieu.[19,51,52]

Membrane-type MMP-14 localizes to the plasma membrane of both resting and activated platelets: resting platelets express the latent form of MT1-MMP on their surface that is then activated during collagen-induced platelet aggregation.[38]

Collagen-induced platelet aggregation was also reported to lead to a reduction of intraplatelet levels of TIMP-4 due to its liberation with the platelet releasate.[50]

Streptococcus sanguinis, a predominant bacterium in the human oral cavity widely associated with the development of infective endocarditis, was found to induce cytokines (SDF-1, VEGF, sCD40L) and MMP-1 release from platelets.[53]

2.5 Receptors Involved in MMP Activity on Platelets

Integrins are ubiquitous transmembrane α/β heterodimers that mediate diverse processes requiring cell–matrix and cell–cell interactions, such as tissue migration during embryogenesis, cell adhesion, cancer metastasis, and lymphocyte helper and killer cell functions. Eighteen integrin α-subunits and 8 integrin β-subunits have been identified in mammals that combine to form 24 different heterodimers that can be grouped into subfamilies depending on the identity of their β subunit. Platelets express three members of the $\beta1$ subfamily ($\alpha_2\beta_1$, $\alpha_v\beta_1$, and $\alpha_{vI}\beta_1$) that support platelet adhesion to the extracellular matrix (ECM) proteins collagen, fibronectin, and laminin, respectively, and both members of the $\beta3$ subfamily ($\alpha_v\beta_3$ and $\alpha_{IIb}\beta_3$). Although $\alpha_v\beta_3$ mediates platelet adhesion to osteopontin and vitronectin in vitro, it is uncertain whether it plays a role in platelet function in vivo. By contrast $\alpha_{IIb}\beta_3$, a receptor for fibrinogen, VWF, fibronectin, and vitronectin, is strictly required for platelet aggregation.[54]

MMP-1 colocalizes with β_3 integrins on the surface of activated platelets at cell to cell contact sites. Integrin $\alpha_2\beta_1$ also binds pro- or active-MMP-1 via the I domain of α_2 that connects to the linker and hemopexin motifs of MMP-1.[55]

Concerning MMP-2, the interaction between an integrin ($\alpha_V\beta_3$) and this protease was first identified on the surface of melanoma cells and in blood vessels during neoangiogenesis and it was shown to be involved in tumor growth and neoangiogenesis.[56] In fact, the inhibition of the formation of the $\alpha_V\beta_3$/MMP-2 complex, by using the MMP-2 C-terminal domain or a small molecule inhibitor (TSRI265), dramatically suppressed angiogenesis in vivo. A direct interaction between $\alpha_v\beta_3$ and MMP-2 was also demonstrated in vitro, showing that they formed a SDS-stable complex that depended on the C-terminus of MMP-2.[56] Furthermore, MMP-2 interacts with integrin $\alpha_{IIb}\beta_3$ on activated platelets and upregulates GPIb receptor expression, thus potentiating the adhesion to VWF. Preincubation of platelets with phenanthroline, that inhibits the activity of MMPs, led to a reduction of platelet adhesion. These results indicate that the release of MMP-2 during platelet adhesion may potentiate this process by upregulating GPIb.[57]

Recent studies have shown that the protease-activated receptor 1 (PAR1), the main thrombin receptor on human platelets, can be activated by matrix metalloproteases (Fig. 1). Activation of PAR1 on different cells is triggered by serine proteases through the enzymatic cleavage of its amino terminal domain, and it has been implicated in numerous biological processes, including hemostasis, inflammation, and cell proliferation.[58]

PAR1 is directly activated on the surface of platelets and breast cancer cells by MMP-1 by its cleavage at site $D_{39} \downarrow P_{40}$, resulting in the formation of a tethered ligand which is two amino acids longer than the thrombin-generated tethered ligand activating platelets. The collagen–MMP-1–PAR1 pathway

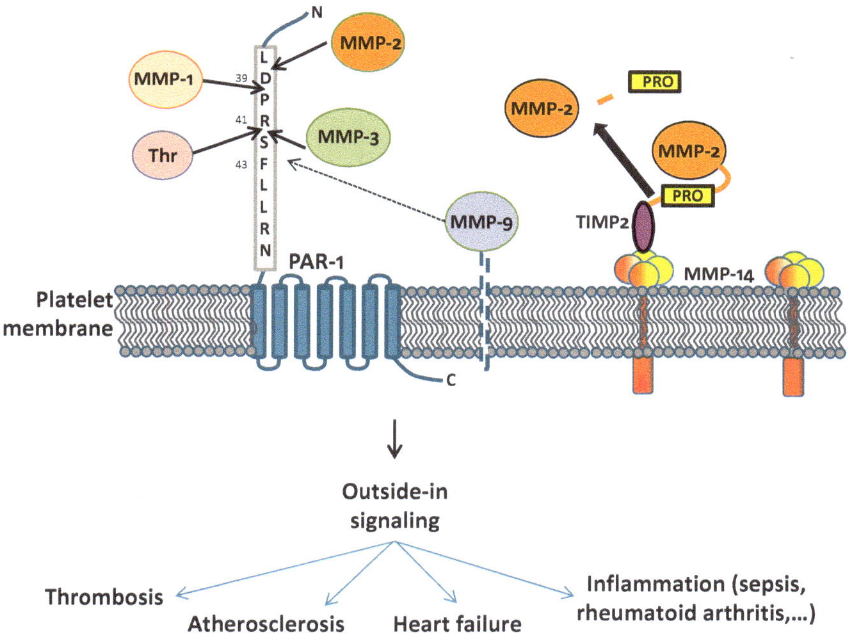

Fig. 1 Activity of different MMPs on platelets: known/unknown receptor involved and functional effects. The N-terminal extracellular domain (exodomain) of PAR1 is cleaved at a canonical site by thrombin and MMP-3 and at noncanonical sites by MMP-1 and MMP-2. Triggered outside-in signaling can lead to platelet-dependent thrombosis, atherosclerosis, in-stent restenosis, heart failure, inflammation. On the platelet surface MMP-14 interacts with TIMP-2 that binds to the hemopexin domain of MMP-2 (thus forming the so-called trimolecular complex) activating it. It has been hypothesized that MMP-9 binds to PAR-1 with low affinity. A specific receptor for MMP-9 on platelets is still not known.

was shown to mediate platelet thrombogenesis and clot retraction, a phenomenon inhibited by PAR1 antagonists.[20] Recently another MMP, MMP-13, was also shown to be able to cleave and activate PAR1 of cardiac fibroblasts and cardiomyocytes.[59] No data, however, are available on the effects of MMP-13 on platelet PAR1.

Emerging evidence suggests that selective proteolytic activation of PAR1 by MMPs, such as MMP-1 and MMP-13, can be an important contributor to the evolution of a variety of disease processes, including thrombus initiation and thrombosis, atherosclerosis and restenosis, sepsis, angiogenesis, heart failure, and cancer.

In particular, studies using human whole blood spiked with either MMP-1- or PAR1-inhibitors, such as a PAR1 pepducin, showed that while primary adhesion of platelets to immobilized collagen fibrils under arterial shear was not affected, the growth rate of platelet aggregate "strings" was significantly attenuated. Blockade of the MMP-1–PAR1 pathway, with the MMP-1 inhibitor FN-439, also greatly curtailed arterial thrombosis in a guinea pig model of ferric chloride-induced injury. These in vitro and in vivo data suggest that the collagen–MMP1–PAR1 pathway may be a point of early intervention in preventing arterial thrombosis.[20]

Moreover, the MMP-13 released from cardiac cells was able to cleave and activate PAR1 on neonatal rat ventricular myocytes and it was shown that either genetic deletion of PAR1 or inhibition of MMP-13 could prevent the deleterious cardiac effects of beta-adrenergic receptor overstimulation. These results indicate that sustained activation of MMP-13–PAR1 in cardiac tissue may be a maladaptive response in heart failure.

Recently, it was also shown that active MMP-2 cleaves PAR1 at $TL_{38}\downarrow D_{39}PR$ on the platelet surface generating a tethered ligand (^{39}DPRSFLLRN) longer than that produced by thrombin. Moreover, integrin $\alpha_{IIb}\beta_3$ is a necessary cofactor for PAR1 cleavage by MMP-2 by binding the MMP-2 hemopexin domain and favoring the interaction of the enzyme with PAR-1.[60]

3. FUNCTIONS OF MMPs IN PLATELETS AND MEGAKARYOCYTES

3.1 Regulation of MK and Platelet Function by MMPs and TIMPs

Over the past few years, MMPs have emerged as a novel system that plays a crucial role in the regulation of platelet function: experiments using blocking antibodies, pharmacological inhibitors, recombinant MMPs, or knock-out mice have shown that MMPs may either activate platelets or potentiate their activation by other agonists.[11]

Mature polyploid human MKs produce and secrete MMP-9 and this metalloproteinase is necessary for the migration of MKs through the basement membrane in response to a chemoattractant stimulus, such as SDF-1. MMP enzymatic activity is also required for subsequent proplatelet formation. Furthermore, administration of a synthetic MMP inhibitor to mice blocked SDF-1-induced platelet increase, demonstrating that MMPs are critical for MK migration out of the bone marrow and for the subsequent platelet production in vivo.[61] These observations show that locomotion of hematopoietic cells requires MMP activity. The exact mechanisms by which MMP-9 or other as yet unknown MMPs regulate platelet release is the subject of ongoing studies.

TIMP-1 and TIMP-2 of MKs exhibited a growth-promoting activity for bone marrow fibroblasts, although TIMP-2 was somewhat less potent.[35]

Concerning platelets, catalytically active MMP-1 enhances tyrosine-phosphorylated proteins in platelets, primes platelets to aggregate in response to submaximal concentrations of thrombin, and clusters β3 integrins on the cell surface.[19]

More recent studies have shown that MMP-1 activates platelets by inducing PAR1-dependent stimulation of $G_{12/13}$-Rho activity and thus eliciting platelet shape change, calcium mobilization, and aggregation. MMP-1 also enhanced phosphorylation of p38MAPK and of its substrate, MAPKAP-K2, a protein involved in cytoskeletal reorganization.[20]

The concentrations of MMP-1 triggering platelet activation (75–150 ng/mL)[19,20] are far above the physiological plasma concentrations (<5 ng/mL).[62] However, serum levels of MMP-1 were found to be highly increased in the culprit coronary artery of patients with acute myocardial infarction (AMI) (~50 ng/mL),[63] suggesting that MMP-1 may contribute to the regulation of platelet function in vivo in some pathologic conditions.

MMP-1 promotes platelet thrombus formation on a collagen-coated surface under arterial flow conditions, a phenomenon blocked by pharmacologic inhibitors of MMP-1 or of PAR1.[20]

At the end of last century, Sawicki and coworkers discovered that low concentrations (0.01–1 ng/mL) of active MMP-2 (but not of its proenzyme) enhanced the aggregation of prestimulated (but not of resting) platelets, while higher concentrations (>100 ng/mL) displayed an inhibitory effect. Neutralization of endogenous MMP-2 with blocking antibodies, recombinant TIMP-2, or pharmacological inhibitors of MMPs reduced collagen-induced platelet aggregation, indicating that platelet-released MMP-2 mediates aggregation. This finding unraveled a novel mechanism of platelet aggregation, although the receptors and signaling events involved were not identified.[51] Later, we showed that active MMP-2 amplifies the platelet aggregation response to a wide range of agonists, besides collagen, acting on different receptors such as thrombin, U46619 (TxA2/PGH2 receptor agonist), and ADP, as well as to agonists acting directly on intracellular signal transduction pathways, such as PMA (a PKC activator) or the calcium ionophore A23187, showing that the effect is mediated by the activation of a common, postreceptorial signaling pathway, that was identified in phosphatidyl-inositol 3-kinase (PI3K). MMP-2 amplifies also platelet granule secretion, calcium fluxes, IP3 formation, and pleckstrin phosphorylation.[52] MMP-2-induced platelet potentiation is resistant to inhibition by aspirin or by ADP receptor antagonists. The concentrations of MMP-2 exerting this priming activity (0.1–50 ng/mL, i.e., 0.0015–0.75 nM) are in the range of those secreted by stimulated platelets in vitro[52] and in vivo in humans at a site of vascular injury (around 0.27 nM).[64] These concentrations may be even higher in the microenvironment of a growing thrombus. In fact, it was reported that the thrombus core, as compared with the shell, provides an environment retaining soluble proteins, such as thrombin.[65] Thus, it can be assumed that MMP-2 concentrations within a growing platelet thrombus at the site of vascular injury easily reach levels potentiating platelet activation.

Intraplatelet MMP-2 was shown to hydrolyze talin, a cytoskeletal protein required for the activation of GPIIb/IIIa in the inside-out signaling pathway. MMP-2 and talin were found associated in resting platelets and to dissociate upon platelet activation. Active MMP-2 was able to hydrolyze talin in vitro within few seconds and this would indicate that intracellular MMP-2 becomes activated, as demonstrated in other cells,[66] and modifies the talin–GPIIb/IIIa complex rapidly enough to participate in the aggregation response to stimuli.[67] It has to be acknowledged that the observations on this mechanism are based essentially on in vitro studies on isolated proteins and with rather high (4 ng/μL) amounts of MMP-2 used to study talin hydrolysis, probably higher than the intracellular concentrations, and thus their functional relevance in vivo awaits confirmation.

The interaction of MMP-2 with integrin $\alpha_{IIb}\beta_3$ is required for the cleavage and release of sCD40L from the surface of activated human platelets, a protein with established roles in inflammation and thrombosis.[68,69]

It was recently shown that active MMP-2 enhances platelet activation by enzymatically cleaving PAR1 at a specific, noncanonical extracellular site with an $\alpha_{IIb}\beta_3$-facilitated mechanism. The cleavage of PAR1 by MMP-2 generates a tethered ligand different from that produced by thrombin that in turn triggers biased PAR1 signaling. In particular, MMP-2 stimulates $G_{12/13}$- and G_q-activation in human platelets, as shown by p38-MAPK phosphorylation, intraplatelet Ca^{+2} increase, and PI3K activation and by the inhibition of MMP-2-priming activity by a Rho-kinase inhibitor and by a phospholipase C inhibitor, but not G_i-signaling. Thus, MMP-2 initiates intraplatelet signaling pathways but in order to generate full activation it requires concomitant G_i-signaling triggered by other agonists, leading to adenylyl cyclase inhibition and full platelet aggregation.[60]

Platelet adhesion to fibrinogen stimulated by thrombin under static conditions is associated with the release of MMP-2 from platelets, and phenanthroline, an aspecific MMPs inhibitor, reduced platelet adhesion, suggesting that the release of MMP-2 promotes platelet adhesion.[70]

Active MMP-2, either exogenously added or released by activated platelets, enhances shear stress-induced platelet activation and potentiates platelet deposition on collagen.[71] Indeed, the exposure of human platelets to high shear stress induces the release of amounts of MMP-2 in the range of those ($\sim$4 ng/10^8 platelets) found to enhance platelet activation. MMP-2 enhances platelet deposition on collagen under flow conditions, an effect due to the potentiation of platelet aggregation and thrombus formation on the initially adhering platelets. MMP-2 potentiated platelet deposition both at low and high shear rates (from 250 to 3000 s^{-1}), suggesting that it does not act by facilitating the interaction between a specific adhesive receptor and collagen, but rather that it acts at a later stage promoting the recruitment of platelets to the growing thrombus.

In real-time microscopy studies, increased deposition of platelets was evident only in the late phases of perfusion and confocal microscopy showed that MMP-2 enhances thrombus volume rather than adhesion.[71]

MMP-2 thus is likely to play a relevant role in thrombus formation at sites of increased shear stress in vivo, like in stenosed atherosclerotic coronary arteries, conditions in which platelet-released or vessel wall-released MMP-2 is enhanced.[72]

The potentiation of platelet activation by MMP-2 may be involved also in platelet-mediated tumor metastasis because it has been reported that some cancer cells aggregate platelets by releasing MMP-2 and in fact the incubation of platelets with a neutralizing anti-MMP-2 antibody reduced the aggregating effects of cancer cells.[73]

In contrast, MMP-9 appears to counteract the platelet-potentiating effects of MMP-2 and to inhibit agonist-induced platelet aggregation.[25] The platelet inhibitory effect of MMP-9 has been ascribed to changes in platelet membrane fluidity and to the reduction of PLC activation followed by inhibition of phosphoinositide breakdown, protein kinase C activation and TxA_2 formation, and intracellular Ca^{2+} mobilization.[74] Activated MMP-9 also increased nitrate production by platelets and thus intraplatelet cyclic GMP, resulting in inhibition of platelet aggregation.[26]

The concentrations of active MMP-9 inhibiting platelet aggregation in vitro (15–90 ng/mL) are in the range of those found in plasma (30–50 ng/mL).[75]

MMP-3 was reported to be devoid of effects on platelets: it did neither induce tyrosine phosphorylation of intracellular proteins nor it potentiated platelet aggregation.[19]

MT1-MMP (MMP-14) participates in the activation of MMP-2 on the platelet surface via the formation of a trimolecular complex involving TIMP-2 (MT1-MMP/TIMP-2/MMP-2).[38] It was also hypothesized that MT1-MMP may contribute to collagen-induced platelet aggregation.[38] In addition to its role as an activator of MMP-2, MMP-14 is a key enzyme in tumor cell migration and invasion.[76,77]

In a recent study comparing the roles of different MMP family members on in vitro thrombus formation and platelet activation on collagen under arterial flow conditions, it was found that pharmacological inhibition of MMP-1 or MMP-2 significantly diminished the surface area covered by platelets, whereas the inhibition of MMP-9 or MMP-14 increased it.[28] This study also showed that MMP-1, MMP-2, MMP-9, and MMP-14 associate with the platelet membrane on a growing thrombus and that, besides modulating platelet activation and thrombus formation, they also degrade the substrate collagen, showing that platelet membrane-associated MMPs exert an enzymatic function on target substrates.[28]

After partial digestion of collagen monomers by MMP-13, static platelet adhesion and thrombus formation in whole flowing blood is diminished, indicating that collagenase activity within an atherosclerotic plaque may reduce the collagen fibril to small components that are unreactive under shear conditions and reduce recruitment of platelet.[78]

Incubation with recombinant TIMP-1 was shown to attenuate phosphatidylserine exposure on thrombin- or calcium ionophore-activated platelets, a negative feed-back signal that may protect from undesired thrombin generation and premature clearance, while TIMP-2, -3, and -4 did not share this effect.[79]

Although TIMPs have historically been thought of as inhibitors of MMPs, it has become clear that in some cases a TIMP may actually activate a MMP. For example, TIMP-2 activates MMP-2 (in complex with active MMP-14) when it binds to the hemopexin-like domain of MMP-2. In contrast, TIMP-2 inhibits MMP-2 when it binds to the catalytic site of MMP-2. TIMPs also affect cell proliferation independent of their inhibitory effects on MMPs. Kasper *et al.*[80] showed that MMPs and TIMPs likely regulate mesenchymal stem cells (MSCs) in response to mechanical force and contribute to osteogenic differentiation. In fact, broad spectrum inhibition of MMPs altered the migration, proliferation, and osteogenic differentiation of MSCs. The balance of MMPs and TIMPs, rather than the individual activity of any single bioactive molecule, is likely the deciding factor.[81]

3.2 Modulation by Platelet MMPs of Other Cell Functions

In addition to regulating platelet function, MMPs secreted during platelet activation exert important effects on surrounding cells, including endothelial cells, monocytes, and tumor cells.

The formation of platelet–leukocyte complexes induced by PAR agonists is associated with increased expression of MMP-1, -2, -3, and -9, although it is not clear if the source of MMPs are platelets or leukocytes (likely both). Moreover, MMP inhibitors reduced the formation of the complexes while the addition of active MMPs promoted them.[82]

Coincubation of platelets with monocytes on immobilized type I collagen greatly increased MMP-9 release from monocytes. MMP-9 synthesis required contact between platelets and monocytes in addition to adhesion of monocytes to collagen, indicating that the synthesis of MMP-9 by monocytes may be spatially restricted by platelets to areas of vascular injury, such as the fibrous cap of atherosclerotic plaques, stenotic coronary lesions, and abdominal aortic aneurysms (AAAs), where type I collagen is abundant.[22]

Thrombin-activated platelets stimulate HUVEC to upregulate mRNA and protein expression of MT1-MMP and to secrete MMP-1, MMP-2, and MMP-9. CD40L blockade and specific GP IIb/IIIa antagonists inhibit MMP-9 and MMP-2 release suggesting that anti-GP IIb/IIIa or anti-CD40L treatments might stabilize plaques.[83]

Platelets stimulate also tumor cells to secrete MMPs thereby facilitating metastasis.[84] Platelets upregulate both the activity and expression of MMP-9 in breast adenocarcinoma MCF7, colon adenocarcinoma Caco-2, and HT-1080 fibrosarcoma cells, leading to increased invasiveness of these cancer cells.[85] Moreover, platelet-derived microvesicles are able to upregulate MT1-MMP and MMP-9 in several lung carcinoma cell lines[86] and to promote invasiveness of prostate cancer cells via the upregulation of MMP-2 production.[87]

Stimulation of human adult dermal fibroblasts with platelet rich plasma results in a marked upregulation of MMP-1 at both the mRNA and protein levels.[88] Moreover, MMP-2 production by synovial fibroblasts was significantly higher in the presence of platelets and this increase was significantly reduced by coincubation with a P-selectin blocking antibody.[89]

3.3 Animal Models

Megakaryocytes and platelets produce and release VEGF-A and other proangiogenic cytokines, including MMPs,[25] thereby promoting angiogenesis. However, the proangiogenic effects of platelets can be counteracted by their capacity to elaborate antiangiogenic factors, including platelet activating factor 4 (PF4) and thrombospondin 1 (TSP1).[90]

The expression and release of TSPs by megakaryocytes and platelets functions as an antiangiogenic switch through the activation of MMP-9 that enhances SDF-1 release and in turn stimulates angiogenesis. TSP deficiency confers a proangiogenic phenotype by an impaired inhibition of proteases, supporting the previous finding that platelets from MMP-9$^{-/-}$ mice show defective SDF-1 release upon stimulation with thrombin.[91] These data suggest that TSP-dependent inhibition of MMP-9 controls SDF-1 release by platelets.[92]

In a model of platelet adhesion to collagen under flow, platelet thrombi were smaller when blood from MMP-2$^{-/-}$ mice was employed as compared with blood from wild type mice. In contrast, perfusion of blood from MMP-9$^{-/-}$ mice resulted in thrombi covering a larger surface area, with platelets expressing higher levels of phosphatidylserine and P-selectin. Blood from MMP-3$^{-/-}$ mice instead did not behave differently from wild type mice for as concerns platelet activation and thrombus formation.[28]

Platelet pulmonary thromboembolism induced by the i.v. injection of collagen + epinephrine- and photochemically induced thrombosis of the femoral artery were reduced in MMP-2$^{-/-}$ mice. To unravel the cellular origin of MMP-2 promoting thrombosis, chimeric mice lacking MMP-2 only in platelets were generated; in these mice thrombus formation was delayed, indicating that it is platelet-derived MMP-2 that facilitates thrombus formation. Finally, platelets activated by a mild vascular damage induced thrombus formation at a downstream arterial injury site by releasing MMP-2 that in turn amplified the platelet response to vessel injury.[93]

Moreover, MMP-$2^{-/-}$ mice showed a mild hemostatic defect, with a prolongation of the bleeding time, a defect resulting from the absence of MMP-2 from platelets as shown by the observation that transfusion of MMP-$2^{-/-}$ platelets into thrombocytopenic mice did not correct the prolonged bleeding time, differently from the transfusion of wild type platelets.[93]

MMP-2 and MMP-9 derived from platelets and macrophages accumulating in the adventitia and media of the aorta have been shown to contribute to the initiation and progression of AAAs by degrading elastin fibers.[94,95] Treatment with aspirin and clopidogrel, inhibiting platelet activation, significantly reduced platelet and macrophage accumulation in the media of the aorta with a parallel reduction of MMPs activity.[96] In agreement with this study, Liu and colleagues demonstrated, that treatment with clopidogrel, a platelet ADP receptor blocker, significantly suppressed aortic aneurysm formation in $ApoE^{-/-}$ mice infused with angiotensin II. Clopidogrel also suppressed elastic lamina degradation, inflammatory cytokine expression and reduced the production of MMPs, particularly of MMP-2, in the aorta.[97]

Platelet-derived CD40L is a potent inducer of lung neutrophil infiltration in abdominal sepsis-induced lung injury. Soluble CD40L in fact induces increased plasma levels of CXC chemokines which are potent stimulators of neutrophils.[98] In turn, neutrophil-derived MMP-9 induces CD40L shedding from platelets.[99] Thus, MMP-9 is crucial for the pathogenic interaction between platelets and neutrophils in sepsis.

Platelets regulate bone formation induced by tumors through the uptake of tumor-derived proteins (i.e., VEGF, TGF-β1, MMP-1, MMP-3, MMP-13, G-CSF, TIMP-1, and TIMP-2), and probably through the secretion of α-granule contents favoring osteoblast differentiation and maturation. In a xenograft tumor model of human prostate cancer (LNCaP-C4-2) implanted in immunocompromised mice, platelet depletion inhibited bone formation in response to tumor growth and in particular MMPs released by platelets modulated bone formation.[100]

Concerning the roles of MMP-1 and MMP-14 in regulating platelet function in vivo and the interactions with other cell, this has not been investigated in mice because murine platelets do not express MMP-1, but the mouse orthologous gene *Mmp-1a*,[101] while MMP-14-deficient mice are not vital.

In a mouse model of sepsis, mouse Mmp-1a was released from the endothelium into the circulation and triggered PAR1-dependent disruption of endothelial barrier function via the Rho pathway. Inhibition of MMP-1 in the early stages of sepsis, by the administration of MMP-1 inhibitor-1, significantly improved the survival of WT mice while the administration of exogenous human MMP-1 caused endothelial barrier dysfunction and increased lung vascular permeability in WT but not in $PAR1^{-/-}$ mice.[102]

3.4 Human Studies

The observation that MMP-2 is released by platelets in vivo in healthy humans during primary hemostasis suggests that MMP-2 plays a physiological role in the regulation of the platelet response to vessel wall damage.[64] MMP-2 concentration was significantly higher in shed blood than in venous blood in healthy volunteers undergoing the measurement of the bleeding time, and increased progressively, consistent with ongoing platelet activation. Active MMP-2 in shed blood was in the range of concentrations (around 1 ng/10^8 platelets) found to potentiate platelet activation. Aspirin does not inhibit this release. The oral intake of 500 mg aspirin, in fact, although resulted in a complete suppression of serum TxB_2 and in a prolongation of the bleeding time, did not affect the surface expression of MMP-2 on platelets recovered from the bleeding time blood and did not significantly modify the amounts of total or active MMP-2 released in shed blood.[64]

4. ROLE OF PLATELET-DERIVED MMPs IN DISEASE

4.1 Atherosclerosis

The activity of MMPs is essential for many of the processes involved in atherosclerotic plaque formation, like infiltration of inflammatory cells, smooth muscle cell migration and proliferation, and angiogenesis. Furthermore, matrix degradation by MMPs causes plaque instability and rupture that lead to unstable angina, myocardial infarction, and stroke.[103]

Therefore, the role of MMPs in atherosclerosis has been extensively evaluated, but only a few studies have explored the role of platelet-derived MMPs. Among MMPs, active MMP-2 recognizes as substrates gelatin, elastin, type IV collagen, fibronectin, laminin-1. The ability of gelatinase A to hydrolyze elastin is especially relevant to its effects on the vasculature, where elastin is an important structural component of the subendothelium of medium- and large-size arteries, and several studies have shown a role of MMP-2 in the vascular remodeling changes associated with atherosclerosis, restenosis, arterial aneurysmal dilation, and plaque rupture.[104]

Increased levels of circulating MMP-2 were found in patients with acute coronary artery syndromes (ACS) relative to control subjects and are considered a marker of plaque rupture or instability.[105]

Interestingly, simultaneous blood sampling from the aorta and the coronary sinus of patients with unstable angina showed that, despite optimal antithrombotic therapy, MMP-2 is released in the coronary circulation concomitantly to the platelet-specific proteins β-TG and PF4, suggesting that coronary MMP-2 derives in large part from activated platelets.[72] This hypothesis is further supported by the observation that the transcardiac gradient of MCP-1, a marker unrelated to platelet activation, was not increased in the ACS group and did not correlate with platelet activation markers or with MMP-2. The release of MMP-2 was found only in the coronaries carrying the culprit lesion, further confirming that it represented platelet activation rather than the expression of generalized coronary inflammation.

Plasma from the coronary sinus of patients with ACS enhanced the expression of P-selectin of platelets from healthy donors, a phenomenon inhibited by preincubation with TIMP-2, stressing the importance of elevated MMP-2 in the pathogenesis of sustained platelet activation in ACS.

Also the transcardiac gradients of MMP-1 are greater in patients with unstable angina and acute myocardial infarction (AMI) than in patients with stable effort angina or control subjects[63,106] and serum MMP-1 is higher in patients with AMI than in patients with stable angina.[107] Recently, elevated baseline plasma levels of MMP-1 have been identified as strong and independent predictors of long-term all-cause mortality in a cohort of patients with known or suspected coronary artery disease.[108] The role of platelet-derived MMP-1 in the increased levels of MMP-1 in patients with ACS has not been explored.

Several studies have analyzed the role of MMPs in plaque stability or progression[109,110] but only one study so far has explored the role of plaque MMPs in modulating platelet activation.[111] Previous studies had shown that atherosclerotic plaques contain thrombogenic substances, such as collagen type I and III and tissue factor, that directly elicit platelet adhesion and stimulate platelet secretion and aggregation.[112–114] Recently, we showed that human carotid plaque extracts promote platelet aggregation due to their content of MMP-2, an effect prevented by three different specific MMP-2 inhibitors (inhibitor II, TIMP-2, moAb anti-MMP-2). The pro-MMP-2/TIMP-2 ratio of plaques potentiating platelet aggregation was significantly higher than that of plaques not potentiating it.

Moreover, an elevated MMP-2 activity in plaques as well as a high aggregation-potentiating effect of plaques were associated with a higher rate of subsequent ischemic cerebrovascular events.[111]

MMP-1, although undetectable in normal arteries, is increased in atherosclerotic plaques[115] and in particular in macrophages, smooth muscle cells, and endothelial cells surrounding the fibrous cap, thus especially in the vulnerable region of the plaque.[115,116] MMP-1 can thus contribute to the destabilization of the plaques and may facilitate thrombus formation on ruptured plaques by its procoagulant function on platelets.

The role of MMPs in the complex pathophysiology of ischemic stroke, in particular of MMP-2 and MMP-9, has been also widely studied in human and in animal models.[117,118] No data, however, are available on a direct involvement of platelet-derived MMPs.

Several observations are emerging about a role of MMPs in peripheral arterial disease (PAD).[119] High circulating levels of MMP-2, -9 and TIMP-1 have been found in patients with PAD, with higher levels in patients with critical limb ischemia as compared with those with intermittent claudication.[120,121] The source of the enhanced MMPs in PAD patients is undefined yet, but it is plausible that part of the raised circulating concentrations may come from platelets, given that platelets release relevant amounts of MMPs upon activation in vivo[64] and that in vivo platelet activation is a hallmark of PAD.[122,123]

4.2 Inflammation

Platelets are innate immune cells and release mediators that strongly contribute to the recruitment and modulation of the activity of other cells that sustain inflammation. The CD40/CD40L pathway is involved in several chronic inflammatory conditions, such as inflammatory bowel disease,[124] rheumatoid arthritis,[125] and atherosclerosis.[126] CD40 ligand (CD40L) is a transmembrane glycoprotein of the tumor necrosis factor family constitutively expressed in platelets and, upon activation, it is cleaved to a soluble form (sCD40L) by MMP-2 with an $\alpha_{IIb}\beta_3$-mediated mechanism.[69] sCD40L displays inflammatory effects through the promotion of platelet-monocyte aggregate formation and the production of reactive oxygen species.[127]

The binding of CD40L to $\alpha_{IIb}\beta_3$ and/or to CD40 facilitates the interaction of platelets with CD40-expressing cells triggering an inflammatory response with the release of several cytokines and of MMPs.[125]

The activation of platelets is also associated with a variety of glomerular diseases with proteinuria. The supernatant of activated platelets and platelet-derived CD40L induced MMP-9 mRNA expression in podocytes, specialized epithelial cells that are pivotal in maintaining the glomerular filtration barrier and its properties.[128]

The interaction with platelets is essential for leukocyte recruitment at inflammatory sites. Platelet–leukocyte aggregates formation induced by PAR agonists is regulated by MMP-1, -2, -3, and -9.[82]

Recently, it was also shown that MMP-1–PAR1 signaling plays an important role also in endothelial barrier function and sepsis outcomes. Sepsis patients had an 18-fold increase in the levels of pro-MMP-1 in their plasma relative to healthy controls.[102]

Platelets act as inflammatory cells in rheumatoid arthritis (RA).[129] The synovial fluid from RA patients contains significantly more activated platelets, platelet–leukocyte and platelet–synoviocyte complexes, and also much more MMP-2 compared with synovial fluid from patients with osteoarthritis. Platelet-released MMP-2 activates synoviocyte PAR1 leading to the release of MMP-2.[130] Moreover, mice selectively depleted of platelet MMP-2 developed significantly less arthritis, and in particular less cartilage damage of the tibio-talar joints, compared with wild-type mice indicating that platelet-derived MMP-2 plays a crucial role in disease progression.[130]

Platelets participate also in the pathogenesis of osteoarthritis by the induction of MMP-2 release by fibroblasts, possibly via P-selectin, thus contributing to cartilage breakdown. Treatment with hyaluronic acid decreases the number of platelets and their level of activation in the synovial fluid and in parallel the concentration of MMP-2. Therefore, the interaction between platelets and synoviocytes leads to platelet activation and MMP-2 release in osteoarthritis possibly contributing to disease progression.[89]

Alzheimer's disease (AD) is the prevalent type of dementia and is characterized by pathological changes in brain with the formation of amyloid-β (Aβ) plaques and neurofibrillar tangles deposition, as well as neuronal death and synaptic loss. MMPs play an important role as inflammatory components in the pathogenesis of AD. Platelets are of particular interest because Aβ peptides are stored in their α-granules and in platelet microparticles (PMPs) that are carriers of sAβ.[131] Impaired clearance of Aβ contributes to the deposition of amyloid plaques. MMPs, in particular MMP-2, participate in the clearance of Aβ, and indeed a reduced content of MMP-2 in platelets has been reported in AD.[132,133]

Recently, it was also shown that MMP-1–PAR1 signaling plays an important role also in endothelial barrier function and sepsis outcomes. Sepsis patients had an 18-fold increase in the levels of pro-MMP-1 in their plasma relative to healthy controls.[102]

4.3 Tumor Growth and Metastasis

Platelets play a fundamental role in hematogenous dissemination of tumor cells. Abundant platelets were detected also in the tumor microenvironment outside blood vessels, thus platelet–tumor cell interaction plays a role also in primary tumor growth.[134]

Over the past 50 years, many studies have contributed to elucidate the molecular mechanisms responsible for mediating tumor cell-induced platelet aggregation (TCIPA) and secretion and how these interactions affect other cells of the tumor microenvironment. Platelets contribute to tumor angiogenesis, immunoevasion, and cancer cell invasion. In particular, platelets form complexes with tumor cells creating emboli that favor tumor cell extravasation to the metastatic niche; the formation of a platelet coat around tumor cells protects them from natural killer (NK) cell cytotoxic activity; platelets release growth factors, proteases, and small molecules that help in tumor growth, invasion, and neoangiogenesis. For most solid tumors, particularly carcinomas, the microenvironment consists of the tumor cells themselves, known as the parenchyma, as well as of the stroma, that consists of nonmalignant mesenchymal cells and of connective tissue that contribute to the structure and survival of the tumor. Platelets flowing through the vascular network of the tumor become part of the tumor microenvironment, thus influencing the parenchyma and tumor-associated stroma.[134]

Activated platelets contribute to degrade structural components of vascular basement membrane either directly, by releasing MMPs, or by favoring tumor- and endothelial cell-production of MMPs. Indeed, platelet depletion reduced metastasis and was associated with decreased ECM degradation and reduced expression of MMP-2, -9 and PAI-1 in the tumor.[134]

MMPs deriving from stromal cells, such as fibroblasts and myofibroblasts, immune cells, and endothelial cells surrounding the tumor are released by nearly all human cancers. The expression of these MMPs, in both the primary tumor and/or metastases, is correlated with tumor progression.

During hematogenous metastasis, cancer cells migrate to the vasculature where they interact with platelets resulting in TCIPA. In particular, cancer cells have the ability to stimulate the release of platelet granules leading to the liberation of proaggregatory agents. ADP contributes to TCIPA induced by SKNMC neuroblastoma,[135] small-cell lung,[136] melanoma M1Do, M3Da, M4Be,[137] breast carcinoma MCF7,[85] and fibroblastoma HT-1080 cells.[73] It has been shown that ADP released during MCF-7-induced TCIPA aggregates platelets via activation of the P2Y12 purinergic receptor.[85] TCIPA is also stimulated by serine proteinases including thrombin, cathepsin B, and MMPs. In fact, it was shown that the release of MMP-2 from platelets as well as from cancer cells is involved in TCIPA induced by HT-1080 and MCF7 cells.[85] Interestingly, increased aggregability of platelets collected from patients with metastatic prostate cancer can be related to enhanced generation of MMP-2.[138] Integrin $\alpha_v\beta_3$ on the surface of invasive angiogenic vascular cells and melanoma cells binds MMP-2 and acts as a receptor for surface-localized metalloproteinase activity. Inhibitors of MMPs and molecules that prevent integrin $\alpha_v\beta_3$ binding to MMP-2, via its hemopexin domain, reduce cellular protrusive activity, and invasive behavior. The fact that $\alpha_v\beta_3$, expressed by tumor cells promotes cell motility while MMP-2, released during platelet activation, potentiates matrix degradation suggests that these proteins function in a cooperative manner promoting tissue remodeling and cancer dissemination.

Recent studies have shown that PAR1 can be activated by MMPs. MMP-1/PAR1-dependent chemotaxis and invasion were demonstrated in ovarian carcinoma cells and were abolished by the PAR1 inhibitor, RWJ-56110. Platelets secrete MMP-1 and express PAR1 that is directly cleaved and activated on their surface and on that of breast cancer cells by matrix metalloprotease-1 (MMP-1). PAR1 activation by MMP-1 provides a link between extracellular proteolytic activities important for remodeling of the matrix, and cell signaling leading to cancer invasion.[139]

Upon activation, platelets release small vesicles encapsulated by plasma membrane, called PMPs. Increased levels of PMPs were demonstrated in the circulation of patients with different cancers. PMP levels are highly correlated with aggressive tumors, elevated number of platelets, and a poor clinical outcome.[140] For example, in gastric cancer, PMP levels are better predictors of metastasis than VEGF, IL-6, and RANTES.[29] PMPs can induce secretion of MMP-2 by prostate cancer cells in vitro, facilitating their passage through collagen, a major component of the extracellular matrix.[87] PMPs may also serve as chemoattractants for several lung cancer cell lines, by activating the expression of membrane type 1-matrix metalloproteinase and by stimulating proliferation and adhesion of cancer cells to fibrinogen and endothelial cells.

5. CONCLUSIONS

The crucial role that platelets play in a vital function such as the arrest of hemorrhage, with the associated need to react in an extremely efficient way to a damage to the vessel wall but simultaneously to avoid an unwanted expansion to thrombosis, explains the extremely sophisticated system of regulation of platelet activation that has evolved.[11,141] The degree of platelet activation is the end result of the platelet response to an array of agonists and inhibitors of platelet activation. Platelet-priming and -potentiating molecules finely tuning the platelet response to stimuli, like MMPs, act as crucial "cofactors" of platelet activation.

A large number of MMP inhibitors that might interfere with the pathologic consequences of the platelet/MMP interaction have been described but the majority of them are nonselective and this limits their potential clinical use enhancing the risk of untoward effects.

On the other hand, activated platelets interact with other cells modulating not only their function but also their ability to produce and release MMPs. Thus, the interaction between MMPs and platelets is at the crossroad between hemostasis, inflammation, and tissue remodeling.

The increasing knowledge of the molecular mechanism regarding the interactions between platelets and MMPs may lead to the development of innovative therapeutic approaches to thrombotic, inflammatory, and neoplastic disorders.

REFERENCES

1. Versteeg HH, Heemskerk JW, Levi M, Reitsma PH. New fundamentals in hemostasis. *Physiol Rev*. 2013;93:327–358.
2. Italiano JE. Megakaryocyte development and platelet production. In: Gresele P, Kleiman NS, Lopez A, Page CP, eds. *Platelet in Thrombotic and Non Thrombotic Disorders—Pathophysiology, Pharmacology and Therapeutics: An Update*. Cham, Switzerland: Springer; 2016:39–54.
3. Harker LA, Hunt P, Marzec UM, et al. Regulation of platelet production and function by megakaryocyte growth and development factor in nonhuman primates. *Blood*. 1996;87:1833–1844.

4. Semple JW, Italiano Jr JE, Freedman J. Platelets and the immune continuum. *Nat Rev Immunol.* 2011;11:264–274.
5. Nurden AT. Platelets, inflammation and tissue regeneration. *Thromb Haemost.* 2011; 105(suppl 1):S13–S33.
6. Rowley JW, Oler AJ, Tolley ND, et al. Genome-wide RNA-seq analysis of human and mouse platelet transcriptomes. *Blood.* 2011;118:e101–e111.
7. Berndt MC, Metharom P, Andrews RK. Primary haemostasis: newer insights. *Haemophilia.* 2014;20:15–22.
8. Andrews RK, Berndt MC. Platelet adhesion: a game of catch and release. *J Clin Invest.* 2008;118:3009–3011.
9. Stalker TJ, Newman DK, Ma P, Wannemacher KM, Brass LF. Platelet signaling. *Handb Exp Pharmacol.* 2012;210:59–85.
10. Vezza R, Roberti R, Nenci GG, Gresele P. Prostaglandin E2 potentiates platelet aggregation by priming protein kinase C. *Blood.* 1993;82:2704–2713.
11. Gresele P, Falcinelli E, Momi S. Potentiation and priming of platelet activation: a potential target for antiplatelet therapy. *Trends Pharmacol Sci.* 2008;29:352–360.
12. Ciferri S, Emiliani C, Guglielmini G, Orlacchio A, Nenci GG, Gresele P. Platelets release their lysosomal content in vivo in humans upon activation. *Thromb Haemost.* 2000;83:157–164.
13. Diamond SL. Systems biology of coagulation. *J Thromb Haemost.* 2013;11:224–232.
14. Panes O, Matus V, Sáez CG, Quiroga T, Pereira J, Mezzano D. Human platelets synthesize and express functional tissue factor. *Blood.* 2007;109:5242–5250.
15. Camera M, Toschi V, Brambilla M, et al. The role of tissue factor in atherothrombosis and coronary artery disease: insights into platelet tissue factor. *Semin Thromb Hemost.* 2015;41:737–746.
16. de Witt SM, Verdoold R, Cosemans JM, Heemskerk JW. Insights into platelet-based control of coagulation. *Thromb Res.* 2014;133:S139–S148.
17. Santos-Martínez MJ, Medina C, Jurasz P, Radomski MW. Role of metalloproteinases in platelet function. *Thromb Res.* 2008;121:535–542.
18. Chesney CM, Harper E, Colman RW. Human platelet collagenase. *J Clin Invest.* 1974;53:1647–1654.
19. Galt SW, Lindemann S, Allen L, et al. Outside-in signals delivered by matrix metalloproteinase-1 regulate platelet function. *Circ Res.* 2002;90:1093–1099.
20. Trivedi V, Boire A, Tchernychev B, et al. Platelet matrix metalloprotease-1 mediates thrombogenesis by activating PAR1 at a cryptic ligand site. *Cell.* 2009;137:332–343.
21. Cecchetti L, Tolley ND, Michetti N, Bury L, Weyrich AS, Gresele P. Megakaryocytes differentially sort mRNAs for matrix metalloproteinases and their inhibitors into platelets: a mechanism for regulating synthetic events. *Blood.* 2011;118:1903–1911.
22. Galt SW, Lindemann S, Medd D, et al. Differential regulation of matrix metalloproteinase-9 by monocytes adherent to collagen and platelets. *Circ Res.* 2001;89:509–516.
23. Kälvegren H, Jönsson S, Jonasson L. Release of matrix metalloproteinases-1 and -2, but not -9, from activated platelets measured by enzyme-linked immunosorbent assay. *Platelets.* 2011;22:572–578.
24. Falcinelli E, Bury L, Tolley N, et al. MMP-9 in platelets: maybe maybe not. *Blood.* 2011;118:6471–6473.
25. Fernandez-Patron C, Martinez-Cuesta MA, Salas E, et al. Differential regulation of platelet aggregation by matrix metalloproteinases-9 and -2. *Thromb Haemost.* 1999;82:1730–1735.
26. Sheu JR, Fong TH, Liu CM, et al. Expression of matrix metalloproteinase-9 in human platelets: regulation of platelet activation in in vitro and in vivo studies. *Br J Pharmacol.* 2004;143:193–201.

27. Wrzyszcz A, Wozniak M. On the origin of matrix metalloproteinase-2 and -9 in blood platelets. *Platelets*. 2012;23:467–474.
28. Mastenbroek TG, Feijge MA, Kremers RM, et al. Platelet-associated matrix metalloproteinases regulate thrombus formation and exert local collagenolytic activity. *Arterioscler Thromb Vasc Biol*. 2015;35:2554–2561.
29. Kim HK, Song KS, Park YS, et al. Elevated levels of circulating platelet microparticles, VEGF, IL-6 and RANTES in patients with gastric cancer: possible role of a metastasis predictor. *Eur J Cancer*. 2003;39:184–191.
30. Malik J, Stulc T, Ceska R. Matrix metalloproteinases in isolated hypercholesterolemia. *Int Angiol*. 2005;24:300–303.
31. Visse R, Nagase H. Matrix metalloproteinases and tissue inhibitors of metalloproteinases: structure, function, and biochemistry. *Circ Res*. 2003;92:827–839.
32. Stetler Stevenson WG. Tissue inhibitors of metalloproteinases in cell signaling: metalloproteinase-independent biological activities. *Sci Signal*. 2008;1:re6.
33. Zhao T, Zhao W, Chen Y, Li VS, Meng W, Sun Y. Platelet-derived growth factor-D promotes fibrogenesis of cardiac fibroblasts. *Am J Physiol Heart Circ Physiol*. 2013;304: H1719–H1726.
34. Eichler W, Friedrichs U, Thies A, Tratz C, Wiedemann P. Modulation of matrix metalloproteinase and TIMP-1 expression by cytokines in human RPE cells. *Invest Ophthalmol Vis Sci*. 2002;43:2767–2773.
35. Murate T, Yamashita K, Isogai C, et al. The production of tissue inhibitors of metalloproteinases (TIMPs) in megakaryopoiesis: possible role of platelet- and megakaryocyte-derived TIMPs in bone marrow fibrosis. *Br J Haematol*. 1997;99:181–189.
36. Gomis-Rüth FX, Maskos K, Betz M, et al. Mechanism of inhibition of the human matrix metalloproteinase stromelysin-1 by TIMP-1. *Nature*. 1997;389:77–78.
37. Overall CM, Tam E, McQuibban GA, et al. Domain interactions in the gelatinase A. TIMP-2.MT1-MMP activation complex. The ectodomain of the 44-kDa form of membrane type-1 matrix metalloproteinase does not modulate gelatinase A activation. *J Biol Chem*. 2000;275:39497–39506.
38. Kazes I, Elalamy I, Sraer JD, Hatmi M, Nguyen G. Platelet release of trimolecular complex components MT1-MMP/TIMP2/MMP2: involvement in MMP2 activation and platelet aggregation. *Blood*. 2000;96:3064–3069.
39. Chang C, Werb Z. The many faces of metalloproteases: cell growth, invasion, angiogenesis and metastasis. *Trends Cell Biol*. 2001;11:S37–S43.
40. Brew K, Nagase H. The tissue inhibitors of metalloproteinases (TIMPs): an ancient family with structural and functional diversity. *Biochim Biophys Acta*. 1803; 2010:55–71.
41. Rowley JW, Schwertz H, Weyrich AS. Platelet mRNA: the meaning behind the message. *Curr Opin Hematol*. 2012;19:385–391.
42. Weis BL, Schleiff E, Zerges W. Protein targeting to subcellular organelles via MRNA localization. *Biochim Biophys Acta*. 1833;2013:260–273.
43. Martin KC, Ephrussi A. mRNA localization: gene expression in the spatial dimension. *Cell*. 2009;136:719–730.
44. Fusco D, Accornero N, Lavoie B, et al. Single mRNA molecules demonstrate probabilistic movement in living mammalian cells. *Curr Biol*. 2003;13:161–167.
45. Lood C, Amisten S, Gullstrand B, et al. Platelet transcriptional profile and protein expression in patients with systemic lupus erythematosus: up-regulation of the type I interferon system is strongly associated with vascular disease. *Blood*. 2010;116:1951–1957.
46. Raghavachari N, Xu X, Harris A, et al. Amplified expression profiling of platelet transcriptome reveals changes in arginine metabolic pathways in patients with sickle cell disease. *Circulation*. 2007;115:1551–1562.

47. Healy AM, Pickard MD, Pradhan AD, et al. Platelet expression profiling and clinical validation of myeloid-related protein-14 as a novel determinant of cardiovascular events. *Circulation*. 2006;113:2278–2284.
48. Villeneuve J, Block A, Le Bousse-Kerdilès MC, et al. Tissue inhibitors of matrix metalloproteinases in platelets and megakaryocytes: a novel organization for these secreted proteins. *Exp Hematol*. 2009;37:849–856.
49. Sawicki G, Sanders EJ, Salas E, Wozniak M, Rodrigo J, Radomski MW. Localization and translocation of MMP-2 during aggregation of human platelets. *Thromb Haemost*. 1998;80:836–839.
50. Radomski A, Jurasz P, Sanders EJ, et al. Identification, regulation and role of tissue inhibitor of metalloproteinases-4 (TIMP-4) in human platelets. *Br J Pharmacol*. 2002;137:1330–1338.
51. Sawicki G, Salas E, Murat J, MisztaLanel H, Radomski MW. Release of gelatinase A during platelet activation mediates aggregation. *Nature*. 1997;386:616–619.
52. Falcinelli E, Guglielmini G, Torti M, Gresele P. Intraplatelet signaling mechanisms of the priming effect of matrix metalloproteinase-2 on platelet aggregation. *J Thromb Haemost*. 2005;3:2526–2535.
53. Cognasse F, Hamzeh-Cognasse H, Chabert A, et al. Streptococcus sanguinis-induced cytokine and matrix metalloproteinase-1 release from platelets. *BMC Immunol*. 2014;15:15.
54. Bennet J. αIIbβ3 (GPIIb/IIIa) structure and function. In: Gresele P, Kleiman NS, Lopez A, Page CP, eds. *Platelet in Thrombotic and Non Thrombotic Disorders—Pathophysiology, Pharmacology and Therapeutics: An Update*. Cham, Switzerland: Springer; 2016:99–112.
55. Stricker TP, Dumin JA, Dickeson SK, et al. Structural analysis of the alpha(2) integrin I domain/procollagenase-1 (matrix metalloproteinase-1) interaction. *J Biol Chem*. 2001;276:29375–29381.
56. Brooks PC, Strömblad S, Sanders LC, et al. Localization of matrix metalloproteinase MMP-2 to the surface of invasive cells by interaction with integrin alpha v beta 3. *Cell*. 1996;85:683–693.
57. Radomski A, Stewart MW, Jurasz P, Radomski MW. Pharmacological characteristics of solid-phase von Willebrand factor in human platelets. *Br J Pharmacol*. 2001;134:1013–1020.
58. Ossovskaya VS, Bunnett NW. Protease-activated receptors: contribution to physiology and disease. *Physiol Rev*. 2004;84:579–621.
59. Jaffré F, Friedman AE, Hu Z, Mackman N, Blaxall BC. β-adrenergic receptor stimulation transactivates protease-activated receptor 1 via matrix metalloproteinase 13 in cardiac cells. *Circulation*. 2012;125:2993–3003.
60. Sebastiano M, Momi S, Falcinelli E, Bury L, Hoylaerts M, Gresele P. MMP-2 mediates PAR1 biased signaling in human platelets: a novel mechanism regulating platelet activation. *Blood*. 2016. [Epub ahead of print].
61. Lane WJ, Dias S, Hattori K, et al. Stromal-derived factor 1-induced megakaryocyte migration and platelet production is dependent on matrix metalloproteinases. *Blood*. 2000;96:4152–4159.
62. Kato R, Momiyama Y, Ohmori R, Taniguchi H, Nakamura H, Ohsuzu F. Levels of matrix metalloproteinase-1 in patients with and without coronary artery disease and relation to complex and noncomplex coronary plaques. *Am J Cardiol*. 2005;95:90–92.
63. Suzuki H, Kusuyama T, Sato R, et al. Elevation of matrix metalloproteinases and interleukin-6 in the culprit coronary artery of myocardial infarction. *Eur J Clin Invest*. 2008;38:166–173.
64. Falcinelli E, Giannini S, Boschetti E, Gresele P. Platelets release active matrix metalloproteinase-2 in vivo in humans at a site of vascular injury: lack of inhibition by aspirin. *Br J Haematol*. 2007;138:221–230.

65. Tomaiuolo M, Stalker TJ, Welsh JD, Diamond SL, Sinno T, Brass LF. A systems approach to hemostasis: 2. Computational analysis of molecular transport in the thrombus microenvironment. *Blood.* 2014;124:1816–1823.
66. Schulz R. Intracellular targets of matrix metalloproteinase-2 in cardiac disease: rationale and therapeutic approaches. *Annu Rev Pharmacol Toxicol.* 2007;47:211–242.
67. Soslau G, Mason C, Lynch S, et al. Intracellular matrix metalloproteinase-2 (MMP-2) regulates human platelet activation via hydrolysis of talin. *Thromb Haemost.* 2014;111:140–153.
68. Reinboldt S, Wenzel F, Rauch BH, et al. Preliminary evidence for a matrix metalloproteinase-2 (MMP-2)-dependent shedding of soluble CD40 ligand (sCD40L) from activated platelets. *Platelets.* 2009;20:441–444.
69. Choi WS, Jeon OH, Kim DS. CD40 ligand shedding is regulated by interaction between matrix metalloproteinase-2 and platelet integrin alpha(IIb)beta(3). *J Thromb Haemost.* 2010;8:1364–1371.
70. Martinez-Cuesta MA, Salas E, Radomski A, Radomski MW. Matrix metalloproteinase-2 in platelet adhesion to fibrinogen: interactions with nitric oxide. *Med Sci Monit.* 2001;7:646–651.
71. Guglielmini G, Appolloni V, Momi S, et al. Matrix metalloproteinase-2 enhances platelet deposition on collagen under flow conditions. *Thromb Haemost.* 2016;115:333–343.
72. Gresele P, Falcinelli E, Loffredo F, et al. Platelets release matrix metalloproteinase-2 in the coronary circulation of patients with acute coronary syndromes: possible role in sustained platelet activation. *Eur Heart J.* 2011;32:316–325.
73. Jurasz P, Sawicki G, Duszyk M, et al. Matrix metalloproteinase 2 in tumor cell-induced platelet aggregation: regulation by nitric oxide. *Cancer Res.* 2001;61:376–382.
74. Lee YM, Lee JJ, Shen MY, Hsiao G, Sheu JR. Inhibitory mechanisms of activated matrix metalloproteinase-9 on platelet activation. *Eur J Pharmacol.* 2006;537:52–58.
75. Nakamura T, Ebihara I, Shimada N, Shoji H, Koide H. Modulation of plasma metalloproteinase-9 concentrations and peripheral blood monocyte mRNA levels in patients with septic shock: effect of fiber-immobilized polymyxin B treatment. *Am J Med Sci.* 1998;316:355–360.
76. Hotary K, Allen E, Punturieri A, Yana I, Weiss SJ. Regulation of cell invasion and morphogenesis in a three-dimensional type I collagen matrix by membrane-type matrix metalloproteinases 1, 2, and 3. *J Cell Biol.* 2000;149:1309–1323.
77. Koshikawa N, Giannelli G, Cirulli V, Miyazaki K, Quaranta V. Role of cell surface metalloprotease MT1-MMP in epithelial cell migration over laminin-5. *J Cell Biol.* 2000;148:615–624.
78. Howes JM, Pugh N, Knäuper V, Farndale RW. Modified platelet deposition on matrix metalloproteinase 13 digested collagen I. *J Thromb Haemost.* 2015;13:2253–2259.
79. Villeneuve J, Belloc F, Hugues M, et al. Tissue inhibitor of matrix metalloproteinase-1 reduces phosphatidylserine exposure on activated and aged platelets. *Br J Haematol.* 2010;149:302–306.
80. Kasper G, Glaeser JD, Geissler S, et al. Matrix metalloprotease activity is an essential link between mechanical stimulus and mesenchymal stem cell behavior. *Stem Cells.* 2007;25:1985–1994.
81. Hire JM, Evanson JL, Johnson PC, et al. Variance of matrix metalloproteinase (MMP) and tissue inhibitor of metalloproteinase (TIMP) concentrations in activated, concentrated platelets from healthy male donors. *J Orthop Surg Res.* 2014;9:29.
82. Chung AW, Radomski A, Alonso-Escolano D, et al. Platelet-leukocyte aggregation induced by PAR agonists: regulation by nitric oxide and matrix metalloproteinases. *Br J Pharmacol.* 2004;143:845–855.
83. May AE, Kälsch T, Massberg S, Herouy Y, Schmidt R, Gawaz M. Engagement of glycoprotein IIb/IIIa (alpha(IIb)beta3) on platelets upregulates CD40L and triggers

CD40L-dependent matrix degradation by endothelial cells. *Circulation*. 2002;106: 2111–2117.

84. Belloc C, Lu H, Soria C, Fridman R, Legrand Y, Menashi S. The effect of platelets on invasiveness and protease production of human mammary tumor cells. *Int J Cancer*. 1995;60:413–417.
85. Alonso-Escolano D, Strongin AY, Chung AW, Deryugina EI, Radomski MW. Membrane type-1 matrix metalloproteinase stimulates tumour cell-induced platelet aggregation: role of receptor glycoproteins. *Br J Pharmacol*. 2004;141:241–252.
86. Janowska-Wieczorek A, Wysoczynski M, Kijowski J, et al. Microvesicles derived from activated platelets induce metastasis and angiogenesis in lung cancer. *Int J Cancer*. 2005;113:752–760.
87. Dashevsky O, Varon D, Brill A. Platelet-derived microparticles promote invasiveness of prostate cancer cells via upregulation of MMP-2 production. *Int J Cancer*. 2009;124:1773–1777.
88. Shin MK, Lee JW, Kim YI, Kim YO, Seok H, Kim NI. The effects of platelet-rich clot releasate on the expression of MMP-1 and type I collagen in human adult dermal fibroblasts: PRP is a stronger MMP-1 stimulator. *Mol Biol Rep*. 2014;41:3–8.
89. Petito E, Alunno A, Falcinelli E, et al. Platelets participate in joint inflammation osteoarthritis: prevention by hyaluronic acid. *Thromb Res*. 2014;134(suppl 2): S79–S80.
90. Maione TE, Gray GS, Petro J, et al. Inhibition of angiogenesis by recombinant human platelet factor-4 and related peptides. *Science*. 1990;247:77–79.
91. Jin DK, Shido K, Kopp HG, et al. Cytokine-mediated deployment of SDF-1 induces revascularization through recruitment of CXCR4(+) hemangiocytes. *Nat Med*. 2006;12:557–567.
92. Kopp HG, Hooper AT, Broekman MJ, et al. Thrombospondins deployed by thrombopoietic cells determine angiogenic switch and extent of revascularization. *J Clin Invest*. 2006;116:3277–3291.
93. Momi S, Falcinelli E, Giannini S, et al. Loss of matrix metalloproteinase 2 in platelets reduces arterial thrombosis in vivo. *J Exp Med*. 2009;206:2365–2379.
94. Pyo R, Lee JK, Shipley JM, et al. Targeted gene disruption of matrix metalloproteinase-9 (gelatinase B) suppresses development of experimental abdominal aortic aneurysms. *J Clin Invest*. 2000;105:1641–1649.
95. Longo GM, Xiong W, Greiner TC, Zhao Y, Fiotti N, Baxter BT. Matrix metalloproteinases 2 and 9 work in concert to produce aortic aneurysms. *J Clin Invest*. 2002;110:625–632.
96. Liu O, Jia L, Liu X, et al. Clopidogrel, a platelet P2Y12 receptor inhibitor, reduces vascular inflammation and angiotensin II induced-abdominal aortic aneurysm progression. *PLoS One*. 2012;7:e51707.
97. Owens 3rd AP, Edwards TL, Antoniak S, et al. Platelet inhibitors reduce rupture in a mouse model of established abdominal aortic aneurysm. *Arterioscler Thromb Vasc Biol*. 2015;35:2032–2041.
98. Rahman M, Zhang S, Chew M, et al. Platelet-derived CD40L (CD154) mediates neutrophil upregulation of Mac-1 and recruitment in septic lung injury. *Ann Surg*. 2009;250:783–790.
99. Rahman M, Zhang S, Chew M, et al. Platelet shedding of CD40L is regulated by matrix metalloproteinase-9 in abdominal sepsis. *J Thromb Haemost*. 2013;11:1385–1398.
100. Kerr BA, McCabe NP, Feng W, Byzova TV. Platelets govern pre-metastatic tumor communication to bone. *Oncogene*. 2013;32:4319–4324.
101. Fanjul-Fernández M, Folgueras AR, Fueyo A, et al. Matrix metalloproteinase Mmp-1a is dispensable for normal growth and fertility in mice and promotes lung cancer progression by modulating inflammatory responses. *J Biol Chem*. 2013;288:14647–14656.

102. Tressel SL, Kaneider NC, Kasuda S, et al. A matrix metalloprotease-PAR1 system regulates vascular integrity, systemic inflammation and death in sepsis. *EMBO Mol Med.* 2011;3:370–384.
103. Newby AC. Metalloproteinases promote plaque rupture and myocardial infarction: A persuasive concept waiting for clinical translation. *Matrix Biol.* 2015;44–46:157–166.
104. Galis ZS, Khatri JJ. Matrix metalloproteinases in vascular remodeling and atherogenesis the good, the bad, and the ugly. *Circ Res.* 2002;90:251–262.
105. Kai H, Ikeda H, Yasukawa H, et al. Peripheral blood levels of matrix metalloproteases-2 and -9 are elevated in patients with acute coronary syndromes. *J Am Coll Cardiol.* 1998;32:368–372.
106. Inoue T, Kato T, Takayanagi K, et al. Circulating matrix metalloproteinase-1 and -3 in patients with an acute coronary syndrome. *Am J Cardiol.* 2003;92:1461–1464.
107. Soejima H, Ogawa H, Sakamoto T, et al. Increased serum matrix metalloproteinase-1 concentration predicts advanced left ventricular remodeling in patients with acute myocardial infarction. *Circ J.* 2003;67(4):301–304.
108. Cavusoglu E, Marmur JD, Hegde S, et al. Relation of baseline plasma MMP-1 levels to long-term all-cause mortality in patients with known or suspected coronary artery disease referred for coronary angiography. *Atherosclerosis.* 2015;239:268–275.
109. Choudhary S, Higgins CL, Chen IY, et al. Quantitation and localization of matrix metalloproteinases and their inhibitors in human carotid endarterectomy tissues. *Arterioscler Thromb Vasc Biol.* 2006;26:2351–2358.
110. Back M, Ketelhuth DF, Agewall S. Matrix metalloproteinases in atherothrombosis. *Prog Cardiovasc Dis.* 2010;52:410–428.
111. Lenti M, Falcinelli E, Pompili M, et al. Matrix metalloproteinase-2 of human carotid atherosclerotic plaques promotes platelet activation. Correlation with ischaemic events. *Thromb Haemost.* 2014;111:1089–1101.
112. Toschi V, Gallo R, Lettino M, et al. Tissue factor modulates the thrombogenicity of human atherosclerotic plaques. *Circulation.* 1997;95:594–599.
113. Penz SM, Reininger AJ, Toth O, et al. Glycoprotein Ibalpha inhibition and ADP receptor antagonists, but not aspirin, reduce platelet thrombus formation in flowing blood exposed to atherosclerotic plaques. *Thromb Haemost.* 2007;97:435–443.
114. Reininger AJ, Bernlochner I, Penz SM, et al. A 2-step mechanism of arterial thrombus formation induced by human atherosclerotic plaques. *J Am Coll Cardiol.* 2010;55:1147–1158.
115. Nikkari ST, O'Brien KD, Ferguson M, et al. Interstitial collagenase (MMP-1) expression in human carotid atherosclerosis. *Circulation.* 1995;92:1393–1398.
116. Sukhova GK, Schönbeck U, Rabkin E, et al. Evidence for increased collagenolysis by interstitial collagenases-1 and -3 in vulnerable human atheromatous plaques. *Circulation.* 1999;99:2503–2509.
117. Kurzepa J, Kurzepa J, Golab P, Czerska S, Bielewicz J. The significance of matrix metalloproteinase (MMP)-2 and MMP-9 in the ischemic stroke. *Int J Neurosci.* 2014;124:707–716.
118. Lenglet S, Montecucco F, Mach F. Role of matrix metalloproteinases in animal models of ischemic stroke. *Curr Vasc Pharmacol.* 2015;13:161–166.
119. Busti C, Falcinelli E, Momi S, Gresele P. Matrix metalloproteinases and peripheral arterial disease. *Intern Emerg Med.* 2010;5:13–25.
120. Tayebjee MH, Tan KT, MacFadyen RJ, Lip GY. Abnormal circulating levels of metalloprotease 9 and its tissue inhibitor 1 in angiographically proven peripheral arterial disease: relationship to disease severity. *J Intern Med.* 2005;257:110–116.
121. Signorelli SS, Anzaldi M, Fiore V, et al. Patients with unrecognized peripheral arterial disease (PAD) assessed by ankle-brachial index (ABI) present a defined profile of proinflammatory markers compared to healthy subjects. *Cytokine.* 2012;59:294–298.

122. Burdess A, Nimmo AF, Campbell N, et al. Perioperative platelet and monocyte activation in patients with critical limb ischemia. *J Vasc Surg*. 2010;52:697–703.
123. Gresele P, Catalano M, Giammarresi C, et al. Platelet activation markers in patients with peripheral arterial disease—a prospective comparison of different platelet function tests. *Thromb Haemost*. 1997;78:1434–1437.
124. Danese S, de la Motte C, Sturm A, et al. Platelets trigger a CD40-dependent inflammatory response in the microvasculature of inflammatory bowel disease patients. *Gastroenterology*. 2003;124:1249–1264.
125. Gotoh H, Kawaguchi Y, Harigai M, et al. Increased CD40 expression on articular chondrocytes from patients with rheumatoid arthritis: contribution to production of cytokines and matrix metalloproteinases. *J Rheumatol*. 2004;31:1506–1512.
126. Büchner K, Henn V, Grafe M, de Boer OJ, Becker AE, Kroczek RA. CD40 ligand is selectively expressed on CD4+ T cells and platelets: implications for CD40-CD40L signalling in atherosclerosis. *J Pathol*. 2003;201:288–295.
127. Henn V, Slupsky JR, Grafe M, et al. CD40 ligand on activated platelets triggers an inflammatory reaction of endothelial cells. *Nature*. 1998;391:591–594.
128. Rigothier C, Daculsi R, Lepreux S, et al. CD154 induces matrix metalloproteinase-9 secretion in human podocytes. *J Cell Biochem*. 2016;117:2737–2747.
129. Morrell CN, Aggrey AA, Chapman LM, Modjeski KL. Emerging roles for platelets as immune and inflammatory cells. *Blood*. 2014;123:2759–2767.
130. Petito E, Alunno A, Falcinelli E, et al. Platelets participate in synovial inflammation in rheumatoid arthritis by increasing MMP-2 formation. *Thromb Res*. 2014;134(suppl 2):S80.
131. Catricala S, Torti M, Ricevuti G. Alzheimer disease and platelets: how's that relevant. *Immun Ageing*. 2012;9:20.
132. Lim NK, Villemagne VL, Soon CP, et al. Investigation of matrix metalloproteinases, MMP-2 and MMP-9, in plasma reveals a decrease of MMP-2 in Alzheimer's disease. *J Alzheimers Dis*. 2011;26:779–786.
133. Hochstrasser T, Ehrlich D, Marksteiner J, Sperner-Unterweger B, Humpel C. Matrix metalloproteinase-2 and epidermal growth factor are decreased in platelets of Alzheimer patients. *Curr Alzheimer Res*. 2012;9:982–989.
134. Li R, Ren M, Chen N, et al. Presence of intratumoral platelets is associated with tumor vessel structure and metastasis. *BMC Cancer*. 2014;14:167.
135. Bastida E, Escolar G, Ordinas A, Jamieson GA. Morphometric evaluation of thrombogenesis by microvesicles from human tumor cell lines with thrombin-dependent (U87MG) and adenosine diphosphate-dependent (SKNMC) platelet-activating mechanisms. *J Lab Clin Med*. 1986;108:622–627.
136. Heinmöller E, Weinel RJ, Heidtmann HH, et al. Studies on tumor-cell-induced platelet aggregation in human lung cancer cell lines. *J Cancer Res Clin Oncol*. 1996;122: 735–744.
137. Boukerche H, Berthier-Vergnes O, Penin F, et al. Human melanoma cell lines differ in their capacity to release ADP and aggregate platelets. *Br J Haematol*. 1994;87:763–772.
138. Jurasz P, North S, Venner P, Radomski MW. Matrix metalloproteinase-2 contributes to increased platelet reactivity in patients with metastatic prostate cancer: a preliminary study. *Thromb Res*. 2003;112:59–64.
139. Foley CJ, Fanjul-Fernández M, Bohm A, et al. Matrix metalloprotease 1a deficiency suppresses tumor growth and angiogenesis. *Oncogene*. 2014;33:2264–2272.
140. Helley D, Banu E, Bouziane A, et al. Platelet microparticles: a potential predictive factor of survival in hormone-refractory prostate cancer patients treated with docetaxel-based chemotherapy. *Eur Urol*. 2009;56:479–484.
141. Momi S, Wiwanitkit V. Phylogeny of blood platelets. In: Gresele P, Kleiman NS, Lopez A, Page CP, eds. *Platelet in Thrombotic and Non Thrombotic Disorders—Pathophysiology, Pharmacology and Therapeutics: An Update*. Cham, Switzerland: Springer; 2016:11–20.

CHAPTER FIVE

Matrix Metalloproteinases and Leukocyte Activation

Kate S. Smigiel, William C. Parks[1]
Women's Guild Lung Institute, Cedars-Sinai Medical Center, Los Angeles, CA, United States
[1]Corresponding author: e-mail address: bill.parks@cshs.org

Contents

Abstract

As their name implies, matrix metalloproteinases (MMPs) are thought to degrade extracellular matrix proteins, a function that is indeed performed by some members. However, regardless of their cell source, matrix degradation is not the only function of these enzymes. Rather, individual MMPs have been shown to regulate specific immune processes, such as leukocyte influx and migration, antimicrobial activity, macrophage activation, and restoration of barrier function, typically by processing a range of nonmatrix protein substrates. Indeed, MMP expression is low under steady-state conditions but is markedly induced during inflammatory processes including infection, wound healing, and cancer. Increasing research is showing that MMPs are not just a downstream consequence of a generalized inflammatory process, but rather are critical factors in the overall regulation of the pattern, type, and duration of immune responses. This chapter outlines the role of leukocytes in tissue remodeling and describes recent progress in our understanding of how MMPs alter leukocyte activity.

Progress in Molecular Biology and Translational Science, Volume 147
ISSN 1877-1173
http://dx.doi.org/10.1016/bs.pmbts.2017.01.003

ABBREVIATIONS

AGTR1 type II angiotensin receptor
DAMPs damage-associated molecular patterns
ECM extracellular matrix
MMP matrix metalloproteinase
NETs neutrophil extracellular traps
PAMPs pathogen-associated molecular patterns

1. INTRODUCTION

Matrix metalloproteinases (MMPs) function in the extracellular space as transmembrane, membrane-anchored, or released enzymes. Consequently, their substrates are also extracellular proteins within with the secretory pathway or present on cell surfaces or within a cellular tissue compartments. A prevailing and long-held concept is that MMPs are responsible for turnover and degradation of extracellular matrix (ECM) proteins. This idea sprung from the initial findings of Gross and Lapiere in 1962 who isolated a neutral proteinase—now called collagenase-4 (MMP-18)—responsible for degrading fibrillar collagens in regressing tadpole tails.[1] Thereafter, over 23 mammalian MMPs were discovered, and most have been shown in vitro to be able degrade or cleave various collagens or other ECM proteins, leading to the addition of the "matrix" modifier.

As a consequence, MMPs were thought to be the extracellular enzymes responsible for turnover of ECM in homeostasis and disease, a concept that remains to this day despite limited supporting evidence. Although defined in vitro degradation assays (i.e., a pure proteinase incubated under ideal conditions with a suspected substrate) have indicated that most MMPs can act on various ECM proteins, such approaches—though quite valuable for validating cleavage sites of candidate physiologic substrates[2]—are limited by themselves to showing what an MMP can do, not what it does do. Indeed, with the emergence of genetically defined animal models,[3] it became clear that, as a family, MMPs do not function in bulk ECM turnover or degradation in vivo. Several reports over the past few decades have demonstrated that MMPs act on a variety of nonmatrix proteins, such as cytokines, chemokines, antimicrobial peptides, and various surface proteins, including receptors, adhesive and junctional proteins, and more.[4–11] About 10% of our genome encodes for proteins with a signal peptide, leading to an extensive array of potential extracellular MMP substrates.

Moreover, unbiased proteomic analyses indicate that, when combined, several MMPs can potentially act on over 600 substrates.[12] Thus, it is not surprising that MMPs evolved to function in a variety of physiologic and disease processes.[9,13–15]

Although some MMPs do act on ECM components, the substrate spectrum of a given enzyme is generally limited to a few matrix proteins. For example, collagenase-3 (MMP-13), MT1-MMP (MMP-14), and MT3-MMP (MMP-15) function as physiologic collagenases acting on fibrillar collagens types I and III,[16–19] and matrilysin (MMP-7) appears to be the potent elastin-degrading proteinase produced by human macrophages.[20] However, when viewed as a family, matrix degradation is neither the only nor central function of MMPs. Depending on the enzyme, cell source, and process involved, a given MMP can act on various proteins and, in turn, affect various processes. Thus, although the presence of a given MMP may be associated with some process, such a cell migration or wound repair, one cannot conclude a priori that these proteinases function by acting on ECM. In particular, and as has been discussed in other reviews, MMPs have emerged as critical effector enzymes controlling a range of immune functions.[8,14,15,21–25] In this chapter, we discuss findings demonstrating that specific MMPs have critical roles in regulating leukocyte function, with an emphasis on neutrophil and macrophage activation.

2. INFLAMMATION AND TISSUE REMODELING

The immune system is a critical player in tissue injury and repair. Upon tissue injury or infection, the rapid release of chemoattractants, histamine, and inflammatory cytokines and the presence of PAMPs (pathogen-associated molecular patterns) and DAMPs (damage-associated molecular patterns) promote the recruitment of inflammatory cells to the affected site. These signals stimulate the ability of leukocytes to migrate across the endothelium, through the vascular wall and interstitium, and for many tissues, across the mucosal/epithelial barrier and into lumenal spaces.[26] Typically, neutrophils are the most abundant leukocytes at the early stages of the response to injury and, together with macrophages, rapidly debride the wound, eliminate infectious organisms, and phagocytize dead or dying host cells. Immune cells have evolved to respond to a variety of PAMPs, such as lipopolysaccharide (LPS) and single-stranded DNA, and DAMPs, such as extracellular ATP, heat-shock proteins, provisional matrix components (e.g., hyaluronan and tenascin C), and uric acid, which are present at sites of sterile injury.[27,28] In response to these signals, innate immune cells secrete a variety of cytokines and chemokines, such as IL-1β, TNF-α, and TGF-β1, that impact the activity and function of resident cells and other leukocytes.

Regardless of the nature of the insult, a common cascade of wound-healing steps is initiated immediately upon injury. The classic model of wound healing is divided into three sequential yet overlapping phases: (1) inflammatory, (2) proliferative, and (3) remodeling.[28] Inflammatory cells are recruited to debride the wound and prevent infection, cytokines produced by these cells stimulate the release of ECM components and the proliferation of myofibroblasts, and then the balance between profibrotic and antifibrotic mediators determines whether the wound will be repaired or whether excess deposition of ECM will case fibrosis. This series of events is shared among target organs, and similar processes dictate tissue remodeling in diseases such as myocardial infarction (MI), cardiac remodeling, pulmonary fibrosis, liver cirrhosis, scleroderma, and Crohn's disease.

The strength of the inflammatory response to injury and the response and duration of leukocyte activity during tissue remodeling often determine the balance between future health (e.g., resolution) and disease (e.g., fibrosis or tissue destruction). Local interactions between immune cells, resident non-immune cells, and ECM dictate the success of tissue remodeling and resolution, and dysregulation of innate or adaptive immune responses is a major contributor to diseases. For example, fibrosis ensues when the tissue damage is severe, the inflammatory response persists, and the repair process becomes dysregulated. The immune response, particularly the balance between type 1 and type 2 responses, largely influences the balance between repair and disease. Type 1 responses, characterized as largely proinflammatory antimicrobial responses involving the cytokines IL-12 and IFN-γ, are critical for effective antimicrobial defenses but also cause immunopathology. In contrast, type 2 responses, characterized as antihelminth and allergic responses involving the cytokines IL-4, IL-5, and IL-13, promote tissue repair.[29] Given the complexity of the signaling pathways and cell types involved in this remodeling process, defects in any stage of the wound repair process can lead to scar formation at the expense of regeneration, which likely explains the complex nature of tissue fibrosis. Below, we discuss briefly the roles of leukocytes in generalized tissue repair and discuss research findings on the mechanisms of MMP-mediated control of leukocyte activity over a range of disease processes.

3. NEUTROPHILS, MACROPHAGES, AND LYMPHOCYTES

3.1 Neutrophils

As stated, neutrophils are the first cells to be recruited to sites of damage and infection, where they contribute to the removal of tissue debris and kill invading bacteria. Neutrophils are a critical component of innate immune defense against invading pathogens, and their antimicrobial role was initially ascribed to direct phagocytosis and the release of toxic components via degranulation. Neutrophil granules contain enzymes and antimicrobial peptides, such as myeloperoxidase, neutrophil elastase, cathepsins, β-defensins, lysozyme, and reactive oxygen species.[30] During activation in response to infection or injury, neutrophils undergo a cell death process termed NETosis, which releases decondensed chromatin fibers coated with histones and granular proteins—termed NETs (neutrophil extracellular traps)—into the extracellular tissue space.[31] NETosis-inducing agents include bacteria, protozoa, fungi, viruses, and host factors, such as GM-CSF and IL-8. While typically associated with infection, NETs are found in models of sterile injury such as mechanical ventilation, transfusion-related acute lung injury, and atherosclerosis.[32–35] Although the recruitment of neutrophils to the injured tissue is important for the wound-healing process, these cells secrete a variety of toxic factors that can be harmful to the nearby host tissue. Indeed, an excess of neutrophils contributes greatly to the tissue damage associated with a variety of conditions, including acute lung injury, severe asthma, and many others.[36–38] Although generally thought of as extremely short-lived cells, with a half-life in humans of 8 h, recent reports have described novel complexity of neutrophil subpopulations, including some with a reported half-life of 5.4 days.[39] Thus, future studies may identify additional roles for long-lived neutrophils in chronic inflammatory settings.

3.2 Macrophages

Macrophages are critical effector leukocytes that reside and function at the intersection of innate and adaptive immunity. Macrophages are either generated from blood monocytes that differentiate into macrophages as they enter tissues or from the local proliferation of long-lived, tissue-resident macrophages that arose from yolk sack hematopoietic tissue in utero. Essentially all tissues have a unique tissue-resident population of macrophages that is typically derived from cells of the embryonic yolk sac. Examples of resident macrophages are Küppfer cells in the liver, alveolar macrophages in the lung, and microglia in the central nervous system.[40]

Macrophages play essential yet distinct roles in both promoting and resolving inflammation and in facilitating tissue repair and contributing to its destruction.[41] That a single-cell type can serve opposing functions may seem counterintuitive, but dramatic phenotypic changes occur when macrophages respond to local stimuli.[41–46] Based on patterns of gene and protein expression and function, macrophages are commonly classified as classically activated (M1) or alternatively activated (M2) cells, as well as sub-M2 types.[41–43,46] The M1 phenotype is induced by infection and proinflammatory Th1 cytokines.[45] M1 macrophages are effective at killing bacteria and release proinflammatory cytokines, such as IL-1β, IL-12, and TNF-α. In contrast, the M2 phenotype is induced by Th2 cytokines IL-4 and IL-13 and other factors.[45,46] M2 macrophages release antiinflammatory factors, such as IL-10 and TGF-β1, are weakly microbicidal, produce arginase-1 that can counter iNOS activity,[47] and promote repair.[45]

Macrophages present early in inflammation are functionally distinct from those at later stages.[46,48–55] Depletion of macrophages in the early phases of wound repair significantly impairs scar formation,[56,57] whereas depletion of macrophages during later stages leads to an inability to resolve scars.[52,58] Hence, early phase macrophages, which are predominately M1-biased cells, contribute to ECM deposition and fibrosis likely by producing profibrotic cytokines that promote the activation of resident fibroblasts and pericytes into ECM-producing myofibroblasts.[46,48–51,59–63] During the later resolution phase, macrophages tend to be alternatively activated, remodeling-competent M2-biased macrophages.[50,61,64]

Although far from being fully understood, resolution of scarring and fibrosis appears to be—not surprisingly—the responsibility of macrophages and, in particular, M2 macrophages.[50,52,65–69] Despite the compelling data in various tissue models with macrophage depletion and direct proteolysis strategies, M2 macrophages—or specific subsets of M2 macrophages—have been considered to be profibrotic[70] and likely for two key reasons. First, M2-like macrophages (i.e., macrophages positive for a few M2 markers) are present in scars and fibrotic tissue.[71,72] However, these are mostly correlative data, whereas functional studies indicate that M2-biased macrophages work to resolve fibrosis, not promote it.[52,66,69,73] Second, M2 macrophages express known or suspected profibrotic factors, particularly TGF-β1 and arginase-1, which stimulates the synthesis of proline, an abundant amino acid in collagens. However, depletion of TGF-β1 or arginase-1 from macrophages does not affect the development nor extent of fibrosis in kidney and lung, respectively.[74,75] In contrast, a recent study concluded that macrophage-derived TGF-β1 is a critical driver of lung fibrosis.[76] However, in this study, the *Tgfb1* gene was conditionally targeted using a Lyz2 Cre driver that, in lung, would also silence the cytokine expression in many airway and alveolar cells,[77] which are an important source of TGF-β1 in lung.

3.3 T Lymphocytes

Akin to M1 and M2 macrophages, the Th1/Th17 populations of CD4+ T cells are classically ascribed antifibrotic functions, whereas Th2 CD4+ T cells are considered to promote repair. The inflammatory cytokines IFN-γ and IL-17 are the prototypic factors produced by Th1 and Th17 cells, respectively. IFN-γ directly inhibits fibroblast proliferation, TGF-β-induced gene expression, and collagen synthesis in activated myofibroblasts. IFN-γ also prevents the differentiation of blood monocytes into fibroblast-like cells called fibrocytes, which are believed to participate in the development of dysregulated wound healing and fibrosis in multiple organ systems.[78,79] Similarly, IL-17 promotes fibrosis by both exacerbating the upstream inflammatory response and regulating the downstream activation of fibroblasts.[47]

Th2-type immunity is a potent driver of tissue remodeling and fibrosis. Th2 cells evolved to combat parasitic infections, which are typically associated with massive tissue damage due to the movement of helminths through host tissues. Accordingly, through the production of IL-4 and IL-13, Th2 cells not only inhibit pathological Th1 immune responses but also promote wound repair. However, if not appropriately regulated, Th2 responses can contribute to the development of lethal fibrotic pathology, which results from overzealous or persistent wound-healing responses.[29]

Regulatory T (Treg) cells expressing the transcription factor Foxp3 are important producers of antiinflammatory cytokines, such as IL-10 and TGF-β1, which function to inhibit inflammation and other T cell responses. Treg cell numbers are expanded in fibrotic diseases, yet reports have demonstrated both profibrotic and antifibrotic functions for these cells. For instance, Treg cells have been found to ameliorate fibrosis in idiopathic pulmonary fibrosis,[80] cardiovascular disease,[81] and liver fibrosis,[82] yet because these cells produce large amounts of TGF-β1, other studies have shown Treg cells to worsen fibrosis.[83]

In the heart, regulatory roles have also been reported for CD8+ T cells following ischemic injury. CD8+ T cells express the type II angiotensin receptor (AGTR1), which promotes antiinflammatory responses. Angiotensin II primarily induces cardiomyocyte hypertrophy and increases vascular tone through AGTR1 signaling, in a process that is amplified by the blockade of AGTR2. Following ischemia–reperfusion injury, AGTR2-expressing CD8+ T cells inhibit inflammation and decrease infarct size by producing antiinflammatory IL-10 in an angiotensin II-dependent manner.[84,85] Thus, T cells and the cytokines they produce contribute not only to the inflammatory phase of wound healing but also to the inhibition of inflammation and tissue remodeling.

4. MMP REGULATION OF LEUKOCYTE ACTIVITY

MMPs are often classified based on the substrates they can cleave or degrade, such as collagens, elastin, and basement membrane components, as determined by in vitro degradation assays. However, this sort of classification can misleading for three reasons: (1) the classification is based only on in vitro data that have seldom predicted or agreed with in vivo functions; (2) the range of potential substrates used is quite limited and biased (i.e., ECM proteins); and (3) only a subset of MMPs have been used in such analyses, and, hence, many are excluded from this classification scheme. As discussed earlier, the validated in vivo substrates of MMPs are quite diverse, yet it is not clear if the substrates for most individual MMPs are confined to functionally similar groups of proteins.

That said, the substrates and, in turn, functions of MMPs often involve the activation of latent signaling molecules, such as cytokines, or processing of proteins to a different functional state, such as ectodomain shedding of syndecan-1 and E-cadherin (discussed later). Hence, one generalization of MMP activity that seems broadly valid is that many of these proteinases function as processing enzymes controlling the activity of a range of effector proteins. Furthermore, another generalization is that many MMP substrates function in some aspect of immunity and inflammation.[15] Among these, MMPs function to shape the migration, activation, proliferation, and function of leukocytes in response to injury or infection. Increasing research demonstrates that MMPs are not just a downstream consequence of a generalized inflammatory process, but rather are critical factors in the overall regulation of the pattern, type, and duration of immune responses.

4.1 Leukocyte Migration

Leukocytes constantly migrate in and out of tissues as part of normal immune surveillance, and this activity is dramatically increased in response to tissue injury or infection. Several MMPs function in migration and invasion and often do so by altering the chemotactic signals received by immune cells. MMP affects chemokine activity by directly modifying the ligand or shedding of accessory proteins that bind, retain, or concentrate chemokines.

MMP-mediated cleavage of chemotactic molecules can lead to enhancement, inactivation, or antagonism of chemokine activity and, hence, can have a range of corresponding effects on infiltrating immune cells and the inflammatory process. MMPs act on both CC- (e.g., CCL7) and CXC motif (e.g., CXCL11, CXCL12, CXCL6, and CXCL8/IL-8) ligands. CXCL11 is a substrate of several MMPs (MMP-1, -2, -3, -9, -13, and -14), and MMP-9 proteolytically alters the function of CXCL5, CXCL6, and mouse CXCL5/LIX. Regarding CC chemokines, MMP-1, -3, -13, and -14 cleave the N-terminus of CCL2 (MCP1), CCL8 (MCP2), and CCL13 (MCP4) to produce antagonist factors.[86] Similarly, MMP-2 acts on CCL7 (MCP3) to convert this chemokine into an antagonistic derivative.[87] Thus, multiple MMPs act on a range of chemokines to collectively shape the inflammatory microenvironment in response to tissue injury.

4.1.1 Neutrophil Influx

As they move from the blood stream into injured or infected tissues, neutrophils go through progressive stages of priming and activation, culminating in the release and production of their cytotoxic/bacteriocidal cargo.[88,89] Although much is known about specific mechanisms controlling neutrophil activation, it is likely that specific checkpoint mechanisms function to moderate or bar the activation of neutrophils as they move through different tissue compartments.

MMP-7 (matrilysin) is an epithelial MMP that is induced in response to a wide range of insults.[90–92] In response to lung or colon injury, the transepithelial migration of neutrophils is halted in *Mmp7*$^{-/-}$ mice,[10,93] and this phenotype is associated with markedly lower levels of CXCL1 in the lumenal space.[10,93] CXCL1, an acute-phase neutrophil chemokine, is the murine functional orthologue of human IL-8/CXCL8. When secreted by injured mucosal epithelium, CXCL1 becomes bound to the glycosaminoglycan chains of syndecan-1 (Scd1), a type I transmembrane proteoglycan on the basolateral surface of epithelial cells.[94,95] In response to epithelial

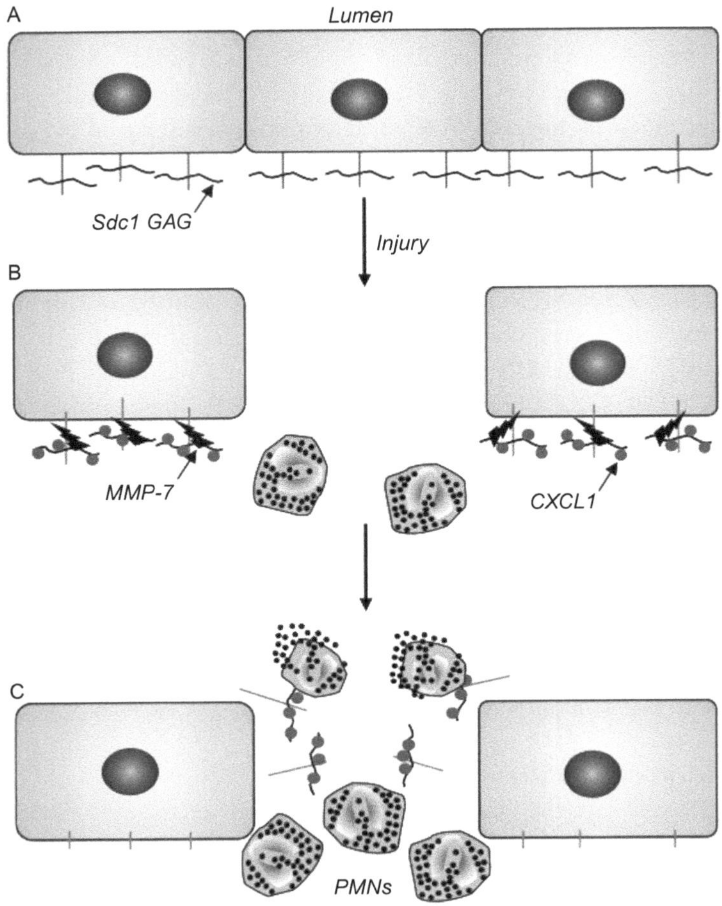

Fig. 1 MMP-7 shedding of epithelial syndecan-1 functions as a checkpoint of neutrophil activation. (A) Syndecan-1 (Sdc1) is present on the basal–lateral surface of intact lung epithelium. (B) In response to injury, MMP-7 (*lightning bolt*) and CXCL1 are induced, and neutrophils begin to infiltrate to the wound site. Secreted CXCL1 accumulates on the GAG chains (*horizontal wavy line*) of syndecan-1, and the complexes are shed by MMP-7. (C) Neutrophils interact with shed syndecan-1/CXCL1 complexes, which promote activation and release of cellular contents. *Reprinted with permission of the American Thoracic Society. Copyright © 2016 American Thoracic Society. Gill SE, Nadler ST, Li Q, et al. Shedding of syndecan-1/CXCL1 complexes by MMP7 functions as an epithelial checkpoint of neutrophil activation.* Am J Respir Cell Mol Biol. *2016;55:243–251.*

damage, MMP-7 sheds syndecan-1 liberating an intact ectodomain with the chemokine cargo,[10] and release of this complex allows both neutrophil migration through the mucosal barrier and their subsequent activation (Fig. 1).[10,93,96] In the absence of shedding of syndecan-1/CXCL1 complexes, *Mmp7*$^{-/-}$ and *Sdc1*$^{-/-}$ mice are protected from the lethal effects of excess neutrophil-mediated lethality in response to acute lung injury.[10]

Although it appears that MMP-7 shedding of syndecan-1/CXCL1 complexes generates a chemokine gradient that neutrophils follow into the lumenal space, the granulocytes would, of course, be moving against the gradient. Thus, MMP-7 shedding of syndecan-1/CXCL1 complexes must control another neutrophil process, which appears to be limiting neutrophil activation at the mucosal interface. In recent studies,[96] our group reported that cell-bound syndecan-1/CXCL1 complexes restrict neutrophil activation, thereby preventing a damaging oxidative burst at the epithelial cell surface, whereas interaction with soluble complexes promotes activation, which would occur ideally at a safer distance from the mucosal layer (Fig. 1). If still intact on the cell surface, syndecan-1/chemokine complexes would support neutrophil binding but would not promote neutrophil activation. This mechanism may have evolved to prevent cell death and host tissue damage due to neutrophil activation occurring too close to a mucosal surface. Thus, the MMP-7 shedding of syndecan-1 functions as a checkpoint to spatially restrict neutrophil activation in proximity to a compromised epithelium.

The mechanism by which the lack of shedding of syndecan-1/CXCL1 complexes (i.e., cell-bound complexes) bars neutrophil activation remains to be determined. As mentioned, several MMPs, including MMP-7, can influence inflammation by direct cleavage of chemokines.[5,86,97–99] Thus, in addition to shedding the syndecan-1, MMP-7 may directly cleave CXCL1 leading to enhanced activity on the released complex. Another possibility is that MMP-7 could control expression of a factor that stimulates neutrophils. In support of this idea, global gene expression analysis between wild-type and *Mmp7*$^{-/-}$ organotypic airway epithelial cultures in response to injury[100] and *Pseudomonas aeruginosa* infection[101] revealed several differences in patterns of gene expression between genotypes. A third possible explanation is that neutrophils ligate to an epithelial surface protein that actively blocks activation. In this mechanism, shedding of syndecan-1/CXCL1 complexes would enable neutrophils to disengage from the repressive interaction providing a green light for activation. While this third mechanism appears attractive, the specific means by which epithelial cells can constrain neutrophil activation remain to be determined.

Neutrophils express MMP-8 (collagenase-2) that cleaves CXCL8 family chemokines to increase their chemotactic effect. In addition to neutrophil chemotaxis, MMP-8 mediates a PMN-controlled feed-forward mechanism to orchestrate the initial inflammatory response and promote responsiveness to LPS stimulation.[102] In contrast, other studies have reported an inhibitory effect of MMP-8 on inflammation. In *Mmp8*-null mice, increased neutrophil recruitment into the alveolar space was observed after intratracheal LPS administration.[103] In a model of murine arthritis, *Mmp8* deficiency exacerbated inflammation, bone erosion, and the accumulation of neutrophils,[104] and in a model of acute lung injury, *Mmp8*-deficient mice showed increased numbers of lung neutrophils and macrophages due to the persistence of the chemokine MIP-1 in the absence of MMP-8 cleavage.[105] Thus, different mechanisms of MMP-8-mediated processing may be disease- or cell type-specific, which holds true for the functions of multiple MMPs in inflammation. In a model of osteoarthritis, MMP-12 produced by macrophages dampens neutrophil influx and NET deposition in the joints, with additional effects on increased phagocytosis and the resolution of inflammation.[106] However, the mechanism for how MMP-12 impacts neutrophil influx remains unresolved.

4.1.2 Macrophage Influx

Regarding macrophage influx, MMP-9, -10, -12, and -28 all appear to regulate the chemotactic signals produced by or received by these cells, and several of these MMPs influence macrophage influx via cell-autonomous means. High levels of MMP-9, or gelatinase-B, are produced by activated macrophages and accumulate on the surface of these cells,[107,108] where this MMP cleaves integrin beta-2 (CD18) to regulate its expression on tissue-infiltrating macrophages.[109] In addition, MMP-12 (macrophage metalloelastase) and MMP-28 (epilysin), both macrophage products, either promote or restrict, respectively, macrophage influx into the lung. Upon cigarette smoke exposure, *Mmp12*-deficient mice did not show increased numbers of macrophages in the lungs and did not develop emphysema,[110] while *Mmp28*-deficient mice demonstrated accelerated macrophage recruitment into *P. aeruginosa*-infected lungs and enhanced bacterial clearance.[111] Thus, macrophages and neutrophils both produce MMPs and respond to the signals processed by MMPs generated from other cellular sources, such as the injured epithelia, and together these MMPs shape the chemotactic milieu.

Recent work from our group indicated that MMP-10 (stromelysin-2) is a critical effector of macrophage biology (Fig. 2), including cellular influx, and hence appears to broadly affect immunity (see later). In a meta-analysis of gene array experiments involving numerous different host–pathogen interactions, MMP-10 was identified as a common host response gene.[112]

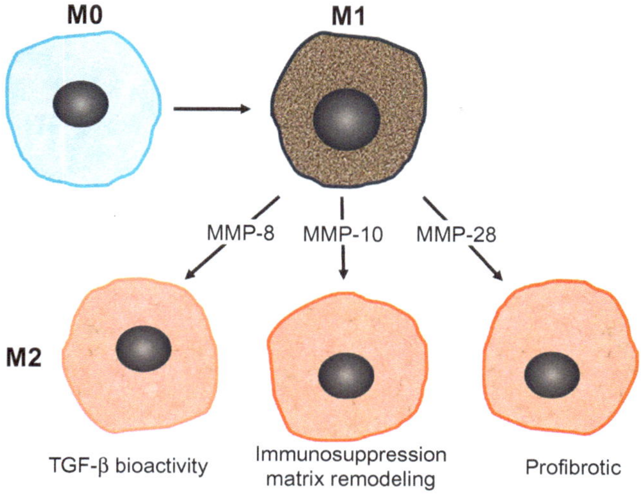

Fig. 2 MMPs control macrophage activation. Monocytes enter tissues and differentiate into macrophages (M0), and in a Th1-rich environment, as in a site of infection or injury, their activation status is biased toward a proinflammatory M1 state. MMP-8, -10, and -28, which all expressed by macrophages, influence macrophage activation by promoting the conversion of M1 cells to M2 cells yet with distinct functional consequences. MMP-8 appears to affect TGF-β activity; MMP-10 drives both the immunosuppressive and ECM remodeling activities of M2 macrophages; and MMP-28 promotes profibrotic activity, a process that is seemingly opposed to the role of MMP-10. Although the mechanism of how these three MMPs function remains unknown, as each are produced by M1 and M2 cells, it is likely that they function via proteolysis of distinct surface proteins critical for macrophage differentiation.

The widespread expression of MMP-10 among tissues suggests that this proteinase serves critical roles in the host response to environmental insults.

In the absence of MMP-10, more macrophages emigrated into the lungs of *Mmp10*$^{-/-}$ mice in response to acute *P. aeruginosa* infection,[113] and a similar phenotype was observed in models of acute liver[114] and colon injury.[115] However, we found no difference in the ability of wild-type and *Mmp10*$^{-/-}$ macrophages to migrate toward serum or wound tissue homogenate,[69] indicating that MMP-10 does not affect the migratory machinery of macrophages. However, compared to wild-type cells in vitro, *Mmp10*$^{-/-}$ macrophages migrate slower over fibronectin and have an impaired ability to invade into Matrigel.[116] Although these data seemingly contradict the increased influx of macrophages seen in vivo in *Mmp10*$^{-/-}$ mice, it is not clear if macrophages migrate over fibronectin on their way into and through the interstitial space. Although fibronectin would be present in this area, it appears that other matrix components, particularly versican and hyaluronan,[117] are the interstitial substrata that many leukocytes migrate on through tissue.[118,119] In addition, Matrigel is a highly dense material that does not mirror the porous architecture of loose connective tissue in the interstitial compartments in lung, liver, and colon.[120,121] Thus, unlike how they move through the interstitium in vivo, macrophages may require proteases to burrow through a thick plug of Matrigel in a culture dish. As we discuss later, MMP-10 activates ECM-degradative pathways in macrophages,[69,122] and hence, *Mmp10*$^{-/-}$ cells may not possess the proteinases needed to invade through or migrate on in vitro substrates. Thus, it appears that MMP-10 does not affect macrophage migration per se in physiologic settings. Rather, MMP-10 may influence the production of macrophage chemokines. In support of this idea, we reported that CCL2 (MCP1) is markedly overexpressed by *Mmp10*$^{-/-}$ alveolar macrophages.[113] CCL2 is a potent macrophage chemokine that is expressed predominantly by resident lung macrophages,[123,124] and overexpression of this factor may account for the excess macrophage influx seen in *Mmp10*$^{-/-}$ mice.

In contrast to the models discussed earlier, we observed no difference in macrophage influx into excisional skin wounds between wild-type and *Mmp10*$^{-/-}$ mice.[69] Although an explanation for these disparate findings is not apparent, it is possible that the mechanisms by which MMP-10 affects macrophage influx are context dependent. In contrast to the ample proinflammatory responses in the liver/colon/lung models, the inflammatory response is less profound in clean skin wounds. Thus, MMP-10-dependent controls on macrophage influx could be modest and, hence, not easily detected in *Mmp10*$^{-/-}$ skin wounds.

4.1.3 T Cells

Under homeostatic conditions, T cells continuously recirculate between secondary lymphoid organs and the blood via the lymphatic system, and targeted migration of different memory T cell populations occurs via the expression of tissue-specific integrins and chemokine receptors. In vitro findings suggest that T cells expressing higher levels of MMP-2 and -9 have increased invasive capacity, with Th1 cells showing the highest migratory capacity.[125,126] In vivo, MMP-2 and -9 were shown to aid in T cell migration into the lung in an allergen-induced airway inflammation model,[127] and due to their degradation of basement membranes, these proteases have also been implicated in T cell migration into the central nervous system during experimental autoimmune encephalitis.[128,129] Thus, because specific populations of memory T cells express unique combinations of chemokine receptors, it is tempting to speculate that the actions of MMPs help establish the tissue specificity of memory cells.

4.1.4 Other Roles for MMPs in Leukocyte Migration

MMPs also play a role in the resolution of inflammation through inflammatory cell trafficking. For example, MMP-2 enables the movement of a variety of inflammatory cells, including eosinophils and macrophages, from the interstitium into the airway lumen in a model of allergic asthma. As part of an IL-13-dependent regulatory loop, MMP-2 establishes the chemotactic gradient required for egression, and $Mmp2^{-/-}$ animals showed increased susceptibility to asphyxiation induced by allergens and accumulation of inflammatory cells in the lung parenchyma.[130] MMP-9 was also shown to contribute to this protective effect through decreased CC chemokines.[131]

In addition to shedding of syndecan-1, MMP-7 also cleaves the ectodomain of E-cadherin from injured lung epithelium.[11] E-cadherin shedding does not begin until several days after injury, by which time shedding of syndecan-1 has ceased, indicating that mechanisms exist to confine MMP-7 activity to specific substrate targets.[132,133] The leukocyte-specific $\alpha_E\beta_7$ integrin (CD103) is expressed on intraepithelial lymphocytes and on specific populations of dendritic cells (DCs), and E-cadherin is the only known CD103 ligand.[134] In the bleomycin toxicity model of lung injury and fibrosis, the influx of CD103+ DCs is reduced in $Mmp7^{-/-}$ mice compared to wild-type animals, and greater fibrosis and persistent neutrophilia are seen in $Cd103^{-/-}$ mice.[135] These findings suggest that MMP-7 shedding of E-cadherin provides a chemotactic signal that promotes the influx CD103+ DCs, which, in turn, serve a beneficial role in immunosuppression and resolution of excess ECM deposition.

4.2 Cytokine Activity

While chemokines are the soluble factors dictating leukocyte migration, cytokines are a form of cell–cell communication that shapes the type of ensuing immune response. Cytokines also serve as a bridge between the activities of different cell types. For instance, in the prototypical type 1 immune response, injured endo/epithelial cells generate proinflammatory cytokines such as IL-1 and TNF-α, which together with PAMPs/DAMPs activate macrophages to produce IL-12, which promotes the differentiation of naïve T cells into Th1 cells that produce IFN-γ and establishes a feed-forward loop to stimulate the proliferation, phagocytic function, and antigen presentation of other macrophages. As a mechanism to limit persistent inflammation, cytokines such as TGF-β1 and IL-10 signal to promote repair. The activity of both proinflammatory and proreparative cytokines has been shown to be impacted by MMPs.

Regarding proinflammatory cytokines, IL-1β is a key mediator of the inflammatory response and has been implicated in the pathology of many conditions, including sepsis, rheumatoid arthritis, atherosclerosis, inflammatory bowel disease, and others. This cytokine is produced as an inactive 31-kDa precursor, termed pro-IL-1β, and must be cleaved intracellularly by caspase-1, also known as IL-1β-converting enzyme, to reach its active form.[136] At least three members of the MMP family, MMP-2, -3, and -9, have been suggested to process pro-IL-1β into its biologically active form; in contrast, MMP-1, -2, -3, and -9 can degrade IL-1β into biologically inactive fragments.[137,138] However, it is clear from numerous studies that inflammasome-associated caspase-1 is the predominant activator of pro-IL-1β.

Similar to IL-1β, TNF-α is another proinflammatory cytokine that requires enzymatic action to reach its active state. TNF-α is expressed as a homotrimer of 26-kDa membrane-bound proproteins (pro-TNF-α). Via proteolysis by ADAM17, also termed TNF-converting enzyme (TACE), the ectodomain is shed to release the soluble active cytokine.[139,140] A number of MMPs have demonstrated TACE activity in vitro, including MMP-1, -2, -3, -7, -9, -12, -14, -15, and -17.[141] In particular, MMP-7 and -12 have been proposed to release active TNF-α from macrophages.[142] However, it is clear that the bulk of active TNF-α is generated by ADAM17.

4.3 Leukocyte Activation and Function

The local cytokine milieu and other factors in the tissue microenvironment drive leukocyte activation down a number of lineages. Most often, these transcriptional programs are plastic and can fluctuate in response to changes in the environment. T cells rely on IL-2 signals for their survival, proliferation, and function. High-affinity IL-2 signaling requires expression of the α chain, or CD25, of the IL-2 receptor, and MMPs, specifically MMP-9, modulate T cell function via cleavage of CD25.[143,144] Coculture experiments with T cells and cancer cells showed that MMP-9 mediates cleavage of CD25 and downregulates the proliferative capacity of cancer-experienced T cells, which suggests a role for MMPs in tumor-mediated immunosuppression.[145] In addition to proliferation,

IL-2 signaling stimulates increased production of IL-2, to promote T cell activity in a cell-intrinsic fashion. Another study showed that mesenchymal stem cell production of MMPs cleaved CD25 from activated T cells and thereby suppressed their production of IL-2.[146]

As discussed, macrophage polarization into the M1 and M2 states is associated with changes in the functions of these cells and hence impacts the outcome of injury, infection, and disease. Macrophage activation is influenced by many factors, both extrinsic and cell autonomous. Several macrophage-expressed MMPs have been reported to influence macrophage polarization. In an in vitro study comparing polarized human macrophages, MMP-1, -3, and -10 are highly expressed in M1 cells, while MMP-12 is strongly expressed in M2 macrophages.[147] Studies with mouse macrophages in vitro found that M1 activation with bacterial LPS or with live *P. aeruginosa* increased the mRNA levels of MMP-8, -13, -14, and -25 and decreased the levels of MMP-19 and TIMP-2, while alternative (M2) activation with IL-4 stimulated expression of MMP-8, -12, -13, and -19.[69,148] In a mouse model of *Helicobacter pylori* infection, MMP-7, which is not expressed by mouse macrophages but by human cells,[20] restrained gastric inflammation and premalignant lesion formation by suppressing M1 polarization, and macrophages isolated from infected *Mmp7*$^{-/-}$ mice expressed significantly higher levels of the macrophage M1 marker IL-1β.[149]

Recent studies have indicated that MMP-8, -10, and -28, which are produced by macrophages, affect macrophage activation; all of these enzymes appear to promote the M1-to-M2 conversion of macrophages but with different functional outcomes (Fig. 2). For example, MMP-8 drives M2 macrophage polarization through activation of TGF-β signaling. *Mmp8*-deficient macrophages have reduced expression of TGF-β1 and lower levels of TGF-β-related signaling molecules, such as pSMAD3.[150]

Postmyocardial infarction, MMP-28, promotes M2 macrophage activation, leading to reduced cardiac dysfunction. In the absence of MMP-28, mice subjected to left ventricular MI demonstrated decreased collagen deposition, fewer myofibroblasts, and less M2 macrophages, leading to aggravated cardiac dysfunction and a defective repair response.[151] In models of lung infection and fibrosis, MMP-28 moderates the M2 polarization of macrophages.[152] In these studies, loss of MMP-28 accentuated proinflammatory macrophage function and reduced M2 polarization, leading to protection from fibrosis. Thus, MMP-28 promotes M2 polarization, although this function can be either beneficial or injurious depending on additional injury-specific factors.

In the heart following MI, the left ventricle undergoes a series of cardiac wound-healing responses that involve both inflammation to clear necrotic myocytes and tissue debris and tissue remodeling with ECM synthesis to generate an infarct scar. MMP-9 is rapidly upregulated in response to MI, with the predominant cellular source being neutrophils and macrophages, and coordinates many aspect of cardiac remodeling. In addition to degrading collagen, fibronectin, and other ECM components, MMP-9 can degrade intracellular proteins such as actin, tubulin, and HMGB1; because these intracellular DAMPs released from necrotic cells perpetuate the inflammatory response, studies have suggested that MMP-9 may serve a protective function to limit the injury caused by dying cells.[153,154] In line with this hypothesis, macrophage-specific transgenic overexpression of MMP-9 was shown to improve post-MI cardiac function by blunting the inflammatory response.[155] However, numerous studies have shown that targeted deletion of MMP-9 also improves cardiac remodeling.[156–158] As a novel in vivo substrate of MMP-9 and multifunctional plasma membrane protein, CD36 was found to link MMP-9 to macrophage function post-MI.[158] Macrophage CD36 recognition and internalization of apoptotic cells inhibit the release of proinflammatory cytokines and initiates the proreparative response mediated by IL-10 and TGF-β. However, MMP-9-mediated degradation of CD36 was shown to decrease macrophage phagocytosis at day 7 post-MI, which implicates MMP-9 in the persistence of inflammation and dysregulated cardiac repair. In neutrophils, MMP-9 signals back to prevent apoptosis through reduced caspase-9 expression, which also perpetuates inflammation, although this pathway is not dependent on CD36 degradation.[158] Because MMP-9 is produced by a range of cell types, including immune cells, fibroblasts, and myocytes, and because its expression shows different temporal patterns following MI injury, the function of MMP-9 and its role in the disease process are likely dictated by the cell source and the presence of substrates available for proteolytic processing.

Recent studies from our group have demonstrated that MMP-10 is a critical regulator of macrophage activation (Fig. 2). In both humans and mice, MMP-10 is induced in numerous tissues in response to injury,[69,114,115,159,160] infection,[101,113] or transformation, especially in lung cancer.[161–167] Compared to wild-type mice, *Mmp10*$^{-/-}$ mice showed increased susceptibility to airway infection with *P. aeruginosa* with no impact on bacterial clearance.[113] In contrast, macrophage numbers were increased in infected *Mmp10*$^{-/-}$ mice, and the expression of several M1 markers was elevated, whereas M2 markers were reduced. These findings were mirrored in cultured bone marrow-derived macrophages (BMDM). Although MMP-10 had little effect on M2-biased macrophages, in M1-activated BMDM, the presence of MMP-10 moderated expression of several proinflammatory markers and stimulated production of M2 factors, including a profound effect on levels of IL-10,[113] an important immunosuppressive cytokine. Furthermore, upon transcriptomic analysis of the genes expressed in wild-type vs *Mmp10*$^{-/-}$ BMDM following M1-biased activation with live *P. aeruginosa*, nearly 4000 of the genes that were differentially expressed between genotypes at 6 h postexposure remained significantly elevated only in *Mmp10*$^{-/-}$ cells at 24 h. These data indicate that MMP-10 functions in a cell-autonomous manner to moderate the proinflammatory activity of M1-biased macrophages.

Furthermore, MMP-10 was also found to promote the expression of collagenolytic MMPs, such as MMP-13, in M2 macrophages and thereby promote the clearance of scar tissue in skin wounds.[69] Our data indicate that MMP-10 does not affect the synthesis and deposition of fibrillar collagens (types I and III) but rather promotes the resolution of fibrosis by controlling collagenolytic MMPs, particularly, MMP-13 (MMP-10 is not a collagenase). Thus, in the absence of MMP-10, fibrosis and scarring are greater and persistent. These findings indicate that MMP-10 controls the ECM remodeling activity of M2 macrophages, and published findings[122] with emphysema support this concept. As part of a multicenter genome-wide association study, *MMP10* was identified via network analysis as a highly connected gene in chronic obstructive pulmonary disease in humans.[122] Using a model of chronic (6 months) exposure to cigarette smoke, we validated its role and found that *Mmp10*$^{-/-}$ mice are fully resistant to the development of emphysema. In support of these findings, MMP-10 is produced by macrophages from human smokers with emphysema[168] and is one of the two genes whose expression correlates with reduced lung function in smokers.[169]

These findings indicate that macrophage MMP-10 contributes to disease progression in emphysema, which is seemingly opposed to the protective/immunosuppressive role of this MMP in acute models, such as in *P. aeruginosa* infection. However, there are important differences between these models, especially with respect to macrophage biology. As discussed earlier, macrophages that function early in inflammation are functionally distinct from those that function late in inflammation or in a persistent inflammatory response, like long-term smoke exposure. Whereas acute infection and injury bias macrophages toward an M1 phenotype,[46] cigarette smoke promotes expansion of M2 macrophages.[170] Macrophages are considered to be the destructive cell in emphysema,[171,172] and findings in wound repair indicate that MMP-10 promotes the matrix-degrading activity of M2 macrophages.[69] Thus, in acute or fibrotic settings, MMP-10 is beneficial by moderating the proinflammatory activity of M1-biased macrophages and by stimulating the ability of M2-biased macrophages to remodel scar tissue. But in a chronic setting, MMP-10-driven matrix remodeling could be excessive and detrimental, as suggested in our smoke exposure studies. Still, the common conclusion among these models is that MMP-10 functions to control macrophage behavior. However, the mechanism by which MMP-10 controls macrophage activation—i.e., the substrate whose processing or degradation by MMP-10 impacts macrophage behavior—remains unidentified.

5. CONCLUDING REMARKS

The inflammatory response comprises the initial detection and migration of immune cells to sites of infection or damage, the proliferation and functional specialization of recruited cells, and the prevention of pathology and resolution of tissue repair. Numerous effector proteins coordinate the activities of both resident and recruited cells, and as proteases, MMPs act beyond the ECM to modify cytokines, chemokines, antimicrobial peptides, surface proteins, receptors, junctional proteins, and more. This activity of MMPs has wide-ranging effects on a variety of leukocytes; because factors in the local environment dictate the functional significance of specific MMPs, many of these functions have been shown to be disease- or cell type-specific. Thus, additional research into precise MMP–substrate interactions will provide much needed information for how to enhance or inhibit specific immune processes by targeting MMPs.

REFERENCES

1. Gross J, Lapiere CM. Collagenolytic activity in amphibian tissues: a tissue culture assay. *Proc Natl Acad Sci USA*. 1962;48:1014–1022.
2. Overall CM, Blobel CP. In search of partners: linking extracellular proteases to substrates. *Nat Rev Mol Cell Biol*. 2007;8:245–257.
3. Gill SE, Kassim SY, Birkland TP, Parks WC. Mouse models of MMP and TIMP function. *Methods Mol Biol*. 2010;622:31–52.
4. Van Lint P, Libert C. Chemokine and cytokine processing by matrix metalloproteinases and its effect on leukocyte migration and inflammation. *J Leukoc Biol*. 2007;82:1375–1381.

5. McQuibban GA, Gong JH, Tam EM, McCulloch CA, Clark-Lewis I, Overall CM. Inflammation dampened by gelatinase a cleavage of monocyte chemoattractant protein-3. *Science*. 2000;289:1202–1206.
6. Wilson CL, Ouellette AJ, Satchell DP, et al. Regulation of intestinal α-defensin activation by the metalloproteinase matrilysin in innate host defense. *Science*. 1999;286:113–117.
7. Levi E, Fridman R, Miao HQ, Ma YS, Yayon A, Vlodavsky I. Matrix metalloproteinase 2 releases active soluble ectodomain of fibroblast growth factor receptor 1. *Proc Natl Acad Sci USA*. 1996;93:7069–7074.
8. Page-McCaw A, Ewald AJ, Werb Z. Matrix metalloproteinases and the regulation of tissue remodelling. *Nat Rev Mol Cell Biol*. 2007;8:221–233.
9. Stamenkovic I. Extracellular matrix remodelling: the role of matrix metalloproteinases. *J Pathol*. 2003;200:448–464.
10. Li Q, Park PW, Wilson CL, Parks WC. Matrilysin shedding of syndecan-1 regulates chemokine mobilization and transepithelial efflux of neutrophils in acute lung injury. *Cell*. 2002;111:635–646.
11. McGuire JK, Li Q, Parks WC. Matrilysin (matrix metalloproteinase-7) mediates E-cadherin ectodomain shedding in injured lung epithelium. *Am J Pathol*. 2003;162:1831–1843.
12. Morrison CJ, Butler GS, Rodriguez D, Overall CM. Matrix metalloproteinase proteomics: substrates, targets, and therapy. *Curr Opin Cell Biol*. 2009;21:645–653.
13. Cauwe B, Van den Steen PE, Opdenakker G. The biochemical, biological, and pathological kaleidoscope of cell surface substrates processed by matrix metalloproteinases. *Crit Rev Biochem Mol Biol*. 2007;42:113–185.
14. McCawley LJ, Matrisian LM. Matrix metalloproteinases: they're not just for matrix anymore! *Curr Opin Cell Biol*. 2001;13:534–540.
15. Parks WC, Wilson CL, Lopez-Boado YS. Matrix metalloproteinases as modulators of inflammation and innate immunity. *Nat Rev Immunol*. 2004;4:617–629.
16. Holmbeck K, Bianco P, Caterina J, et al. MT1-MMP-deficient mice develop dwarfism, osteopenia, arthritis, and connective tissue disease due to inadequate collagen turnover. *Cell*. 1999;99:81–92.
17. Hotary K, Allen E, Punturieri A, Yana I, Weiss SJ. Regulation of cell invasion and morphogenesis in a three-dimensional type I collagen matrix by membrane-type matrix metalloproteinases 1, 2, and 3. *J Cell Biol*. 2000;149:1309–1323.
18. Hotary KB, Allen ED, Brooks PC, Datta NS, Long MW, Weiss SJ. Membrane type I matrix metalloproteinase usurps tumor growth control imposed by the three-dimensional extracellular matrix. *Cell*. 2003;114:33–45.
19. Shi J, Son MY, Yamada S, et al. Membrane-type MMPs enable extracellular matrix permissiveness and mesenchymal cell proliferation during embryogenesis. *Dev Biol*. 2008;313:196–209.
20. Filippov S, Caras I, Murray R, et al. Matrilysin-dependent elastolysis by human macrophages. *J Exp Med*. 2003;198:925–935.
21. Chen P, Parks WC. Role of matrix metalloproteinases in epithelial migration. *J Cell Biochem*. 2009;108:1233–1243.
22. Egeblad M, Werb Z. New functions for the matrix metalloproteinases in cancer progression. *Nat Rev Cancer*. 2002;2:161–174.
23. Gill SE, Parks WC. Metalloproteinases and their inhibitors: regulators of wound healing. *Int J Biochem Cell Biol*. 2008;40:1334–1347.
24. Greenlee KJ, Werb Z, Kheradmand F. Matrix metalloproteinases in lung: multiple, multifarious, and multifaceted. *Physiol Rev*. 2007;87:69–98.
25. Dufour A, Overall CM. Missing the target: matrix metalloproteinase antitargets in inflammation and cancer. *Trends Pharmacol Sci*. 2013;34:233–242.

26. Nourshargh S, Alon R. Leukocyte migration into inflamed tissues. *Immunity*. 2014;41:694–707.
27. Medzhitov R. Origin and physiological roles of inflammation. *Nature*. 2008;454: 428–435.
28. Atala A, Irvine DJ, Moses M, Shaunak S. Wound healing versus regeneration: role of the tissue environment in regenerative medicine. *MRS Bull*. 2010;35:597–606.
29. Gause WC, Wynn TA, Allen JE. Type 2 immunity and wound healing: evolutionary refinement of adaptive immunity by helminths. *Nat Rev Immunol*. 2013;13:607–614.
30. Borregaard N, Sorensen OE, Theilgaard-Monch K. Neutrophil granules: a library of innate immunity proteins. *Trends Immunol*. 2007;28:340–345.
31. Brinkmann V, Reichard U, Goosmann C, et al. Neutrophil extracellular traps kill bacteria. *Science*. 2004;303:1532–1535.
32. Warnatsch A, Ioannou M, Wang Q, Papayannopoulos V. Inflammation. Neutrophil extracellular traps license macrophages for cytokine production in atherosclerosis. *Science*. 2015;349:316–320.
33. Yildiz C, Palaniyar N, Otulakowski G, et al. Mechanical ventilation induces neutrophil extracellular trap formation. *Anesthesiology*. 2015;122:864–875.
34. Caudrillier A, Kessenbrock K, Gilliss BM, et al. Platelets induce neutrophil extracellular traps in transfusion-related acute lung injury. *J Clin Invest*. 2012;122:2661–2671.
35. Rossaint J, Herter JM, Van Aken H, et al. Synchronized integrin engagement and chemokine activation is crucial in neutrophil extracellular trap-mediated sterile inflammation. *Blood*. 2014;123:2573–2584.
36. Slebos DJ, Postma DS, Koeter GH, Van Der Bij W, Boezen M, Kauffman HF. Bronchoalveolar lavage fluid characteristics in acute and chronic lung transplant rejection. *J Heart Lung Transplant*. 2004;23:532–540.
37. Ordonez CL, Shaughnessy TE, Matthay MA, Fahy JV. Increased neutrophil numbers and IL-8 levels in airway secretions in acute severe asthma: clinical and biologic significance. *Am J Respir Crit Care Med*. 2000;161:1185–1190.
38. Silliman CC, McLaughlin NJ. Transfusion-related acute lung injury. *Blood Rev*. 2006;20:139–159.
39. Kolaczkowska E, Kubes P. Neutrophil recruitment and function in health and inflammation. *Nat Rev Immunol*. 2013;13:159–175.
40. Haldar M, Murphy KM. Origin, development, and homeostasis of tissue-resident macrophages. *Immunol Rev*. 2014;262:25–35.
41. Gordon S, Taylor PR. Monocyte and macrophage heterogeneity. *Nat Rev Immunol*. 2005;5:953–964.
42. Lumeng CN, Bodzin JL, Saltiel AR. Obesity induces a phenotypic switch in adipose tissue macrophage polarization. *J Clin Invest*. 2007;117:175–184.
43. Laskin DL, Weinberger B, Laskin JD. Functional heterogeneity in liver and lung macrophages. *J Leukoc Biol*. 2001;70:163–170.
44. Mantovani A, Sica A, Locati M. Macrophage polarization comes of age. *Immunity*. 2005;23:344–346.
45. Gordon S. Alternative activation of macrophages. *Nat Rev Immunol*. 2003;3:23–35.
46. Benoit M, Desnues B, Mege JL. Macrophage polarization in bacterial infections. *J Immunol*. 2008;181:3733–3739.
47. Wynn TA, Ramalingam TR. Mechanisms of fibrosis: therapeutic translation for fibrotic disease. *Nat Med*. 2012;18:1028–1040.
48. Porcheray F, Viaud S, Rimaniol AC, et al. Macrophage activation switching: an asset for the resolution of inflammation. *Clin Exp Immunol*. 2005;142:481–489.
49. Stout RD, Jiang C, Matta B, Tietzel I, Watkins SK, Suttles J. Macrophages sequentially change their functional phenotype in response to changes in microenvironmental influences. *J Immunol*. 2005;175:342–349.

50. Lichtnekert J, Kawakami T, Parks WC, Duffield JS. Changes in macrophage phenotype as the immune response evolves. *Curr Opin Pharmacol.* 2013;13:555–564.
51. Aggarwal NR, King LS, D'Alessio FR. Diverse macrophage populations mediate acute lung inflammation and resolution. *Am J Physiol Lung Cell Mol Physiol.* 2014;306: L709–L725.
52. Duffield JS, Forbes SJ, Constandinou CM, et al. Selective depletion of macrophages reveals distinct, opposing roles during liver injury and repair. *J Clin Invest.* 2005;115:56–65.
53. Tabas I, Bornfeldt KE. Macrophage phenotype and function in different stages of atherosclerosis. *Circ Res.* 2016;118:653–667.
54. Nahrendorf M, Swirski FK, Aikawa E, et al. The healing myocardium sequentially mobilizes two monocyte subsets with divergent and complementary functions. *J Exp Med.* 2007;204:3037–3047.
55. Ricardo SD, van Goor H, Eddy AA. Macrophage diversity in renal injury and repair. *J Clin Invest.* 2008;118:3522–3530.
56. Lucas T, Waisman A, Ranjan R, et al. Differential roles of macrophages in diverse phases of skin repair. *J Immunol.* 2010;184:3964–3977.
57. Goren I, Allmann N, Yogev N, et al. A transgenic mouse model of inducible macrophage depletion: effects of diphtheria toxin-driven lysozyme M-specific cell lineage ablation on wound inflammatory, angiogenic, and contractive processes. *Am J Pathol.* 2009;175:132–147.
58. Ramachandran P, Pellicoro A, Vernon MA, et al. Differential Ly-6C expression identifies the recruited macrophage phenotype, which orchestrates the regression of murine liver fibrosis. *Proc Natl Acad Sci USA.* 2012;109:E3186–E3195.
59. Song E, Ouyang N, Horbelt M, Antus B, Wang M, Exton MS. Influence of alternatively and classically activated macrophages on fibrogenic activities of human fibroblasts. *Cell Immunol.* 2000;204:19–28.
60. Duffield JS. The inflammatory macrophage: a story of Jekyll and Hyde. *Clin Sci (Lond).* 2003;104:27–38.
61. Wynn TA, Barron L. Macrophages: master regulators of inflammation and fibrosis. *Semin Liver Dis.* 2010;30:245–257.
62. Mahdavian Delavary B, van der Veer WM, van Egmond M, Niessen FB, Beelen RH. Macrophages in skin injury and repair. *Immunobiology.* 2011;216:753–762.
63. Pellicoro A, Ramachandran P, Iredale JP, Fallowfield JA. Liver fibrosis and repair: immune regulation of wound healing in a solid organ. *Nat Rev Immunol.* 2014;14:181–194.
64. Mosser DM, Edwards JP. Exploring the full spectrum of macrophage activation. *Nat Rev Immunol.* 2008;8:958–969.
65. Atabai K, Jame S, Azhar N, et al. Mfge8 diminishes the severity of tissue fibrosis in mice by binding and targeting collagen for uptake by macrophages. *J Clin Invest.* 2009;119:3713–3722.
66. Madsen DH, Leonard D, Masedunskas A, et al. M2-like macrophages are responsible for collagen degradation through a mannose receptor-mediated pathway. *J Cell Biol.* 2013;202:951–966.
67. Iredale JP, Bataller R. Identifying molecular factors that contribute to resolution of liver fibrosis. *Gastroenterology.* 2014;146:1160–1164.
68. Vannella KM, Barron L, Borthwick LA, et al. Incomplete deletion of IL-4Ralpha by LysM(Cre) reveals distinct subsets of M2 macrophages controlling inflammation and fibrosis in chronic schistosomiasis. *PLoS Pathog.* 2014;10:e1004372.
69. Rohani MG, McMahan RS, Razumova MV, et al. MMP-10 regulates collagenolytic activity of alternatively activated resident macrophages. *J Invest Dermatol.* 2015;135:2377–2384.

70. Duffield JS, Lupher M, Thannickal VJ, Wynn TA. Host responses in tissue repair and fibrosis. *Annu Rev Pathol.* 2013;8:241–276.
71. Prasse A, Pechkovsky DV, Toews GB, et al. A vicious circle of alveolar macrophages and fibroblasts perpetuates pulmonary fibrosis via CCL18. *Am J Respir Crit Care Med.* 2006;173:781–792.
72. Gibbons MA, MacKinnon AC, Ramachandran P, et al. Ly6Chi monocytes direct alternatively activated profibrotic macrophage regulation of lung fibrosis. *Am J Respir Crit Care Med.* 2011;184:569–581.
73. Madsen DH, Ingvarsen S, Jurgensen HJ, et al. The non-phagocytic route of collagen uptake: a distinct degradation pathway. *J Biol Chem.* 2011;286:26996–27010.
74. Barron L, Smith AM, El Kasmi KC, et al. Role of arginase 1 from myeloid cells in Th2-dominated lung inflammation. *PLoS One.* 2013;8:e61961.
75. Huen SC, Moeckel GW, Cantley LG. Macrophage-specific deletion of transforming growth factor-beta1 does not prevent renal fibrosis after severe ischemia-reperfusion or obstructive injury. *Am J Physiol Renal Physiol.* 2013;305:F477–F484.
76. Larson-Casey JL, Deshane JS, Ryan AJ, Thannickal VJ, Carter AB. Macrophage Akt1 kinase-mediated mitophagy modulates apoptosis resistance and pulmonary fibrosis. *Immunity.* 2016;44:582–596.
77. Desai TJ, Brownfield DG, Krasnow MA. Alveolar progenitor and stem cells in lung development, renewal and cancer. *Nature.* 2014;507:190–194.
78. Shao DD, Suresh R, Vakil V, Gomer RH, Pilling D. Pivotal advance: Th-1 cytokines inhibit, and Th-2 cytokines promote fibrocyte differentiation. *J Leukoc Biol.* 2008;83:1323–1333.
79. Loomis-King H, Moore BB. Fibrocytes in the pathogenesis of chronic fibrotic lung disease. *Curr Respir Med Rev.* 2013;9:34–41.
80. Kotsianidis I, Nakou E, Bouchliou I, et al. Global impairment of CD4+CD25+ FOXP3+ regulatory T cells in idiopathic pulmonary fibrosis. *Am J Respir Crit Care Med.* 2009;179:1121–1130.
81. Kanellakis P, Dinh TN, Agrotis A, Bobik A. CD4(+)CD25(+)Foxp3(+) regulatory T cells suppress cardiac fibrosis in the hypertensive heart. *J Hypertens.* 2011;29:1820–1828.
82. Claassen MA, de Knegt RJ, Tilanus HW, Janssen HL, Boonstra A. Abundant numbers of regulatory T cells localize to the liver of chronic hepatitis C infected patients and limit the extent of fibrosis. *J Hepatol.* 2010;52:315–321.
83. Liu F, Liu J, Weng D, et al. CD4+CD25+Foxp3+ regulatory T cells depletion may attenuate the development of silica-induced lung fibrosis in mice. *PLoS One.* 2010;5: e15404.
84. Epelman S, Liu PP, Mann DL. Role of innate and adaptive immune mechanisms in cardiac injury and repair. *Nat Rev Immunol.* 2015;15:117–129.
85. Curato C, Slavic S, Dong J, et al. Identification of noncytotoxic and IL-10-producing CD8+AT2R+ T cell population in response to ischemic heart injury. *J Immunol.* 2010;185:6286–6293.
86. McQuibban GA, Gong JH, Wong JP, Wallace JL, Clark-Lewis I, Overall CM. Matrix metalloproteinase processing of monocyte chemoattractant proteins generates CC chemokine receptor antagonists with anti-inflammatory properties in vivo. *Blood.* 2002;100:1160–1167.
87. McQuibban GA, Butler GS, Gong JH, et al. Matrix metalloproteinase activity inactivates the CXC chemokine stromal cell-derived factor-1. *J Biol Chem.* 2001;276:43503–43508.
88. Lee WL, Downey GP. Neutrophil activation and acute lung injury. *Curr Opin Crit Care.* 2001;7:1–7.
89. Ellis TN, Beaman BL. Interferon-gamma activation of polymorphonuclear neutrophil function. *Immunology.* 2004;112:2–12.

90. Dunsmore SE, Saarialho-Kere UK, Roby JD, et al. Matrilysin expression and function in airway epithelium. *J Clin Invest.* 1998;102:1321–1331.
91. López-Boado YS, Wilson CL, Hooper LV, Gordon JI, Hultgren SJ, Parks WC. Bacterial exposure induces and activates matrilysin in mucosal epithelial cells. *J Cell Biol.* 2000;148:1305–1315.
92. Goswami S, Angkasekwinai P, Shan M, et al. Divergent functions for airway epithelial matrix metalloproteinase 7 and retinoic acid in experimental asthma. *Nat Immunol.* 2009;10:496–503.
93. Swee M, Wilson CL, Wang Y, McGuire JK, Parks WC. Matrix metalloproteinase-7 (matrilysin) controls neutrophil egress by generating chemokine gradients. *J Leukoc Biol.* 2008;83:1404–1412.
94. Teng YH, Aquino RS, Park PW. Molecular functions of syndecan-1 in disease. *Matrix Biol.* 2012;31:3–16.
95. Choi Y, Chung H, Jung H, Couchman JR, Oh ES. Syndecans as cell surface receptors: unique structure equates with functional diversity. *Matrix Biol.* 2011;30:93–99.
96. Gill SE, Nadler ST, Li Q, et al. Shedding of syndecan-1/CXCL1 complexes by MMP7 functions as an epithelial checkpoint of neutrophil activation. *Am J Respir Cell Mol Biol.* 2016;55:243–251.
97. Van den Steen PE, Proost P, Wuyts A, Van Damme J, Opdenakker G. Neutrophil gelatinase B potentiates interleukin-8 tenfold by aminoterminal processing, whereas it degrades CTAP-III, PF-4, and GRO-alpha and leaves RANTES and MCP-2 intact. *Blood.* 2000;96:2673–2681.
98. Van den Steen PE, Wuyts A, Husson SJ, Proost P, Van Damme J, Opdenakker G. Gelatinase B/MMP-9 and neutrophil collagenase/MMP-8 process the chemokines human GCP-2/CXCL6, ENA-78/CXCL5 and mouse GCP-2/LIX and modulate their physiological activities. *Eur J Biochem.* 2003;270:3739–3749.
99. Kruidenier L, MacDonald TT, Collins JE, Pender SL, Sanderson IR. Myofibroblast matrix metalloproteinases activate the neutrophil chemoattractant CXCL7 from intestinal epithelial cells. *Gastroenterology.* 2006;130:127–136.
100. Gharib SA, Altemeier WA, Van Winkle LS, et al. Matrix metalloproteinase-7 coordinates airway epithelial injury response and differentiation of ciliated cells. *Am J Respir Cell Mol Biol.* 2013;48:390–396.
101. Kassim SY, Gharib SA, Mecham BH, Birkland TP, Parks WC, McGuire JK. Individual matrix metalloproteinases control distinct transcriptional responses in airway epithelial cells infected with *Pseudomonas aeruginosa. Infect Immun.* 2007;75:5640–5650.
102. Tester AM, Cox JH, Connor AR, et al. LPS responsiveness and neutrophil chemotaxis in vivo require PMN MMP-8 activity. *PLoS One.* 2007;2:e312.
103. Owen CA, Hu Z, Lopez-Otin C, Shapiro SD. Membrane-bound matrix metalloproteinase-8 on activated polymorphonuclear cells is a potent, tissue inhibitor of metalloproteinase-resistant collagenase and serpinase. *J Immunol.* 2004;172:7791–7803.
104. Garcia S, Forteza J, Lopez-Otin C, Gomez-Reino JJ, Gonzalez A, Conde C. Matrix metalloproteinase-8 deficiency increases joint inflammation and bone erosion in the K/BxN serum-transfer arthritis model. *Arthritis Res Ther.* 2010;12:R224.
105. Quintero PA, Knolle MD, Cala LF, Zhuang Y, Owen CA. Matrix metalloproteinase-8 inactivates macrophage inflammatory protein-1 alpha to reduce acute lung inflammation and injury in mice. *J Immunol.* 2010;184:1575–1588.
106. Bellac CL, Dufour A, Krisinger MJ, et al. Macrophage matrix metalloproteinase-12 dampens inflammation and neutrophil influx in arthritis. *Cell Rep.* 2014;9:618–632.
107. Owen CA, Hu Z, Barrick B, Shapiro SD. Inducible expression of tissue inhibitor of metalloproteinases-resistant matrix metalloproteinase-9 on the cell surface of neutrophils. *Am J Respir Cell Mol Biol.* 2003;29:283–294.
108. La Fleur M, Underwood JL, Rappolee DA, Werb Z. Basement membrane and repair of injury to peripheral nerve: defining a potential role for macrophages, matrix

metalloproteinases, and tissue inhibitor of metalloproteinases-1. *J Exp Med.* 1996;184:2311–2326.
109. Vaisar T, Kassim SY, Gomez IG, et al. MMP-9 sheds the beta 2 integrin subunit (CD18) from macrophages. *Mol Cell Proteomics.* 2009;8:1044–1060.
110. Hautamaki RD, Kobayashi DK, Senior RM, Shapiro SD. Requirement for macrophage elastase for cigarette smoke-induced emphysema in mice. *Science.* 1997;277:2002–2004.
111. Manicone AM, Birkland TP, Lin M, et al. Epilysin (MMP-28) restrains early macrophage recruitment in *Pseudomonas aeruginosa* pneumonia. *J Immunol.* 2009;182:3866–3876.
112. Jenner RG, Young RA. Insights into host responses against pathogens from transcriptional profiling. *Nat Rev Microbiol.* 2005;3:281–294.
113. McMahan RS, Birkland TP, Smigiel KS, et al. Stromelysin-2 (MMP10) moderates inflammation by controlling macrophage activation. *J Immunol.* 2016;197:899–909.
114. Garcia-Irigoyen O, Carotti S, Latasa MU, et al. Matrix metalloproteinase-10 expression is induced during hepatic injury and plays a fundamental role in liver tissue repair. *Liver Int.* 2014;34:e257–e270.
115. Koller FL, Dozier EA, Nam KT, et al. Lack of MMP10 exacerbates experimental colitis and promotes development of inflammation-associated colonic dysplasia. *Lab Invest.* 2012;92:1749–1759.
116. Murray MY, Birkland TP, Howe JD, et al. Macrophage migration and invasion is regulated by MMP10 expression. *PLoS One.* 2013;8:e63555.
117. Chang MY, Tanino Y, Vidova V, et al. A rapid increase in macrophage-derived versican and hyaluronan in infectious lung disease. *Matrix Biol.* 2014;34:1–12.
118. de La Motte CA, Hascall VC, Calabro A, Yen-Lieberman B, Strong SA. Mononuclear leukocytes preferentially bind via CD44 to hyaluronan on human intestinal mucosal smooth muscle cells after virus infection or treatment with poly(I.C). *J Biol Chem.* 1999;274:30747–30755.
119. de la Motte CA, Hascall VC, Drazba J, Bandyopadhyay SK, Strong SA. Mononuclear leukocytes bind to specific hyaluronan structures on colon mucosal smooth muscle cells treated with polyinosinic acid:polycytidylic acid: inter-alpha-trypsin inhibitor is crucial to structure and function. *Am J Pathol.* 2003;163:121–133.
120. Rowe RG, Weiss SJ. Breaching the basement membrane: who, when and how? *Trends Cell Biol.* 2008;18:560–574.
121. Sabeh F, Shimizu-Hirota R, Weiss SJ. Protease-dependent versus -independent cancer cell invasion programs: three-dimensional amoeboid movement revisited. *J Cell Biol.* 2009;185:11–19.
122. Gharib SA, Loth DW, Soler Artigas M, et al. Integrative pathway genomics of lung function and airflow obstruction. *Hum Mol Genet.* 2015;24:6836–6848.
123. Belperio JA, Keane MP, Burdick MD, et al. Critical role for the chemokine MCP-1/CCR2 in the pathogenesis of bronchiolitis obliterans syndrome. *J Clin Invest.* 2001;108:547–556.
124. Sakanashi Y, Takeya M, Yoshimura T, Feng L, Morioka T, Takahashi K. Kinetics of macrophage subpopulations and expression of monocyte chemoattractant protein-1 (MCP-1) in bleomycin-induced lung injury of rats studied by a novel monoclonal antibody against rat MCP-1. *J Leukoc Biol.* 1994;56:741–750.
125. Abraham M, Shapiro S, Karni A, Weiner HL, Miller A. Gelatinases (MMP-2 and MMP-9) are preferentially expressed by Th1 vs. Th2 cells. *J Neuroimmunol.* 2005;163:157–164.
126. Ivanoff A, Ivanoff J, Hultenby K, Sundqvist KG. Infiltrative capacity of T leukemia cell lines: a distinct functional property coupled to expression of matrix metalloproteinase-9

(MMP-9) and tissue inhibitor of metalloproteinases-1 (TIMP-1). *Clin Exp Metastasis.* 1999;17:695–711.
127. Kumagai K, Ohno I, Okada S, et al. Inhibition of matrix metalloproteinases prevents allergen-induced airway inflammation in a murine model of asthma. *J Immunol.* 1999;162:4212–4219.
128. Sato W, Tomita A, Ichikawa D, et al. CCR2(+)CCR5(+) T cells produce matrix metalloproteinase-9 and osteopontin in the pathogenesis of multiple sclerosis. *J Immunol.* 2012;189:5057–5065.
129. Sixt M, Engelhardt B, Pausch F, Hallmann R, Wendler O, Sorokin LM. Endothelial cell laminin isoforms, laminins 8 and 10, play decisive roles in T cell recruitment across the blood-brain barrier in experimental autoimmune encephalomyelitis. *J Cell Biol.* 2001;153:933–946.
130. Corry DB, Rishi K, Kanellis J, et al. Decreased allergic lung inflammatory cell egression and increased susceptibility to asphyxiation in MMP2-deficiency. *Nat Immunol.* 2002;3:347–353.
131. Corry DB, Kiss A, Song LZ, et al. Overlapping and independent contributions of MMP2 and MMP9 to lung allergic inflammatory cell egression through decreased CC chemokines. *FASEB J.* 2004;18:995–997.
132. Ra HJ, Parks WC. Control of matrix metalloproteinase catalytic activity. *Matrix Biol.* 2007;26:587–596.
133. Tocchi A, Parks WC. Functional interactions between matrix metalloproteinases and glycosaminoglycans. *FEBS J.* 2013;280:2332–2341.
134. Taraszka KS, Higgins JM, Tan K, Mandelbrot DA, Wang JH, Brenner MB. Molecular basis for leukocyte integrin alpha(E)beta(7) adhesion to epithelial (E)-cadherin. *J Exp Med.* 2000;191:1555–1567.
135. Manicone AM, Huizar I, McGuire JK. Matrilysin (Matrix Metalloproteinase-7) regulates anti-inflammatory and antifibrotic pulmonary dendritic cells that express CD103 (alpha(E)beta(7)-integrin). *Am J Pathol.* 2009;175:2319–2331.
136. Guo H, Callaway JB, Ting JP. Inflammasomes: mechanism of action, role in disease, and therapeutics. *Nat Med.* 2015;21:677–687.
137. Ito A, Mukaiyama A, Itoh Y, et al. Degradation of interleukin 1beta by matrix metalloproteinases. *J Biol Chem.* 1996;271:14657–14660.
138. Schonbeck U, Mach F, Libby P. Generation of biologically active IL-1 beta by matrix metalloproteinases: a novel caspase-1-independent pathway of IL-1 beta processing. *J Immunol.* 1998;161:3340–3346.
139. Moss ML, Jin SL, Milla ME, et al. Cloning of a disintegrin metalloproteinase that processes precursor tumour-necrosis factor-alpha. *Nature.* 1997;385:733–736.
140. Black RA, Rauch CT, Kozlosky CJ, et al. A metalloproteinase disintegrin that releases tumour-necrosis factor-alpha from cells. *Nature.* 1997;385:729–733.
141. Mohan MJ, Seaton T, Mitchell J, et al. The tumor necrosis factor-alpha converting enzyme (TACE): a unique metalloproteinase with highly defined substrate selectivity. *Biochemistry.* 2002;41:9462–9469.
142. Manicone AM, McGuire JK. Matrix metalloproteinases as modulators of inflammation. *Semin Cell Dev Biol.* 2008;19:34–41.
143. Schulz O, Sewell HF, Shakib F. Proteolytic cleavage of CD25, the alpha subunit of the human T cell interleukin 2 receptor, by Der p 1, a major mite allergen with cysteine protease activity. *J Exp Med.* 1998;187:271–275.
144. Bank U, Reinhold D, Schneemilch C, Kunz D, Synowitz HJ, Ansorge S. Selective proteolytic cleavage of IL-2 receptor and IL-6 receptor ligand binding chains by neutrophil-derived serine proteases at foci of inflammation. *J Interferon Cytokine Res.* 1999;19:1277–1287.

145. Sheu BC, Hsu SM, Ho HN, Lien HC, Huang SC, Lin RH. A novel role of metalloproteinase in cancer-mediated immunosuppression. *Cancer Res.* 2001; 61:237–242.
146. Park MJ, Shin JS, Kim YH, et al. Murine mesenchymal stem cells suppress T lymphocyte activation through IL-2 receptor alpha (CD25) cleavage by producing matrix metalloproteinases. *Stem Cell Rev.* 2011;7:381–393.
147. Roch T, Akymenko O, Kruger A, Jung F, Ma N, Lendlein A. Expression pattern analysis and activity determination of matrix metalloproteinase derived from human macrophage subsets. *Clin Hemorheol Microcirc.* 2014;58:147–158.
148. Hayes EM, Tsaousi A, Di Gregoli K, et al. Classical and alternative activation and metalloproteinase expression occurs in foam cell macrophages in male and female ApoE null mice in the absence of T and B lymphocytes. *Front Immunol.* 2014;5:537.
149. Krakowiak MS, Noto JM, Piazuelo MB, et al. Matrix metalloproteinase 7 restrains Helicobacter pylori-induced gastric inflammation and premalignant lesions in the stomach by altering macrophage polarization. *Oncogene.* 2015;34:1865–1871.
150. Wen G, Zhang C, Chen Q, et al. A novel role of matrix metalloproteinase-8 in macrophage differentiation and polarization. *J Biol Chem.* 2015;290:19158–19172.
151. Ma Y, Halade GV, Zhang J, et al. Matrix metalloproteinase-28 deletion exacerbates cardiac dysfunction and rupture after myocardial infarction in mice by inhibiting M2 macrophage activation. *Circ Res.* 2013;112:675–688.
152. Gharib SA, Johnston LK, Huizar I, et al. MMP28 promotes macrophage polarization toward M2 cells and augments pulmonary fibrosis. *J Leukoc Biol.* 2014;95:9–18.
153. Iyer RP, Jung M, Lindsey ML. MMP-9 signaling in the left ventricle following myocardial infarction. *Am J Physiol Heart Circ Physiol.* 2016;311:H190–H198.
154. Cauwe B, Martens E, Proost P, Opdenakker G. Multidimensional degradomics identifies systemic autoantigens and intracellular matrix proteins as novel gelatinase B/MMP-9 substrates. *Integr Biol (Camb).* 2009;1:404–426.
155. Zamilpa R, Ibarra J, de Castro Bras LE, et al. Transgenic overexpression of matrix metalloproteinase-9 in macrophages attenuates the inflammatory response and improves left ventricular function post-myocardial infarction. *J Mol Cell Cardiol.* 2012;53:599–608.
156. Lindsey ML, Escobar GP, Dobrucki LW, et al. Matrix metalloproteinase-9 gene deletion facilitates angiogenesis after myocardial infarction. *Am J Physiol Heart Circ Physiol.* 2006;290:H232–H239.
157. Heymans S, Luttun A, Nuyens D, et al. Inhibition of plasminogen activators or matrix metalloproteinases prevents cardiac rupture but impairs therapeutic angiogenesis and causes cardiac failure. *Nat Med.* 1999;5:1135–1142.
158. DeLeon-Pennell KY, Tian Y, Zhang B, et al. CD36 is a matrix metalloproteinase-9 substrate that stimulates neutrophil apoptosis and removal during cardiac remodeling. *Circ Cardiovasc Genet.* 2016;9:14–25.
159. Saarialho-Kere UK, Kovacs SO, Pentland AP, Parks WC, Welgus HG. Distinct populations of keratinocytes express stromelysin-1 and -2 in chronic wounds. *J Clin Invest.* 1994;94:79–88.
160. Krampert M, Bloch W, Sasaki T, et al. Activities of the matrix metalloproteinase stromelysin-2 (MMP-10) in matrix degradation and keratinocyte organization in wounded skin. *Mol Biol Cell.* 2004;15:5242–5254.
161. Kren L, Goncharuk VN, Krenova Z, et al. Expression of matrix metalloproteinases 3, 10 and 11 (stromelysins 1, 2 and 3) and matrix metalloproteinase 7 (matrilysin) by cancer cells in non-small cell lung neoplasms. *Cesk Patol.* 2006;42:16–19.
162. Gill JH, Kirwan IG, Seargent JM, et al. MMP-10 is overexpressed, proteolytically active, and a potential target for therapeutic intervention in human lung carcinomas. *Neoplasia.* 2004;6:777–785.

163. Lin TS, Chiou SH, Wang LS, et al. Expression spectra of matrix metalloproteinases in metastatic non-small cell lung cancer. *Oncol Rep.* 2004;12:717–723.
164. Bodey B, Bodey Jr B, Groger AM, Siegel SE, Kaiser HE. Invasion and metastasis: the expression and significance of matrix metalloproteinases in carcinomas of the lung. *In Vivo.* 2001;15:175–180.
165. Cho NH, Hong KP, Hong SH, Kang S, Chung KY, Cho SH. MMP expression profiling in recurred stage IB lung cancer. *Oncogene.* 2004;23:845–851.
166. Aung PP, Oue N, Mitani Y, et al. Systematic search for gastric cancer-specific genes based on SAGE data: melanoma inhibitory activity and matrix metalloproteinase-10 are novel prognostic factors in patients with gastric cancer. *Oncogene.* 2006;25:2546–2557.
167. Frederick LA, Matthews JA, Jamieson L, et al. Matrix metalloproteinase-10 is a critical effector of protein kinase Ciota-Par6alpha-mediated lung cancer. *Oncogene.* 2008;27:4841–4853.
168. Kaner R, Santiago F, Crystal R. Up-regulation of alveolar macrophage matrix metalloproteinases in HIV1+ smokers with early emphysema. *J Leukoc Biol.* 2009; 86:913–922.
169. Gosselink JV, Hayashi S, Elliott WM, et al. Differential expression of tissue repair genes in the pathogenesis of COPD. *Am J Respir Crit Care Med.* 2010;181:1329–1335.
170. Shaykhiev R, Krause A, Salit J, et al. Smoking-dependent reprogramming of alveolar macrophage polarization: implication for pathogenesis of chronic obstructive pulmonary disease. *J Immunol.* 2009;183:2867–2883.
171. Shapiro SD. The macrophage in chronic obstructive pulmonary disease. *Am J Respir Crit Care Med.* 1999;160:S29–S32.
172. Tetley TD. Macrophages and the pathogenesis of COPD. *Chest.* 2002;121:156S–159S.

CHAPTER SIX

Evidence for the Involvement of Matrix-Degrading Metalloproteinases (MMPs) in Atherosclerosis

Bethan A. Brown, Helen Williams, Sarah J. George[1,2]
School of Clinical Sciences, University of Bristol, Bristol, United Kingdom
[2]Corresponding author: e-mail address: s.j.george@bristol.ac.uk

Contents

[1] Senior author.

Progress in Molecular Biology and Translational Science, Volume 147
ISSN 1877-1173
http://dx.doi.org/10.1016/bs.pmbts.2017.01.004

Abstract

Atherosclerosis leads to blockage of arteries, culminating in myocardial infarction, and stroke. The involvement of matrix-degrading metalloproteinases (MMPs) in atherosclerosis is established and many studies have highlighted the importance of various MMPs in this process. MMPs were first implicated in atherosclerosis due to their ability to degrade extracellular matrix components, which can lead to increased plaque instability. However, more recent work has highlighted a multitude of roles for MMPs in addition to breakdown of extracellular matrix proteins. MMPs are now known to be involved in various stages of plaque progression: from initial macrophage infiltration to plaque rupture. This chapter summarizes the development and progression of atherosclerotic plaques and the contribution of MMPs. We provide data from human studies showing the effect of MMP polymorphisms and the expression of MMPs in both the atherosclerotic plaque and within plasma. We also discuss work in animal models of atherosclerosis that show the effect of gain or loss of function of MMPs. Together, the data provided from these studies illustrate that MMPs are ideal targets as both biomarkers and potential drug therapies for atherosclerosis.

1. ATHEROSCLEROSIS AND CARDIOVASCULAR DISEASE

Atherosclerosis is the underlying cause of cardiovascular diseases including coronary artery disease and stroke, conditions, which were responsible for 73,000 and 41,000 deaths in the United Kingdom in 2012, respectively.[1] In addition, it is estimated that 2.3 million people in the United Kingdom live with coronary artery disease.[2] Accumulating evidence suggests that matrix metalloproteinases (MMPs) play a role in atherosclerosis development and progression, and hence may represent a potential target for cardiovascular disease therapy.

1.1 Healthy Arterial Anatomy

The arterial wall consists of three layers: the tunica intima, the tunica media, and the tunica adventitia,[3,4] see Fig. 1. The tunica intima is the layer which is closest to the lumen. This is the thinnest layer of the vascular wall consisting of a single sheet of endothelial cells resting on a basement membrane and a thin subendothelial extracellular matrix (ECM) composed of collagen and elastin. The endothelium not only acts as a physical barrier to separate the blood from surrounding tissue but also participates in the regulation of coagulation, inflammation, and vessel tone. The tunica media is the middle layer and is composed of organized layers of vascular smooth muscle cells (VSMCs) embedded in a collagenous ECM containing structural

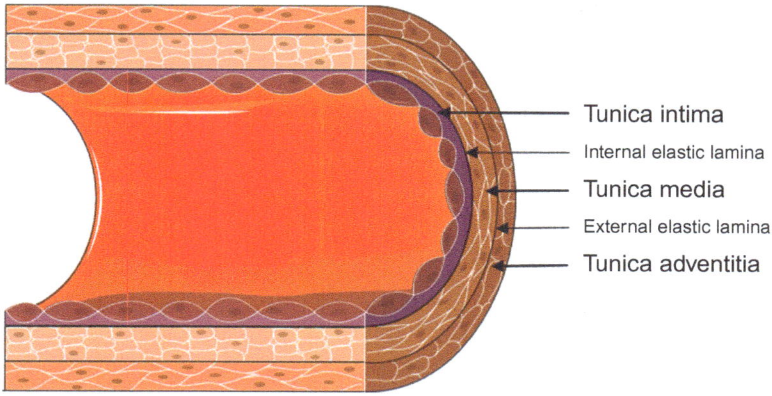

Fig. 1 Structure of the healthy arterial wall. The arterial wall is composed of three layers: an inner tunica intima consisting of endothelial cells and subendothelial extracellular matrix (ECM), a thick muscular tunica media composed of vascular smooth muscle cells (VSMCs), and an outermost tunica adventitia composed of collagenous ECM and containing a network of finely lined blood vessels called the vasa vasorum. The internal elastic lamina separates the tunica intima and media, and the external elastic lamina separates the tunica media and adventitia. Images adapted from www.servier.com.

glycoproteins and proteoglycans such as hyaluronan and decorin, and delineated by elastic laminae. This muscular layer is responsible for adjusting oscillations in blood flow and maintaining vascular tone. The outermost tunica adventitia is composed of collagenous connective tissue populated by fibroblasts. The adventitia can also contain the lymphatic and nerve plexi and a network of finely lined blood vessels called the vaso vasorum. The tunica intima and media, and the tunica media and adventitia are separated by the internal elastic lamina and the external elastic lamina, respectively.

1.2 Atherosclerosis

Atherosclerotic lesions can start to develop as early as adolescence and progress over a period of decades.[5] Risk factors for atherosclerosis include a family history of cardiovascular disease, hypertension, obesity, diabetes, smoking, and hypercholesterolemia (see review by Lusis[6]). Atherosclerotic lesions tend to form in distinct regions of the vasculature such as curvatures, branches, and bifurcations where the hemodynamic shear stress is disturbed evoking a dysfunctional endothelial phenotype (see review by Heo *et al.*[7]). Additionally, atherosclerosis develops at sites of intimal hyperplasia and restenosis, which occurs following surgical interventions to treat coronary artery disease such as balloon angioplasty, intracoronary stent implantation, or coronary artery bypass graft surgery (see reviews by Schwartz *et al.*[8] and Wallitt *et al.*[9]). Initiation of hyperplasia at these sites is thought to be due to damage of the protective endothelium, leading to migration of underlying VSMCs into the subendothelial space where these cells proliferate and synthesize ECM to create a thickened intima[10] and reviews by Schwartz *et al.*[8] and Wallitt *et al.*[9] This involves a change in the VSMC phenotype from the contractile phenotype observed in the tunica media to a synthetic phenotype with increased ability to migrate, proliferate, and synthesize ECM (see review by Owens *et al.*[11] and Lacolley *et al.*[12]). This thickened intima, sometimes termed a neointima, can act as soil for atherosclerosis development promoting further restenosis of the vessel (reviewed by Schwartz *et al.*[8] and Wallitt *et al.*[9]). In addition, intimal thickening also occurs naturally with aging, which renders the artery more susceptible to atherosclerosis (see reviews by Kovacic *et al.*,[13] Najjar *et al.*,[14] and Wang and Bennett[15]).

The progression of atherosclerosis is shown in Fig. 2 and reviewed extensively by McLaren *et al.*,[16] Libby,[17] and Tabas *et al.*[18] Atherosclerosis is initiated in regions of thickened intima or dysfunctional endothelial cells by the accumulation of lipoproteins, such as low-density lipoprotein (LDL), in the subendothelial matrix of the artery. Once in the vessel wall, lipoproteins are modified by oxidation and activate the overlying endothelium. Activated endothelial cells release inflammatory cytokines and chemokines, for example, monocyte chemoattractant protein-1 (MCP-1), which promote the recruitment of circulating immune cells such as monocytes and T-cells. Simultaneously, activated endothelial cells express adhesion molecules such as P-selectin and vascular cell adhesion protein-1 (VCAM-1), to which monocytes and T-cells then adhere, facilitating migrating through the endothelium into the subendothelial intima. Once here monocytes differentiate into macrophages and engulf modified lipoproteins, transforming them into lipid-rich foam cells, which are the characteristic feature of early-stage atherosclerosis—the fatty streak. As atherosclerosis progresses, inflammatory- and growth-stimulating factors released from activated endothelial cells and the infiltrating immune cells stimulate the migration of VSMCs from the tunica media over the top of the fatty streak. As in intimal hyperplasia, this involves a change in the VSMC phenotype from a contractile to a synthetic phenotype with increased ability to migrate, proliferate, and synthesize ECM. Once overlying the fatty lesion, synthetic VSMCs secrete ECM proteins to form the fibrous cap, which is the characteristic feature of the atheroma stage of atherosclerosis. The fibrous cap

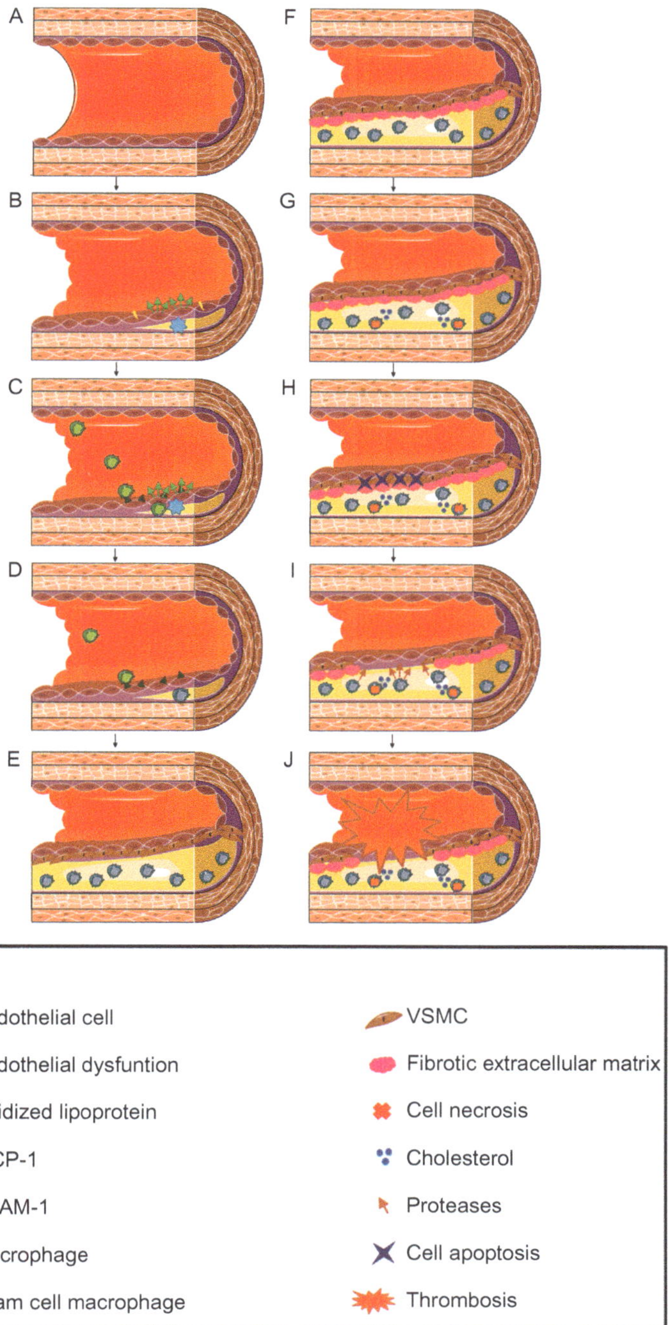

Fig. 2 See legend on next page.

confers plaque stability by preventing contact between circulating platelets and the thrombogenic fatty core of the atheroma. As atherosclerosis progresses, lipid-laden macrophages within the lesion die, leading to deposition of extracellular cholesterol to form cholesterol clefts, calcification, and further inflammation. In the late stages of atherosclerosis, thinning of the fibrous cap occurs due to a combination of VSMC apoptosis and ECM degradation by proteases. Thinning of the fibrous cap predisposes plaque rupture, platelet activation, and subsequent thrombosis, reviewed by Clarke and Bennett[19] and Bennett *et al.*[20] In some cases, plaque rupture can be silent, resulting in no immediate symptoms, and the fibrous cap can heal as a result of VSMC proliferation; this results in the formation of buried layers, increased plaque size, and greater luminal stenosis. Alternatively, due to the location of the plaque rupture and extent of the thrombosis and subsequent arterial occlusion, it can lead to myocardial or cerebral ischemia. Such catastrophic plaque rupture usually occurs after multiple rounds of previous rupture and healing.[21]

Fig. 2 Atherosclerotic plaque formation and progression. (A) The healthy arterial wall consists of three layers, the tunica intima, tunica media, and the tunica adventitia. (B) Endothelial cell dysfunction and entry of lipoproteins into the subendothelial intima occurs. Trapped lipoproteins can become modified by oxidation and promote activation of the overlying endothelium. Once activated, endothelial cells secrete inflammatory cytokines and chemokines such as monocyte chemoattractant protein-1 (MCP-1). (C) MCP-1 recruits circulating monocytes to the area. Activated endothelial cells also express cell adhesion molecules such as vascular cell adhesion protein-1 (VCAM-1), which promote monocyte adherence followed by migration through the endothelial cell layer. (D) Once in the subendothelial space, monocytes differentiate into macrophages and engulf the modified lipoproteins transforming them into foam cell macrophages. These lipid-laden cells characterize the early stage of atherosclerosis—the fatty streak. (E) As atherosclerosis progresses, monocyte recruitment and foam cell formation continue and the fatty streak increases in size. Inflammatory and growth-regulating factors secreted from activated endothelial cells and infiltrating immune cells stimulate migration of VSMCs over the fatty streak. (F) VSMCs secrete ECM to cover the fatty lesion, termed the fibrous cap. This cap typifies the atheroma stage of atherosclerosis. (G) Foam cells trapped in the plaque core undergo cell death. These dying cells deposit cholesterol extracellularly forming cholesterol clefts. (H) VSMCs in the overlying fibrous cap undergo apoptosis. (I) In addition to the reduced number of VSMCs secreting ECM within the fibrous cap, macrophages secrete proteases, which degrade matrix components. Together, this leads to thinning of the fibrous cap. (J) Eventually thinning of the fibrous cap can lead to plaque rupture and thrombosis. In some cases, plaque rupture is clinically silent and the plaque can heal by VSMC proliferation. Alternatively, plaque rupture and thrombosis can lead to tissue ischemia presenting as a myocardial infarction or a stroke. Images adapted from www.servier.com.

2. INTRODUCTION TO MMPs

The MMPs are a family of 21 zinc-dependent endopeptidases, which mediate degradation or remodeling of the ECM (Table 1). Together, the MMP family can degrade all of the components of the blood vessel wall, and therefore play a major role in both physiological and pathological events that involve the degradation of ECM (Table 2). Additionally, MMPs have

Table 1 MMP Pseudonyms

MMP	Pseudonyms
MMP-1	Collagenase-1
MMP-2	Gelatinase-A
MMP-3	Stromelysin-1
MMP-7	Matrilysin-1, PUMP-1, uterine metalloproteinase
MMP-8	Collagenase-2, neutrophil collagenase, microbial collagenase
MMP-9	Gelatinase-B, macrophage gelatinase, type IV collagenase
MMP-10	Stromelysin-2, transin-2
MMP-11	Stromelysin-3
MMP-12	Macrophage metalloelastase, macrophage elastase
MMP-13	Collagenase-3
MMP-14	MT1-MMP
MMP-15	MT2-MMP
MMP-16	MT3-MMP
MMP-17	MT4-MMP
MMP-19	RASI-1
MMP-20	Enamelysin
MMP-23	CA-MMP (cysteine array)
MMP-24	MT5-MMP
MMP-25	MT6-MMP, leukolysin
MMP-26	Matrilysin-2, endometase
MMP-28	Epilysin

Table 2 MMP Substrates

MMP	ECM Substrates	Alternative Substrates
MMP-1	Collagen type I, II, III, VII, VIII, X Gelatin, aggrecan, casein, serpins, versican, perlecan, proteoglycan link protein, tenascin-C	Pro-TNFα, FGF/perlecan complex, IL-1β, MCP-3, IGFBPs, CC1q, α1-ACT, α2-MG, α1-PI, fibrin, fibrinogen
MMP-2	Collagen type I, IV, V, VII, X, XI, XIV Gelatin, elastin, aggrecan, fibronectin, versican, laminin, decorin, proteoglycan, proteoglycan link protein	Pro-TGFβ, pro-TNFα, IL-1β, MCP-3, IGFBPs, CC1q, α1-ACT, α1-PI, fibrin, fibrinogen
MMP-3	Collagen type II, IV, IX, X Gelatin, elastin, aggrecan, fibronectin, versican, casein, decorin, laminin, perlecan, proteoglycan, proteoglycan link protein	Pro-TGFβ, pro-TNFα, FGF/ perlecan complex, IL-1β, MCP-3, pro-HBEGF, IGFBPs, plasminogen, CC1q, E-cad, α1-ACT, α2-MG, α1-PI, fibrin, fibrinogen
MMP-7	Collagen type I, II, III, IV, V, VI, VIII, X Elastin, aggrecan, casein, laminin, entactin, proteoglycan, proteoglycan link protein	Pro-TGFβ, pro-TNFα, β4int, plasminogen, E-cad, N-cad, Fas-L, α1-PI, fibrinogen, ApoA-IV
MMP-8	Collagen type I, II, III, V, VII, VIII, X Gelatin, aggrecan, laminin	CC1q, α2-MG, α1-PI, fibrinogen, Ang-I, ADAM10, ApoA-1
MMP-9	Collagen type VI, V, VII, X, XIV Fibronectin, laminin, versican, proteoglycan link protein	Pro-TGFβ, pro-TNFα, IL-1β, IL2Rα, PEGF, plasminogen, CC1q, α2-MG, α1-PI, fibrin, fibrinogen, N-cad
MMP-10	Collagen type VI, V, VII, X, XIV Gelatin, fibronectin, laminin	Fibrinogen
MMP-11	Laminin, serpins	IGFBPs, α2-MG, α1-PI
MMP-12	Elastin	Plasminogen, factor XII, α2-MG, α1-PI, fibrinogen, N-cad
MMP-13	Collagen type I, II, III, IV, V, IX, X, XI Gelatin, aggrecan, fibronectin, laminin, perlecan, tenascin	FGF/perlecan complex, MCP-3, factor XII, CC1q, α2-MG, fibrinogen, ICAM-1

Table 2 MMP Substrates—cont'd

MMP	ECM Substrates	Alternative Substrates
MMP-14	Collagen type I, II, III Gelatin, aggrecan, fibronectin, laminin, perlecan, tenascin, vitronectin, dermatan sulfate proteoglycan	Pro-TNFα, MCP-3, TTG, factor XII, CD44, α2-MG, α1-PI, fibrin, fibrinogen, ApoA-IV
MMP-15	Collagen type I, II, III Gelatin, aggrecan, fibronectin, laminin, perlecan, vitronectin	TTG
MMP-16	Collagen type I, III Gelatin, aggrecan, fibronectin, laminin, casein, tenascin	TTG
MMP-17	Gelatin, fibronectin	Fibrin
MMP-19	Collagen type I, IV, gelatin, aggrecan, casein, fibronectin, laminin, tenascin	
MMP-20	Aggrecan, amelogenin, cartilage oligomeric protein	
MMP-23	Gelatin, fibronectin, chondroitin sulfate, dermatan sulfate	
MMP-24	Gelatin, fibronectin	Fibrin
MMP-25	Collagen type IV, gelatin, fibronectin, casein	Fibrinogen
MMP-26	Collagen type IV, gelatin, casein	α1-PI, fibrinogen

Abbreviations: *ADAM10*, a disintegrin and metalloproteinase domain 10; *Ang-I*, angiotensin I; *Apo*, apolipoprotein; *E-cad*, E-cadherin; *Fas-L*, Fas ligand; *FGF*, fibroblast growth factor; *HBEGF*, heparin-binding EGF-like growth factor; *IGFBPs*, insulin-like growth factor-binding proteins; *IL-1β*, interleukin-1β; *IL2Rα*, interleukin-2 receptor-α; *MCP-3*, monocyte-chemotactic protein 3; *N-cad*, N-cadherin; *PEGF*, provascular endothelial growth factor; *pro-HBEGF*, proheparin-binding epidermal growth factor; *TGFβ*, transforming growth factor-β; *TNFα*, tumour necrosis factor-α; *TTG*, tissue transglutaminase; *α1-ACT*, α1-antichymotrypsin; *α1-PI*, α1-proteinase inhibitor; *α2-MG*, α2-macroglobulin; *β4int*, β4 integrin.

Table has been adapted and updated from a review by Johnson JL. Matrix metalloproteinases: influence on smooth muscle cells and atherosclerotic plaque stability. *Expert Rev Cardiovasc Ther.* 2007;5(2):265–282.

non-ECM substrates (Table 2). Since the MMPs can modulate the cell–ECM and cell–cell interactions that control cell behavior, their activity affects processes as diverse as cellular differentiation, migration, proliferation, and apoptosis.[22]

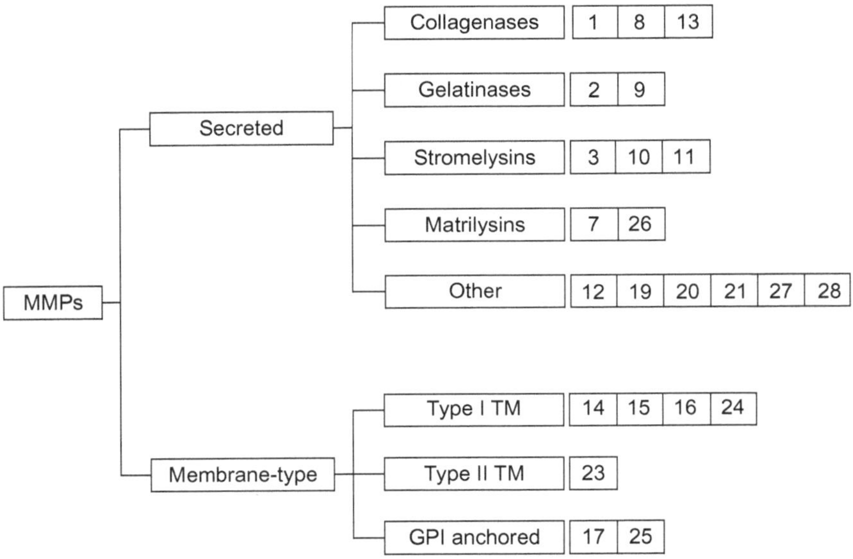

Fig. 3 The matrix metalloproteinase family tree.

The MMP family is divided into subfamilies, based on their function and structure as shown in Fig. 3. Although each MMP is a product of a different gene, there is a high degree of sequence and structural domain homology between the MMPs. All MMPs have a short signal sequence and a pro-peptide region at the N-terminus, containing a cysteine residue that ligates with zinc at the catalytic domain and maintains the enzyme in the inactive or pro-form. The C-terminus of all MMPs, except MMP-7 contains a region that has a high level of homology with the hemopexin family and confers the substrate-binding and degradation specificity. The gelatinases also contain a fibronectin type II-like region that can also confer substrate specificity. The C- and N-termini are connected by a hinge region, which varies in length between the MMP groups. Although the MT-MMPs are attached to the cell surface by a membrane domain at the C-terminus, it has been recently demonstrated that some MT-MMPs also exist as soluble proteases, which may add greater flexibility to their function.[23]

All MMPs are expressed as inactive zymogens requiring extracellular proteolytic processing to expose the active catalytic site, although MMP-11 is activated by furin intracellularly.[24] MT-MMPs also possess a furin recognition motif, and therefore they can also be activated intracellularly by furin.[25] In the secreted latent, zymogen form, the prodomain folds over and shields the catalytic site. Thiol interactions between cysteine residues in the prodomain and the zinc atom present in the catalytic site of all MMPs maintains this folding. Activation of the proenzymes occurs in two stages. Partial activation occurs when the cysteine–zinc interaction is disrupted allowing partial cleavage of the prodomain by other proteases such as plasmin, trypsin, kallikrein, tryptase, chymase, and some MMPs or by nonproteolytic compounds such as thiol reactive agents and denaturants or by heat treatment.[26,27] Partial activation causes conformational changes rendering the enzyme susceptible to autocatalytic or exogenous cleavage of the entire propeptide region by proteases, including other MMPs permitting complete activation.

MMP activity is counterbalanced by the family of four endogenous inhibitors, the tissue inhibitors of metalloproteinases (TIMPs). TIMPs efficiently inhibit MMPs and, while there are fundamental variations in the affinity of different TIMPs for individual MMP enzymes, each TIMP can inhibit multiple MMPs. The preferential TIMP–MMP interactions and tissue-restricted TIMP expression suggests that each TIMP has defined functions.[28] For example, TIMP-1 binds preferentially to pro-MMP-9, whereas TIMP-2 possesses a higher affinity for pro-MMP-2.[29,30] TIMPs bind to MMPs at two sites. TIMPs bind to pro-MMPs within the C-terminal region of the MMP, which stabilizes MMP activity in the extracellular space and further serves to delay pro-MMP activation, hence limiting the activity of MMPs. Once activated the active MMPs may still be inhibited by the binding of TIMPs at the active site in a 1:1 stoichiometric ratio leading to inhibition of MMP-mediated ECM degradation.[28] For the purpose of delineating MMP involvement in disease processes the levels of TIMPs within the extracellular space as well as the level of latent and active MMPs, therefore, requires detailed analysis.

There has been extensive interest in the role of MMPs in the pathogenesis of atherosclerosis over the last three decades. In this chapter, we have reviewed evidence of the association of MMP expression and activity with atherosclerosis in human and animal models. Our greater understanding of the involvement of MMPs in atherosclerosis has led to the interest in developing MMP specific inhibitors to retard plaque progression and rupture.

2.1 MMP-1

Human data implicate the association of MMP-1 in atherosclerosis. Levels of MMP-1 mRNA were increased in carotid plaques compared to healthy arteries and were also significantly increased in vulnerable compared to stable plaques.[31] In addition, augmented MMP-1 mRNA and protein were detected in VSMCs from carotid plaques from symptomatic patients compared to asymptomatic patients.[32,33] Nuclear factor-κB (NFκB)/protein 38 (P38)/phosphatidylinositol 3-kinase (PI3K), or Jun N-terminal kinase (JNK) inhibitors reduced the expression of MMP-1 in these cells from symptomatic patients in response to TNFα,[32] suggesting the involvement of these signaling molecules in the enhanced expression. Plasma levels of MMP-1 were also found to correlate to plaque burden as assessed by CT-angiography[34] or MRI scanning,[35] associated with high levels of C-reactive protein (CRP),[36] and an independent predictor of mortality over 5 years in male patients.[37] However, these associations do not demonstrate whether MMP-1 has a causal role. An insertion polymorphism in the MMP-1 gene, G-1607GG, was found to show a weak but not significant association with the rate of clinical events in patients with coronary artery disease.[38] However, a larger study revealed that this polymorphism significantly correlated to smaller plaques and thicker fibrous caps, as well as better clinical outcomes after an acute vascular event, in a Tunisian cohort of patients with heart disease,[39] suggestive of a protective/stabilizing role.

Mice do not have a homologous gene to human MMP-1; therefore, to investigate the effect of MMP-1 in a mouse model Lemaitre *et al.* used an ApoE knockout ($ApoE^{-/-}$) mouse expressing human MMP-1 in the macrophages to investigate the effect of overexpression of human MMP-1 on atherosclerosis in the mouse aorta. The resultant atherosclerotic plaques were less extensive and less mature lesions compared to control with less collagen,[40] which supports the proposed protective role of MMP-1.

2.2 MMP-2

In comparison to other MMPs, the gelatinases, MMP-2 and MMP-9, have been extensively studied in vascular disease. Multiple studies have reported increased circulating MMP-2 levels in patients with atherosclerosis,[41,42] peripheral arterial disease in combination with type II diabetes[43] or acute MI.[44,45] In addition, MMP-2 expression or activity by peripheral blood mononuclear cells has been found to be higher in patients with atherosclerotic disease.[44,46] A handful of studies have reported increased MMP-2 levels with disease progression and plaque instability. Kai *et al.* reported increased

MMP-2 in serum taken from patients with acute coronary syndrome (ACS), including unstable angina and myocardial infarction, compared to stable angina.[41] Similarly in a cohort of patients with carotid atherosclerosis,[47] augmented serum MMP-2 levels in patients with prior neurological ischemia compared to asymptomatic controls was reported. Although interestingly, this difference was not observed when patients with unstable and stable carotid lesions were compared.[47] Furthermore, Fiotti *et al.* detected no significant difference in plasma MMP-2 levels when comparing patients with ACS and stable angina.[42]

Accumulating evidence suggests that polymorphisms in the MMP-2 gene may be associated with the prevalence or severity of atherosclerotic disease. An early study by Price *et al.*[48] identified multiple single-nucleotide polymorphisms (SNPs) within the human MMP-2 gene, and most intriguingly, the MMP-2 promoter.[48] The authors showed that one such promoter polymorphism, –1306C/T, impaired binding of the transcription factor stimulating protein-1 (Sp1) and blunted MMP-2 promoter activation.[48] More recently, T variant carriers have been reported to have reduced cardiovascular risk in a study of Polish Caucasian patients with type II diabetes, implying that low MMP-2 expression may be protective for atherosclerosis.[49] That said, cytosine to thymine variation at this site was not associated with differences in plasma levels of MMP-2 in an Iranian cohort.[44] Another promoter polymorphism, –790T/G, has also been studied in relation to cardiovascular disease. Vasku *et al.*[50] observed a greater than twofold increased risk of severe atherosclerosis, defined as stenosis in three coronary vessels: in T-allele carriers with differential binding of the transcription factors S8, gut-enriched Kruppel-like factor, and ectopic viral integration site 1-encoded factor, at this site in the MMP-2 promoter.[50] A more recent study reaffirmed the association between the –790T/G and cardiovascular risk in patients with angina pectoris.[51] It is tempting to speculate that this mutation may increase MMP-2 levels; however, Vasku *et al.*[50] did not investigate MMP-2 expression and previously Price and coworkers[48] observed no difference in MMP-2 promoter activity in in vitro transfection assays employing –790T or –790G constructs.[48,50] Therefore, although multiple polymorphisms have been identified in the MMP-2 gene, the exact effect of some variations on MMP-2 expression and their causal role requires further investigation.

In the human vasculature, luminal endothelial cells and medial VSMCs constitutively secrete MMP-2 in an inactive complex with TIMP-2.[52–54]

Expression of vascular MMP-2 is increased with age, especially in the intima.[55] Increased MMP-2 expression and activity are also observed following vascular injury and intimal thickening.[54] Similarly, MMP-2 expression and activity is increased in human atherosclerosis.[31,53,56] Within plaques, MMP-2 expression by luminal endothelial cells and medial VSMCs was maintained; however, high levels of MMP-2 were also expressed by macrophages and VSMCs within the diseased intima, as well as endothelial cell lining plaque microvessels.[31,53,57] Increased gelatinolytic activity has been observed in the core and shoulder of human atherosclerotic plaques.[53] As fibrous cap thinning at the shoulder region predisposes plaque rupture, this finding implies a role for the gelatinases in plaque instability.[53] Interestingly, Heo and colleagues identified a significant correlation between the level of MMP-2 protein in human carotid artery atherosclerotic lesions and contributors to plaque vulnerability, including ulceration, intraplaque hemorrhage, and fibrous cap thinning and rupture.[57] It said, this study could not detect differences in MMP-2 expression between plaques causing ischemic symptoms and asymptomatic plaques.[57] Similarly, Fiotti *et al.*[42] observed no difference in MMP-2 in fragments of coronary atherosclerotic plaques from patients with unstable angina or acute myocardial infarction (MI) compared to stable angina.[42] On the other hand, Sluijter and coworkers reported higher MMP-2 levels in fibrous VSMC-rich carotid plaques compared to less fibrous lesions,[58] while a study by Choudhary and colleagues observed more MMP-2 protein in carotid artery fatty streaks compared to more advanced plaques.[59] Therefore, the relationship between circulating or vascular MMP-2 levels and plaque stability remains controversial.

In human endothelial cells, Levkau and colleagues found that apoptosis following growth factor withdrawal was accompanied by translocation of MMP-2 to focal adhesion sites, binding of MMP-2 to α_v and β_1 integrins, and MMP-2 activity.[60] The authors showed that cell death could be rescued by inhibition of MMP activity or integrin–MMP binding, and hypothesized that once bound to integrins, MMP-2-mediated cleavage of the pericellular ECM may reduce prosurvival integrin–matrix signaling.[60] Meanwhile in VSMCs, many studies have investigated the role of MMP-2 in migration. In order to migrate into the intima and contribute to intimal thickening or atherosclerosis, the basket of basement membrane type IV collagen which surrounds medial VSMCs must first be degraded.[61] In vitro studies found that MMP-2 inhibition or deficiency could impair VSMC invasion through a reconstituted basement membrane, implying a role for MMP-2 in basement membrane degradation and VSMC migration.[62–64]

Similar to the aforementioned effect of injury in human vein,[54] multiple animal studies have reported increased MMP-2 expression or activity following vascular injury, especially in the neointima.[64–73] Following balloon catheter injury of the rat carotid artery or rabbit aorta, this increase in MMP-2 expression was temporally accompanied by increased MMP-14 expression.[68,69] As in vitro evidence has demonstrated a role for MMP-14 in MMP-2 activation at the cell surface,[29,74–77] these results imply that MMP-14 may be responsible for increased MMP-2 activity following injury.[68,69] Meanwhile, a causal role for MMP-2 in intimal thickening has been demonstrated by loss-of-function studies in vivo. Tsukioka and colleagues showed that MMP-2 knockdown, using ribozyme gene transfer, could inhibit coronary artery intimal thickening in mice following cardiac transplant.[78] While, two other groups described smaller neointimas with fewer intima cells following carotid artery ligation in mice with MMP-2 deficiency compared to wild types.[64,79] Crucially, inhibition of intimal thickening occurred in the absence of changes to neointimal or medial proliferation or leukocyte infiltration.[64] Together with the aforementioned in vitro studies, these reports suggest that MMP-2 promotes intimal thickening via basement membrane degradation and VSMC migration.

Studies in animal models have also shed light on the role of MMP-2 in atherosclerosis. MMP-2 was increased in animal models of atherosclerosis[80–82] and correlated with plaque instability.[81,83] Kuzuya and coworkers reported smaller aortic plaques, with thinner fibrous caps containing fewer VSMCs and less collagen, in high-fat diet-fed ApoE$^{-/-}$ mice with MMP-2 deficiency compared to MMP-2$^{+/+}$ controls.[84] These findings suggest that MMP-2 promotes VSMC migration into atherosclerotic plaques leading to formation of, and collagenase deposition within, the fibrous cap, hence promoting plaque stability.[84] Interestingly, the authors also reported reduced macrophage accumulation within the aortic sinus, but not the aortic arch, perhaps implying that MMP-2 may also affect macrophage infiltration in some areas of the vascular tree.[84] Furthermore, a role for MMP-2 in plaque calcification has also been proposed. Evidence suggests MMP-2 is expressed by procalcifying chondrocyte-like VSMCs in calcified aortic atherosclerotic lesions in mice.[85] Additionally, MMP-2$^{-/-}$ mice have reduced calcification and expression of the calcification-related proteins osteocalcin and bone morphogenetic protein-2 in aortic plaques compared to MMP-2 homozygous controls.[85] Together, these studies suggest that MMP-2 is involved in multiple aspects of atherosclerotic plaque progression and stability, including fibrous cap formation, collagen deposition, and calcification.

2.3 MMP-3

Elevated plasma levels of MMP-3 were with prior history of cardiovascular disease, aging, and in patients with increased carotid plaque burden, measured by MRI scan,[35] and patients with peripheral artery disease.[86] Moreover, high levels of active and total MMP-3 were found in the blood of patients with a high carotid plaque score.[87] In this study, there was a dose-dependent correlation between plasma MMP-3 and carotid plaque score, and was associated with a common MMP-3 polymorphism, −1612 6A6A, in this Taiwanese cohort.[87] In another study, plasma and peripheral blood mononuclear cells isolated from the blood of patients following acute MI and compared to controls with no MI and the level of active MMP-3 measured by ELISA and zymography was significantly elevated in acute MI patients.[44]

Studying the MMP-3 promoter polymorphism −1171 5A/6A in the Chinese population, Huang *et al.*[88] found increased propensity to ischemic stroke, making the presence of this polymorphism an independent risk factor. An association was also demonstrated between the frequency of this MMP-3 polymorphism and internal carotid artery stenosis.[89] However, a large study looking at several different SNPs in the MMP-3 gene failed to find any correlations with MI in a German population.[90] It is possible that differences in the outcome of polymorphism studies may be influenced by SNP location, ethnicity, or by other environmental or risk factors, thus explaining the discrepancies between these studies.[91]

Evidence from knockout mice reveals the involvement of MMP-3 in intimal thickening. Intimal thickening after carotid artery ligation model was significantly attenuated in MMP-3-deficient mice compared to wild-type controls.[92] Interestingly, deficiency of MMP-3 significantly reduced MMP-9 activation and cell migration both in vivo and in VSMCs in vitro, implying the effect of MMP-3 on intimal thickening was mediated via MMP-9 activity.[92]

In a mouse model of atherosclerosis, MMP-3-deficient mice were crossed with $ApoE^{-/-}$ mice fed a high-fat diet and atherosclerotic plaques were examined in the brachiocephalic artery. MMP-3 deficiency resulted in fourfold larger plaques, more buried layers, and less VSMCs within the plaques.[93] Silence and colleagues reported similar findings of larger lesions containing more collagen in the thoracic aorta of the same knockout mice. While in their control $ApoE^{-/-}$ MMP-3 wild-type mice, the authors observed colocalization of MMP-3 and uPA with macrophages in the vulnerable areas of plaque and suggested that MMP-3 was produced by macrophages and activated by uPA.[94] MMP-3 was also implicated in plaque progression in rabbits, as rabbits fed a high-fat diet had increased levels of plaque MMP-3.[95] Together, these studies suggest that MMP-3 protects against plaque growth; however, further research may be needed to clarify the effect of this MMP on plaque stability.

2.4 MMP-7

Elevated plasma MMP-7 levels were observed in patients with cardiovascular disease[35,96] and confirmed as a predictor of future mortality.[96] Plasma MMP-7 levels could predict the likelihood of a future cardiovascular event,[97] in a study measuring levels of MMPs in the plasma of diabetic patients. Within the atherosclerotic plaque, there are multiple targets for MMP-7 including MMPs 1, 2, and 9.[98,99] MMP-7 can be activated by MMP-3,[98] which is also present in atherosclerotic plaques.[100] MMP-7 activity was found in atherosclerotic plaques from patients, but not in healthy human arteries.[101] MMP-7 was augmented with carotid plaque burden and plaque calcification, as measured using MRI, and was also increased with age, hypertension, and diabetes.[35] Particularly, high levels of MMP-7 were seen in the macrophage-rich and collagen-poor areas of plaques from patients with carotid atherosclerosis who had experienced recent symptoms.[96] In an epigenetic study investigating a polymorphism in the MMP-7 gene, −181A>G, an association was found between A allele carriers and the presence of atherosclerotic plaque in the femoral artery, but not the carotid artery. This association was stronger in men than in women.[102]

Deficiency of MMP-7 in ApoE$^{-/-}$ mice resulted increased prevalence of VSMCs in brachiocephalic artery plaques, but without a change in plaque size.[93] Interestingly, MMP-7 has been shown to cleave N-cadherin, resulting in VSMC apoptosis, which may explain the increased VSMC number in the atherosclerotic plaques from MMP-7$^{-/-}$ApoE$^{-/-}$ mice.[101] MMP-7 can also cleave apolipoprotein A-IV (ApoA-IV), and therefore eliminate the antioxidant effect[103] and antiatherogenic effect of ApoA-IV; therefore, MMP-7 may be proatherogenic via a number of mechanisms.

2.5 MMP-8

Clinical studies have found that plasma MMP-8 correlates with cardiovascular disease.[104–106] Plasma MMP-8 is also increased with age, body mass index (BMI), and CRP, and decreased with high-density lipoprotein

(HDL) cholesterol.[35] Levels of MMP-8 in the plasma also correlated with a number of markers of inflammation and disease, such as CRP, urea, AST, and creatinine.[106] MMP-8 is minimally expressed in healthy arterial vessels, but significantly upregulated in human atherosclerotic plaques in VSMCs, macrophages, and endothelial cells, and particularly in the vulnerable shoulder regions of the plaques.[107] Interestingly, intraplaque MMP-8 levels have been shown to correlate with plaque progression,[108,109] while a number of statin drugs have been shown to reduce the amount of active MMP-8 in human atherosclerotic plaques.[110] Epigenetic studies have highlighted the potential importance of a MMP-8 polymorphism on the incidence of cardiovascular disease.[111]

MMP-8 has a number of actions that could affect atherosclerotic plaque progression and stability. The canonical function of MMP-8, degrading fibrillar collagens, will contribute to plaque destabilization by weakening the plaque fibrous cap. In fact, MMP-8 cleaves collagen I three times more effectively than MMP-1 or MMP-13.[106] In addition to this, MMP-8 can also act to convert angiotensin I (Ang-I) to angiotensin II (Ang-II), leading to downstream effects such as VCAM expression, macrophage recruitment, and angiogenesis. MMP-8 also cleaves ADAM 10, an enzyme that cleaves N-cadherin to activate β-catenin, resulting in smooth muscle cell migration and proliferation.[112] Additionally, MMP-8 can prevent cholesterol efflux from macrophages by cleavage of ApoA1, thus promoting cholesterol accumulation in the blood vessels.[113] Therefore, depletion of MMP-8 may facilitate cholesterol efflux, and therefore reduce the size of the necrotic core. Depletion of MMP-8 in mice was shown to decrease blood pressure via reducing the conversion of Ang-I to Ang-II[111] and mice exhibited lesions that had increased collagen, as expected with the loss of the collagenase. The lesions were also found to be significantly smaller than controls, due to reduced VSMC migration and proliferation, and had a reduced macrophage content.[111] Together, this implicates a proatherogenic role for MMP-8.

2.6 MMP-9

Circulating MMP-9 levels were increased in patients with coronary atherosclerosis,[41,42,114] acute MI,[44,45,115] and ischemic heart disease in combination with type II diabetes.[116] Additionally, MMP-9 expression and activity was also increased in peripheral blood mononuclear cells from patients with atherosclerosis.[45,46] There is also substantial evidence for a link between heightened MMP-9 and plaque vulnerability. For instance, increased circulating MMP-9 levels have been observed in patients with ACS compared to stable angina,[42,114] patients with previous neurological ischemic symptoms compared to asymptomatic controls,[47] and in patients with unstable compared to stable lesions.[47,117] Importantly, MMP-9 levels have even been reported to be indicative of the presence of a vulnerable plaque,[47] while serum levels of MMP-9 have been associated with increased risk of MI or stroke,[118] advanced plaque phenotype, or vulnerability.[119] In addition, Blankenberg and coworkers reported increased plasma MMP-9 in patients that went on to have a fatal cardiac event, compared to those who did not.[120] Therefore, a multitude of evidence suggests that MMP-9 is heightened in unstable vascular disease, and some authors have proposed that MMP-9 may be employed as a biomarker to identify patients at risk of plaque rupture.[117,119,121]

A −1562C/T SNP in the MMP-9 promoter has been widely studied in relation to vascular disease. Initial experiments by Zhang and colleagues found that transition to a T nucleotide at this site increased promoter activity and were associated with the severity of atherosclerotic disease, quantified by the number of coronary arteries with severe stenosis.[122] Later studies reported increased frequency of the T-allele or TT genotype in patients with atherosclerosis compared to undiseased controls,[89,123] although this was not observed when patients with MI were compared to controls.[44,122] As an extension of these findings, Morgan and colleagues calculated a 1.5-fold increased risk of coronary atherosclerosis in T-allele carriers,[123] while Biscetti and coworkers reported a greater than three times increased risk for unstable carotid atherosclerosis for TT homozygotes.[89]

Inactive pro-MMP-9 is expressed in very low amounts in undiseased blood vessels, mainly by medial VSMCs, luminal endothelial cells, and microvascular endothelial cells.[53,54] However, evidence suggests that MMP-9 is upregulated by surgical preparation, intimal thickening,[54] aging,[55] and in atherosclerosis.[31,53] In plaques, MMP-9 is expressed by multiple cell types including endothelia, macrophages, and VSMCs.[31,53,57] Galis and coworkers reported that VSMC-derived MMP-9 expression was induced by the inflammatory cytokines IL-1 and TNFα in vitro and colocalized to these cytokines in human atherosclerosis in vivo.[52,53] As previously mentioned, the authors also described gelatinolytic activity in the rupture-prone shoulder of human plaques, which was mirrored by MMP-9 protein in this region, supporting a role for MMP-9 in plaque instability.[53] In further support of this role enhanced MMP-9 was reported in plaque fragments from patients with ACS compared to stable angina,[42] in stable compared to unstable human carotid artery lesions,[31] as well as plaques with more macrophages and fewer VSMCs and reduced collagen deposition,[58] and even correlated with contributors to plaque vulnerability, including cap rupture, lipid core size, and macroscopic ulceration.[57] That said, when MMP-9 expression in carotid atherosclerotic lesions from symptomatic and asymptomatic patients was compared, findings of both no difference[57] and increased MMP-9 in symptomatic patients have been reported.[32] In the latter study, the authors identified colocalization between MMP-9 protein and triggering receptor expressed on myeloid cells-1 (TREM-1) in symptom-causing plaques.[32] This study went onto show that TNFα-mediated MMP-9 expression in primary VSMCs involved TREM-1 and multiple signaling pathways including p38 mitogen-activated protein kinase, JNK, PI3K, and NFκB, suggesting a role for these factors in plaque instability.[32]

Evidence suggests that MMP-9 plays a vital role in VSMC migration. Mason and colleagues found that rat VSMCs overexpressing MMP-9 had augmented migration through a collagen gel,[124] whereas Cho and Reidy, and Johnson and Galis reported diminished migration of primary VSMCs from MMP-9-deficient mice in a transwell and scratch wound assay, respectively.[73,79] Interestingly, VSMCs from these MMP-9 mice also exhibited reduced ability to assemble collagen monomers and adhere to gelatin.[79] The authors went onto show that MMP-9, in combination with the hyaluronan receptor CD44, mediated attachment of VSMCs to the ECM, thus permitting collagen assembly and contraction.[79] Evidence suggests that MMP-9 also promotes VSMC proliferation via cleavage of N-cadherin, thus activating β-catenin signaling and subsequent upregulation of the proproliferative gene cyclin-D1.[125,126] However, no difference in VSMC proliferation was observed by Johnson and colleagues following stimulation of cells with active MMP-9 or inhibition with MMP-9 siRNA.[92] Intriguingly, MMP-9 has also been shown to bind to HDL in vitro and impair the ability of this lipoprotein to inhibit LDL oxidation.[127] As MMP-9 is present in fractions of HDL from patients with coronary artery disease, but not healthy controls, further investigation into the relationship between MMP-9 and HDL may shed light onto the role of these MMPs in atherosclerosis development.[127]

To further understand the role of MMP-9 in vascular disease, animal studies of intimal thickening and atherosclerosis have been employed. MMP-9 expression and activity is increased in animal models of intimal thickening,[65,66,71–73,128] similar to human vessels.[54] Reports have demonstrated impaired intimal thickening and reduced neointimal cells in MMP-9-deficient mice, compared to wild-type controls, in response to catheter-induced denudation of carotid arteries[73] or carotid artery ligation.[79,92,129] Furthermore, reduced injury-induced intimal proliferation and cyclin-D1 protein upregulation has also been reported in these MMP-9-deficient mice.[73]

In atherosclerosis-prone ApoE$^{-/-}$ mice fed a high-fat diet, Luttun *et al.* described multiple effects of MMP-9 deficiency on plaque composition and stability, including reduced plaque amount, size, macrophage content, and fibrillar collagen deposition by VSMCs.[130] However, the exact effect of MMP-9 deficiency appeared to be dependent on the region of the aorta studied.[130] Similarly, Choi and coworkers reported reduced plaque volume with reduced foam cell macrophage, VSMC, and collagen content in carotid atherosclerotic lesions produced in MMP-9$^{-/-}$ApoE$^{-/-}$ mice by temporary carotid artery ligation and a western diet.[131] In these studies, bone marrow transplant experiments provided conflicting evidence whether MMP-9 production by bone marrow-derived cells or resident vascular cells is responsible for the observed effects of MMP-9 deficiency on limiting plaque growth.[130,131] On the other hand, Johnson *et al.* reported increased plaque area in brachiocephalic arteries from high-fat-fed MMP-9$^{-/-}$ApoE$^{-/-}$ mice with reduced VSMC content, increased macrophage content, and increased buried layers.[93] Thus, although the precise effect of lifelong MMP-9 deficiency on plaque size and macrophage content appears to depend on the vessel examined, MMP-9 deficiency generally resulted in reduced VSMC and collagen content.[93,130,131] That said, an altogether different role for MMP-9 has been described in mice with established atherosclerosis. Jin and colleagues reported that MMP-9 silencing, using lentiviral transfer of MMP-9 shRNA, in ApoE$^{-/-}$ mice previously fed a high-fat diet for 20 weeks, increased VSMC content and fibrous cap thickness in already established lesions, suggesting that MMP-9 may actually promote plaque rupture in late atherosclerosis.[132] Together, these studies suggest that lifelong MMP-9 deficiency may produce differing results compared to MMP-9 silencing in established disease. Similar findings have been reported in MMP-9 gain-of-function experiments. Lemaitre *et al.* observed increased aortic plaque collagen content in high-fat-fed ApoE$^{-/-}$ mice with macrophage-specific MMP-9 overexpression, but no effect on VSMC content or fibrous cap thickness or rupture compared to wild types.[133] Whereas Gough and coworkers found that bone marrow transplant of cells overexpressing autoactivating MMP-9 into ApoE$^{-/-}$ mice with established vascular disease promoted multiple characteristics of plaque instability including hemorrhage of plaque microvessels, fibrous cap disintegration, and even fibrin deposition.[134] Similarly, de Nooijer and colleagues found that in ApoE$^{-/-}$ mice with established carotid artery atherosclerosis intraluminal addition of an adenovirus to overexpress MMP-9 reduced cap thickness and increased prevalence of intraplaque hemorrhage.[135] Thus, together, these animal studies suggest that although MMP-9 may promote VSMC accumulation and collagen deposition in early disease, this enzyme is detrimental to plaque stability in late atherosclerosis.

Interestingly, adenoviral MMP-9 overexpression also induced outward vascular remodeling around intermediate atherosclerotic lesions, a phenomenon known to occur to limit luminal occlusion.[135,136] A similar increase in outward remodeling has been observed in balloon-injured rat carotid arteries following luminal seeding of rat VSMCs overexpressing MMP-9.[124] Furthermore, Lessner and colleagues described reduced outward remodeling following carotid artery ligation and high-fat diet in MMP-9$^{-/-}$ApoE$^{-/-}$ mice compared to MMP-9$^{+/+}$ApoE$^{-/-}$ controls.[137] Together, these studies show that MMP-9 has multiple roles in vascular pathology, which may differ throughout the stages of plaque progression.

2.7 MMP-10

Plasma MMP-10 was elevated in patients with peripheral arterial disease and acted as a predictor of both overall and cardiovascular disease related mortality.[138] In chronic kidney disease patients, atherosclerosis severity correlated to levels of MMP-10.[139] While in a different study employing patients with subclinical atherosclerosis, plasma MMP-10 correlated to carotid plaque size as well as levels of inflammatory markers such as CRP.[36,140] MMP-10 is known to be expressed in endothelial cells, monocytes, and macrophages,[36,141,142] and can activate MMP 1, 7, 8, and 9.[143] It colocalizes with COX-2 and NFκB in plaques, has a fibrinolytic role, and can be induced by thrombin.[144] Consequently, it is proposed that MMP-10 is proatherogenic.

2.8 MMP-11

MMP-11 is released in its active form and is expressed in human carotid atherosclerotic plaques in smooth muscle cells, macrophages, and endothelial cells, while healthy arteries and fatty streaks showed no expression. MMP-11 was induced by CD40 ligand in human smooth muscle cells, macrophages, and endothelial cells grown in culture.[145] MMP-11 has also been detected in all cell types in atherosclerotic lesions in the aortic arch of proatherosclerotic LDL receptor (LDLR)-deficient mice fed a high-fat diet. Treatment with anti-CD40L antibody reduced MMP-11 and atherosclerotic lesion area.[145,146] Meanwhile, in a model of intimal thickening, MMP-11-deficient mice exhibited significantly enhanced intimal thickening compared to wild-type controls following electric injury to the femoral artery. This was associated with increased cell number in the intima and increased numbers of VSMCs implying increased cell migration from the media, as there was no detectable proliferation in this study. Arteries from these MMP-11-deficient mice also had increased elastin degradation in the vessel wall.[147] This in vivo finding was surprising as in vitro studies have shown MMP-11 exhibits proteolytic activity,[147] suggesting this enzyme has other overriding effects in vivo. MMP-11 does not cleave any collagens, but can inactivate serine proteinase inhibitors (serpins), which regulate cellular functions involved in atherosclerosis[145,148] and insulin-like growth factor-binding proteins (IGFBPs).[149] However, further work is necessary to establish the exact role of this MMP in atherosclerosis.

2.9 MMP-12

Using a proteomics approach, it has been demonstrated in patients with carotid artery plaques there is an association with plasma levels of MMP-12.[150] Moreover, plasma samples of diabetic patients revealed that MMP-12 was higher in type II diabetics and was independently associated with cardiovascular disease. The study also found that plasma MMP-12 levels could predict the likelihood of a future cardiovascular event.[97] In addition, an epigenetic study by Panayiotou and coworkers[102] found an association between the MMP-12 polymorphism –82A>G and the presence of atherosclerotic plaque in the femoral arteries of women but not men. Additionally, MMP-12 mRNA was enhanced in human carotid plaques compared to healthy arteries and was also significantly increased in vulnerable plaques compared to stable plaques.[31]

No significant effect on carotid artery ligation-induced intimal thickening was observed in mice with MMP-12 deficiency[92]; however, in MMP-$12^{-/-}$ApoE$^{-/-}$ mice fed a high-fat diet, deficiency in MMP-12 resulted in smaller brachiocephalic artery plaques. These plaques also showed increased signs of stability such as less buried fibrous layers, more VSMCs, and less macrophages.[93] Use of an MMP-12 inhibitor confirmed this effect as it also reduced atherosclerotic plaque development and increased plaque stability in a fat fed ApoE$^{-/-}$ mouse model. These effects appeared to be mainly mediated through changes in macrophages, thus explaining the contrast in effect compared to that seen in the neointimal formation model. Inhibition of MMP-12 reduced the invasive capacity of macrophages as well as decreasing apoptosis.[151,152] In a similar model examining plaques in the thoracic aorta, Luttun and colleagues did not observe a difference in lesion size in MMP-12-deficient mice, while in control ApoE$^{-/-}$ mice, MMP-12 was upregulated in the atherosclerotic plaques compared to healthy arteries. Deletion of MMP-12 also had no effect on macrophage or collagen content in this study, but did protect against the loss of elastic lamellae in the region of the plaque.[130]

In hypercholesterolemic transgenic rabbits overexpressing human MMP-12 in tissue macrophages, atherosclerosis in the aorta, and coronary arteries was enhanced and lesions exhibited increased macrophage infiltration that was associated with accelerated degradation of medial elastic laminae in advanced atherosclerosis. The increased MMP-12 also led to augmented expression of MMP-3[153,154] and activation of both MMP-2 and MMP-3.[155] Together, this data highlights the proatherogenic role of MMP-12 and identifies it as a target for selective inhibition.

2.10 MMP-13

In human carotid artery plaques, no connection was found between levels of MMP-13 and the level of atherosclerotic disease or patients symptoms. MMP-13 was also expressed at considerably lower levels in the plaques compared to the levels of MMPs 1 and 8, and TIMPs 1 and 2.[109]

As MMP-13 is a member of the collagenase family, inhibition of MMP-13 increases the amount of collagen present in mouse plaques.[144] MMP-13 was present in all cell types[156] and was secreted by macrophages in mouse atherosclerotic plaques.[157] This discovery led to studies by Quillard *et al.*[157,158] to elucidate the influence of MMP-13 on atherosclerotic plaque development and progression in ApoE$^{-/-}$ mice fed a high-fat diet. They found that when mice were given an MMP-13 inhibitor, this effectively reduced MMP-13 activity in plaques and increased plaque interstitial collagen; however, somewhat surprisingly, the inhibitor did not affect plaque size or macrophage accumulation. Improvements in plaque stability were observed through increased collagen in the fibrous cap of the plaques, resulting in larger and thicker fibrous caps.[157] Later, both Quillard *et al.* and Deguchi *et al.* used a similar $ApoE^{-/-}$ mouse model, but this time crossed these mice with MMP-13-deficient mice. The lesions in the aortic roots had, as expected, increased fibrillar collagen compared to $ApoE^{-/-}$ controls. Further analysis showed that there was no collagenolytic activity, measured by in situ zymography, in these plaques. However, no difference in plaque size was seen in the aortic root plaques or those found in the descending aorta or the brachiocephalic arteries. In addition, no effect was seen on macrophage accumulation in this study, however, a reduction in necrotic core size was reported in the MMP-13-deficient mice.[158,159] MMP-13 has been implicated as a mediator of the ability of uPA to accelerate atherosclerosis. In $ApoE^{-/-}$ mice, MMP inhibition could reverse the proatherosclerotic effect of uPA overexpression in mouse macrophages and the effect was found to be predominantly via MMP-13.[160] MMP-13 is also known to have a number of noncanonical downstream actions, including the cleavage of ICAM-1 in the vascular endothelium[161] and stimulation of VSMC migration via Akt/ERK.[162] Consequently the human and mouse data are contradictory and highlights the need for caution in extrapolating animal results to the human.

2.11 MT-MMPs

MT-MMPs are anchored to the cell membrane rather than being soluble proteases that are released. As with all MMPs, MT-MMPs are inhibited by TIMPs. However, transmembrane MT-MMPs are not inhibited by TIMP-1, whereas GPI anchored are inhibited by all TIMPs.[163]

2.12 MMP-14

MMP-14 (also known as MT1-MMP) is inhibited by TIMP-2, but not TIMP-1, and activates MMP-2 and MMP-13.[29,74–77,164]

MMP-14 mRNA was increased in carotid plaques compared to healthy arteries and was also significantly increased in vulnerable compared to stable looking plaques[31] and found within the vulnerable shoulder regions of these plaques.[165,166] MMP-14 is the dominant MT-MMP in both monocytes and macrophages and promotes monocyte invasion and recruitment.[167,168] Interestingly, carrying at least one allele of +7096T>C polymorphism in the MMP-14 gene has been associated with lower risk of a vulnerable plaque in the carotid artery, implying a role for this MMP in plaque stability.[169]

MMP-14-deficient mice die by 3 weeks of age, so in order to study the effect of MMP-14 loss in macrophages Schneider *et al.*[170] created a mouse model using LDL-deficient mice that were irradiated and the bone marrow repopulated with either normal or MMP-14-deficient bone marrow. Deletion of macrophage MMP-14 in this way did not alter aortic root plaque size after 16 weeks of high-fat feeding. It did, however, increase plaque interstitial collagen. The deficient macrophages were found to have less collagenase activity compared to those with MMP-14. The aortic plaques with MMP-14-deficient macrophages showed less activation of MMP-13, but there was no change in activation of MMP-2 or MMP-8. On the other hand, increasing the amount of MMP-14 in the macrophages using anti-microRNA-24 promoted macrophage invasion and increased plaque size and markers of instability.[171]

Meanwhile, in hypercholesterolemic rabbit, aortic atherosclerotic lesions the amount of MMP-14 detected were correlated to the severity of the atherosclerotic plaques present.[81,164] Although conversely Liu and coworkers found that MMP-14 decreased as aortic atherosclerotic lesions developed in rabbits,[95] MMP-14 was found in vulnerable, macrophage-rich areas of the atherosclerotic plaques, where it correlated with levels of MMP-2 and COX-2.[81] MMP-14 interacts with LOX-1 to activate signaling pathways in the presence of oxidized LDL. MMP-14 can act via a wide range of downstream pathways, including RhoA/Rac1, ROS generation, and Akt signaling.[172] MMP-14 can cleave ApoA-IV, and therefore remove its antioxidant effect.[103] Together, these studies imply a proatherogenic role for MMP-14.

2.13 MMP-16

MMP-16 (MT3-MMP) is abundantly expressed in VSMCs of normal healthy arteries. In human atherosclerotic plaques obtained from autopsy samples, MMP-16 was found to colocalize with both smooth VSMCs and macrophages. In addition, in vitro human monocyte-derived macrophages have been shown to contain active MMP-16 protein. MMP-16 may be involved in atherosclerosis but currently supportive data are lacking.

3. TIMPs

3.1 TIMP-1

TIMP-1 is increased with age, BMI, hypertension, and CRP, as well as in patients with increased plaque burden and increased lipid core size, as measured by MRI.[35] TIMP-1 mRNA was also increased in samples of plaque debris obtained from patients with both ACS and stable angina compared to mRNA from healthy internal mammary arteries. Intriguingly, the highest levels of TIMP-1 were found in the debris from patients with ACS, suggesting that unstable plaques exhibit higher TIMP-1 compared to more stable lesions.[42] Another study showed that hypertensive patients had significantly less TIMP-1 mRNA in the plasma compared to normotensive controls, but this did not change in hypertensive patients with atherosclerosis.[173]

Deletion of TIMP-1 in ApoE$^{-/-}$ mice fed a high-fat diet did not affect brachiocephalic atherosclerotic plaque size compared to control ApoE$^{-/-}$ mice after 8 weeks of high-fat diet. However, deletion of TIMP-1 resulted in reduced VSMC numbers within the plaque compared to wild-type controls. TIMP-1-deficient mice had no other significant plaque changes compared to wild types.[167] Similarly, Lemaitre *et al.* observed no difference in aortic lesion size, macrophage content, or the amount of collagen in the lesions in TIMP-1$^{-/-}$ApoE$^{-/-}$ mice compared to wild-type controls. However, TIMP-1-deficient mice had increased propensity for degradation of the elastic lamellae in the aorta.[174] Conversely, using the same mouse model of atherosclerosis, Silence and coworkers found that deletion of TIMP-1 reduced the size of lesions in the thoracic aorta, which also exhibited more lipid staining and enhanced MMP activity compared to controls.[175] These data may appear counterintuitive in the light of the prior data, which highlights the involvement of MMPs in atherosclerosis, but may be due to compensation by other TIMPs.

3.2 TIMP-2

Overexpression of TIMP-2 reduced brachiocephalic lesion area in ApoE$^{-/-}$ mice on high-fat diet. TIMP-2 also stabilized plaques. These effects were thought to be via inhibition of macrophage migration and apoptosis.[176] While a similar study by Rouis *et al.* found that overexpression of TIMP-1 did in fact reduce lesion size in the aortae of mice with overexpression of TIMP-1.[177] The difference in outcome of the two studies may illustrate the differences seen by looking at different sites of atherosclerosis or the more dominant role of TIMP-2 in the inhibition of proatherogenic MMPs.

Deletion of TIMP-2 in ApoE$^{-/-}$ mice fed a high-fat diet did not significantly affect brachiocephalic atherosclerotic plaque size compared to controls, but did lead to increased markers of instability including a higher macrophage/VSMC ratio, less VSMCs, increased necrotic core size, less collagen, and increased apoptosis. TIMP-2 specifically decreased MMP-14-dependent monocyte/macrophage infiltration into atherosclerotic lesions. The absence of TIMP-2 resulted in more invasive macrophages due to increased action of MMP-14.[167] Clearly, TIMP-2 is a powerful inhibitor of MMPs that promote atherogenesis.

3.3 TIMP-3

TIMP-3 was increased in atherosclerotic plaques compared to healthy arteries in human endarterectomy samples; however, levels of this inhibitor were decreased in plaques described as vulnerable on visual inspection compared to stable plaques.[31]

Overexpression of TIMP-3 in mouse macrophages of LDLR-deficient mice resulted in smaller aortic atherosclerotic plaques compared to controls. Macrophage-specific overexpression of TIMP-3 resulted in a more stable phenotype, with more intraplaque collagen, smaller necrotic cores, and fewer T-cells and macrophages. They also had reduced signs of oxidative stress.[178] In contrast, the deletion of TIMP-3 in an $ApoE^{-/-}$ mouse model increased atherosclerosis and plaque macrophages.[179] Together, this implicates TIMP-3 as an inhibitor of atherosclerosis.

4. CONCLUSION

This review has discussed the evidence for multiple roles of the MMP family in atherosclerotic plaque development, progression, and rupture (summarized in Table 3 and Fig. 4). It is evident that these roles may change temporally and may act in concert or opposition with other MMP members.

Based on this knowledge, future studies could investigate the possibility of utilizing MMPs as biomarkers of disease. Many of the MMPs discussed here, including MMPs 1, 2, 3, 7, 8, 9, 10, and 12, are increased in the circulation of patients with cardiovascular disease compared to in healthy controls, and hence may be utilized as potential biomarkers in the future. Alternatively, some MMPs, such as MMP-9, have been proposed to identify patients at risk of plaque rupture and subsequent ischemia.[117,119,121] Alternatively, levels of MMPs could be imaged in plaques themselves. For instance, a recent study by Qin and colleagues has described a method of visualizing MMP-2, as a marker of plaque severity and instability, in patients using photoacoustic imaging combined with gold nanorods linked to anti-MMP-2 antibodies.[180]

Table 3 The Effect of Loss of Function and Gain of Function of MMPs and TIMPs in Animal Models of Atherosclerosis

MMP	Effect of Loss of Function	Effect of Gain of Function
MMP-1	—	↓ size = stability[40]
MMP-2	↓ size ↓ stability[84] ↓ size ↓ stability[85]	—
MMP-3	↑ size ↓ stability[93] ↑ size ↑ stability[94]	—
MMP-7	= size ↑ stability[93]	—
MMP-8	↓ size ↑ stability[111] = size = stability[158]	—
MMP-9	↓ size = stability[130]	= size ↓ stability[134]
	↓ size = stability[131]	= size ↓ stability[135]
	↑ size ↓ stability[93] ? size ↑ stability[132]	
MMP-12	↓ size ↑ stability[93]	↑ size = stability[153]
	↓ size ↑ stability[151]	↑ size = stability[154]
	= size = stability[130]	
MMP-13	= size ↑ stability[157] = size ↑ stability[158] = size ↑ stability[159]	
MMP-14	= size ↑ stability[170]	↑ size ↓ stability[171]
TIMP-1	= size ↓ stability[167]	= size = stability[176]
	= size = stability[174] ↓ size ↓ stability[175]	
TIMP-2	= size ↓ stability[167]	↓ size ↑ stability[176]
TIMP-3	↑ size ↓ stability[179]	↓ size ↑ stability[178]

Alternatively, targeting MMPs could be therapeutic in cardiovascular disease. For reviews on MMP inhibitors in cardiovascular disease, see Newby.[181] However, briefly, evidence suggests that due to the complex and sometimes opposing roles of MMP members, broad spectrum MMP inhibitors do not beneficially affect restenosis, atherosclerosis development, or plaque stability in animal models[182–184] and are hindered with side effects in patients.[185] Thus, more specific methods of targeting the activity of

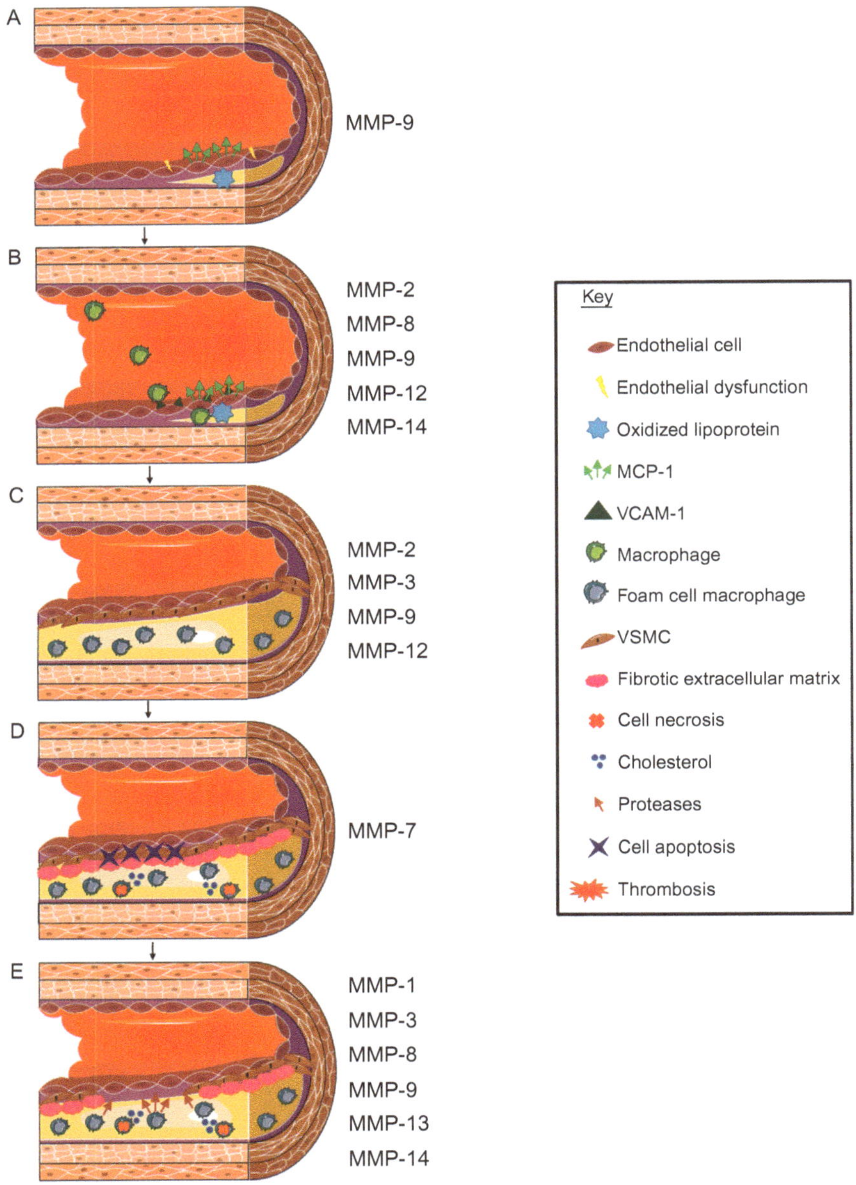

Fig. 4 Involvement of MMPs in multiple stages of atherosclerotic plaque progression. Images adapted from www.servier.com.

detrimental MMPs is necessary.[93,181] For instance, MMP-12 inhibition has been shown to inhibit lesion development and plaque stability in ApoE$^{-/-}$ mice[151]; however, MMP-13 inhibition did not affect plaque size, but did increase plaque collagen levels.[157] In addition to inhibitors, MMP expression may be reduced by nanoparticles carrying specific small interfering RNAs to target specific MMPs.[186,187] Alternatively, as exercise has been shown to reduce aortic plaque gelatinase activity in a murine model of atherosclerosis, life style changes may also represent a route to target MMP expression and activity in patients with vascular disease.[188] Alternatively, endogenous MMPs could be employed to cleave therapeutic drugs, such as antiproliferative drugs, from stents to specifically target MMP-expressing cells in restenosis.[189] Thus, the studies discussed here have prepared the ground for exciting new research into the use of MMPs as tools for atherosclerosis diagnosis, prognosis, and treatment.

REFERENCES

1. Townsend N, Williams J, Bhatnagar P, Wickramasinghe K, Rayner M. *Cardiovascular Disease Statistics 2014*. London: British Heart Foundation; 2014.
2. BHF. *BHF Cardiovascular Disease Statistics UK Factsheet 2015*. Oxford: British Heart Foundation; 2015.
3. Pugsley MK, Tabrizchi R. The vascular system. An overview of structure and function. *J Pharmacol Toxicol Methods*. 2000;44(2):333–340.
4. Jacob MP. Extracellular matrix remodeling and matrix metalloproteinases in the vascular wall during aging and in pathological conditions. *Biomed Pharmacother*. 2003;57(5–6):195–202.
5. Strong JP, Malcom GT, McMahan CA, et al. Prevalence and extent of atherosclerosis in adolescents and young adults: implications for prevention from the Pathobiological Determinants of Atherosclerosis in Youth Study. *JAMA*. 1999;281(8):727–735.
6. Lusis AJ. Atherosclerosis. *Nature*. 2000;407(6801):233–241.
7. Heo KS, Fujiwara K, Abe J. Shear stress and atherosclerosis. *Mol Cells*. 2014;37(6):435–440.
8. Schwartz SM, deBlois D, O'Brien ER. The intima. Soil for atherosclerosis and restenosis. *Circ Res*. 1995;77(3):445–465.
9. Wallitt EJ, Jevon M, Hornick PI. Therapeutics of vein graft intimal hyperplasia: 100 years on. *Ann Thorac Surg*. 2007;84(1):317–323.
10. Thyberg J, Blomgren K, Hedin U, Dryjski M. Phenotypic modulation of smooth muscle cells during the formation of neointimal thickenings in the rat carotid artery after balloon injury: an electron-microscopic and stereological study. *Cell Tissue Res*. 1995;281(3):421–433.
11. Owens GK, Kumar MS, Wamhoff BR. Molecular regulation of vascular smooth muscle cell differentiation in development and disease. *Physiol Rev*. 2004;84(3):767–801.
12. Lacolley P, Regnault V, Nicoletti A, Li Z, Michel J-B. The vascular smooth muscle cell in arterial pathology: a cell that can take on multiple roles. *Cardiovasc Res*. 2012;95(2):194–204.
13. Kovacic JC, Moreno P, Nabel EG, Hachinski V, Fuster V. Cellular senescence, vascular disease, and aging: part 2 of a 2-part review: clinical vascular disease in the elderly. *Circulation*. 2011;123(17):1900–1910.
14. Najjar SS, Scuteri A, Lakatta EG. Arterial aging: is it an immutable cardiovascular risk factor? *Hypertension*. 2005;46(3):454–462.
15. Wang JC, Bennett M. Aging and atherosclerosis: mechanisms, functional consequences, and potential therapeutics for cellular senescence. *Circ Res*. 2012;111(2):245–259.
16. McLaren JE, Michael DR, Ashlin TG, Ramji DP. Cytokines, macrophage lipid metabolism and foam cells: implications for cardiovascular disease therapy. *Prog Lipid Res*. 2011;50(4):331–347.

17. Libby P. Inflammation in atherosclerosis. *Arterioscler Thromb Vasc Biol.* 2012;32(9):2045–2051.
18. Tabas I, Garcia-Cardena G, Owens GK. Recent insights into the cellular biology of atherosclerosis. *J Cell Biol.* 2015;209(1):13–22.
19. Clarke M, Bennett M. The emerging role of vascular smooth muscle cell apoptosis in atherosclerosis and plaque stability. *Am J Nephrol.* 2006;26(6):531–535.
20. Bennett MR, Sinha S, Owens GK. Vascular smooth muscle cells in atherosclerosis. *Circ Res.* 2016;118(4):692–702.
21. Burke AP, Kolodgie FD, Farb A, et al. Healed plaque ruptures and sudden coronary death: evidence that subclinical rupture has a role in plaque progression. *Circulation.* 2001;103(7):934–940.
22. Newby AC. Dual role of matrix metalloproteinases (matrixins) in intimal thickening and atherosclerotic plaque rupture. *Physiol Rev.* 2005;85(1):1–31.
23. Shofuda K, Nagashima Y, Kawahara K, Yasumitsu H, Miki K, Miyazaki K. Elevated expression of membrane-type 1 and 3 matrix metalloproteinases in rat vascular smooth muscle cells activated by arterial injury. *Lab Investig.* 1998;78(8):915–923.
24. Pei D, Weiss SJ. Furin-dependent intracellular activation of the human stromelysin-3 zymogen. *Nature.* 1995;375:244–247.
25. Sato H, Kinoshita T, Takino T, Nakayama K, Seiki M. Activation of a recombinant membrane type 1-matrix metalloproteinase (MT1-MMP) by furin and its interaction with tissue inhibitor of metalloproteinases (TIMP)-2. *FEBS Lett.* 1996;393(1):101–104.
26. Nagase H. Activation mechanisms of matrix metalloproteinases. *Biol Chem.* 1997;378:151–160.
27. Van Wart HE, Birkedal-Hansen H. The cysteine switch: a principle of regulation of metalloproteinase activity with potential applicability to the entire matrix metalloproteinase gene family. *Proc Natl Acad Sci USA.* 1990;87:5578–5582.
28. Willenbrock F, Murphy G. Structure-function relationships in the tissue inhibitors of metalloproteinases. *Am J Respir Crit Care Med.* 1994;150:S165–S170.
29. Strongin AY, Collier I, Bannikov G, Marmer BL, Grant GA, Goldberg GI. Mechanism of cell-surface activation of 72 kDa type IV collagenase—isolation of the activated form of membrane metalloprotease. *J Biol Chem.* 1995;270(10):5331–5339.
30. Murphy G, Willenbrock F, Ward RV, Cockett MI, Eaton D, Docherty AJP. The C-terminal domain of 72 kDa gelatinase A is not required for catalysis, but is essential for membrane activation and modulates interactions with tissue inhibitors of metalloproteinases. *Biochem J.* 1992;283:636–641.
31. Müller A, Krämer SD, Meletta R, et al. Gene expression levels of matrix metalloproteinases in human atherosclerotic plaques and evaluation of radiolabeled inhibitors as imaging agents for plaque vulnerability. *Nucl Med Biol.* 2014;41(7):562–569.
32. Rao VH, Rai V, Stoupa S, Subramanian S, Agrawal DK. Tumor necrosis factor-α regulates triggering receptor expressed on myeloid cells-1-dependent matrix metalloproteinases in the carotid plaques of symptomatic patients with carotid stenosis. *Atherosclerosis.* 2016;248:160–169.
33. Rao VH, Kansal V, Stoupa S, Agrawal DK. MMP-1 and MMP-9 regulate epidermal growth factor-dependent collagen loss in human carotid plaque smooth muscle cells. *Physiol Rep.* 2014;2(2):e00224.
34. Lehrke M, Greif M, Broedl UC, et al. MMP-1 serum levels predict coronary atherosclerosis in humans. *Cardiovasc Diabetol.* 2009;8:50.
35. Gaubatz JW, Ballantyne CM, Wasserman BA, et al. Association of circulating matrix metalloproteinases with carotid artery characteristics: the Atherosclerosis Risk in Communities Carotid MRI Study. *Arterioscler Thromb Vasc Biol.* 2010;30(5):1034–1042.

36. Montero I, Orbe J, Varo N, et al. C-reactive protein induces matrix metalloproteinase-1 and -10 in human endothelial cells: implications for clinical and subclinical atherosclerosis. *J Am Coll Cardiol*. 2006;47(7):1369–1378.
37. Cavusoglu E, Marmur JD, Hegde S, et al. Relation of baseline plasma MMP-1 levels to long-term all-cause mortality in patients with known or suspected coronary artery disease referred for coronary angiography. *Atherosclerosis*. 2015;239(1):268–275.
38. Jguirim-Souissi I, Jelassi A, Slimani A, et al. Matrix metalloproteinase-1 and matrix metalloproteinase-12 gene polymorphisms and the outcome of coronary artery disease. *Coron Artery Dis*. 2011;22(6):388–393.
39. Fiotti N, Moretti ME, Bussani R, et al. Features of vulnerable plaques and clinical outcome of UA/NSTEMI: relationship with matrix metalloproteinase functional polymorphisms. *Atherosclerosis*. 2011;215(1):153–159.
40. Lemaitre V, O'Byrne TK, Borczuk AC, Okada Y, Tall AR, D'Armiento J. ApoE knockout mice expressing human matrix metalloproteinase-1 in macrophages have less advanced atherosclerosis. *J Clin Invest*. 2001;107(10):1227–1234.
41. Kai H, Ikeda H, Yasukawa H, et al. Peripheral blood levels of matrix metalloproteases-2 and -9 are elevated in patients with acute coronary syndromes. *J Am Coll Cardiol*. 1998;32(2):368–372.
42. Fiotti N, Altamura N, Orlando C, et al. Metalloproteinases-2, -9 and TIMP-1 expression in stable and unstable coronary plaques undergoing PCI. *Int J Cardiol*. 2008;127(3):350–357.
43. Signorelli SS, Malaponte G, Libra M, et al. Plasma levels and zymographic activities of matrix metalloproteinases 2 and 9 in type II diabetics with peripheral arterial disease. *Vasc Med*. 2005;10(1):1–6.
44. Ghaderian SM, Akbarzadeh Najar R, Tabatabaei Panah AS. Genetic polymorphisms and plasma levels of matrix metalloproteinases and their relationships with developing acute myocardial infarction. *Coron Artery Dis*. 2010;21(6):330–335.
45. Ghaderian SM, Najar RA, Tabatabaei Panah AS, et al. Matrix metalloproteinase: investigation from gene to protein as effective factor in myocardial infarction. *J Thromb Thrombolysis*. 2010;30(4):404–411.
46. Dabek J, Glogowska-Ligus J, Szadorska B. Transcription activity of MMP-2 and MMP-9 metalloproteinase genes and their tissue inhibitor (TIMP-2) in acute coronary syndrome patients. *J Postgrad Med*. 2013;59(2):115–120.
47. Alvarez B, Ruiz C, Chacon P, Alvarez-Sabin J, Matas M. Serum values of metalloproteinase-2 and metalloproteinase-9 as related to unstable plaque and inflammatory cells in patients with greater than 70% carotid artery stenosis. *J Vasc Surg*. 2004;40(3):469–475.
48. Price SJ, Greaves DR, Watkins H. Identification of novel, functional genetic variants in the human matrix metalloproteinase-2 gene: role of Sp1 in allele-specific transcriptional regulation. *J Biol Chem*. 2001;276(10):7549–7558.
49. Buraczynska M, Dragan M, Buraczynska K, Orlowska-Kowalik G, Ksiazek A. Matrix metalloproteinase-2 (MMP-2) gene polymorphism and cardiovascular comorbidity in type 2 diabetes patients. *J Diabetes Complicat*. 2015;29(6):829–833.
50. Vasku A, Goldbergova M, Izakovicova Holla L, et al. A haplotype constituted of four MMP-2 promoter polymorphisms (-1575G/A, -1306C/T, -790T/G and -735C/T) is associated with coronary triple-vessel disease. *Matrix Biol*. 2004;22(7):585–591.
51. Vasku A, Bienertova-Vasku J, Parenica J, et al. Central pulse pressure and variability in matrix metalloproteinases genes and their inhibitors in patients with ischemic heart disease. *Physiol Res*. 2014;63(suppl 4):S497–S507.
52. Galis ZS, Muszynski M, Sukhova GK, et al. Cytokine-stimulated human vascular smooth muscle cells synthesize a complement of enzymes required for extracellular matrix digestion. *Circ Res*. 1994;75(1):181–189.

53. Galis ZS, Sukhova GK, Lark MW, Libby P. Increased expression of matrix metalloproteinases and matrix degrading activity in vulnerable regions of human atherosclerotic plaques. *J Clin Invest.* 1994;94(6):2493–2503.
54. George SJ, Zaltsman AB, Newby AC. Surgical preparative injury and neointima formation increase MMP-9 expression and MMP-2 activation in human saphenous vein. *Cardiovasc Res.* 1997;33(2):447–459.
55. Wang M, Zhang J, Jiang LQ, et al. Proinflammatory profile within the grossly normal aged human aortic wall. *Hypertension.* 2007;50(1):219–227.
56. Li Z, Li L, Zielke HR, et al. Increased expression of 72-kd type IV collagenase (MMP-2) in human aortic atherosclerotic lesions. *Am J Pathol.* 1996;148(1):121–128.
57. Heo SH, Cho CH, Kim HO, et al. Plaque rupture is a determinant of vascular events in carotid artery atherosclerotic disease: involvement of matrix metalloproteinases 2 and 9. *J Clin Neurol.* 2011;7(2):69–76.
58. Sluijter JP, Pulskens WP, Schoneveld AH, et al. Matrix metalloproteinase 2 is associated with stable and matrix metalloproteinases 8 and 9 with vulnerable carotid atherosclerotic lesions: a study in human endarterectomy specimen pointing to a role for different extracellular matrix metalloproteinase inducer glycosylation forms. *Stroke.* 2006;37(1): 235–239.
59. Choudhary S, Higgins CL, Chen IY, et al. Quantitation and localization of matrix metalloproteinases and their inhibitors in human carotid endarterectomy tissues. *Arterioscler Thromb Vasc Biol.* 2006;26(10):2351–2358.
60. Levkau B, Kenagy RD, Karsan A, et al. Activation of metalloproteinases and their association with integrins: an auxiliary apoptotic pathway in human endothelial cells. *Cell Death Differ.* 2002;9(12):1360–1367.
61. Aguilera CM, George SJ, Johnson JL, Newby AC. Relationship between type IV collagen degradation, metalloproteinase activity and smooth muscle cell migration and proliferation in cultured human saphenous vein. *Cardiovasc Res.* 2003;58(3): 679–688.
62. Pauly RR, Passaniti A, Bilato C, et al. Migration of cultured vascular smooth muscle cells through a basement membrane barrier requires type IV collagenase activity and is inhibited by cellular differentiation. *Circ Res.* 1994;75(1):41–54.
63. Cheng L, Mantile G, Pauly R, et al. Adenovirus-mediated gene transfer of the human tissue inhibitor of metalloproteinase-2 blocks vascular smooth muscle cell invasiveness in vitro and modulates neointimal development in vivo. *Circulation.* 1998;98(20): 2195–2201.
64. Kuzuya M, Kanda S, Sasaki T, et al. Deficiency of gelatinase a suppresses smooth muscle cell invasion and development of experimental intimal hyperplasia. *Circulation.* 2003;108(11):1375–1381.
65. Bendeck MP, Zempo N, Clowes AW, Galardy RE, Reidy MA. Smooth muscle cell migration and matrix metalloproteinase expression after arterial injury in the rat. *Circ Res.* 1994;75(3):539–545.
66. Zempo N, Kenagy RD, Au YP, et al. Matrix metalloproteinases of vascular wall cells are increased in balloon-injured rat carotid artery. *J Vasc Surg.* 1994;20(2):209–217.
67. Strauss BH, Robinson R, Batchelor WB, et al. In vivo collagen turnover following experimental balloon angioplasty injury and the role of matrix metalloproteinases. *Circ Res.* 1996;79(3):541–550.
68. Jenkins GM, Crow MT, Bilato C, et al. Increased expression of membrane-type matrix metalloproteinase and preferential localization of matrix metalloproteinase-2 to the neointima of balloon-injured rat carotid arteries. *Circulation.* 1998;97(1):82–90.
69. Wang H, Keiser JA. Expression of membrane-type matrix metalloproteinase in rabbit neointimal tissue and its correlation with matrix-metalloproteinase-2 activation. *J Vasc Res.* 1998;35(1):45–54.

70. Lijnen HR, Lupu F, Moons L, Carmeliet P, Goulding D, Collen D. Temporal and topographic matrix metalloproteinase expression after vascular injury in mice. *Thromb Haemost.* 1999;81(5):799–807.
71. Southgate KM, Fisher M, Banning AP, et al. Upregulation of basement membrane-degrading metalloproteinase secretion after balloon injury of pig carotid arteries. *Circ Res.* 1996;79(6):1177–1187.
72. Southgate KM, Mehta D, Izzat MB, Newby AC, Angelini GD. Increased secretion of basement membrane-degrading metalloproteinases in pig saphenous vein into carotid artery interposition grafts. *Arterioscler Thromb Vasc Biol.* 1999;19(7):1640–1649.
73. Cho A, Reidy MA. Matrix metalloproteinase-9 is necessary for the regulation of smooth muscle cell replication and migration after arterial injury. *Circ Res.* 2002;91(9):845–851.
74. Sato H, Takino T, Okada Y, et al. A matrix metalloproteinase expressed on the surface of invasive tumour cells. *Nature.* 1994;370(6484):61–65.
75. Will H, Atkinson SJ, Butler GS, Smith B, Murphy G. The soluble catalytic domain of membrane type 1 matrix metalloproteinase cleaves the propeptide of progelatinase A and initiates autoproteolytic activation. Regulation by TIMP-2 and TIMP-3. *J Biol Chem.* 1996;271(29):17119–17123.
76. Kinoshita T, Sato H, Okada A, et al. TIMP-2 promotes activation of progelatinase A by membrane-type 1 matrix metalloproteinase immobilized on agarose beads. *J Biol Chem.* 1998;273(26):16098–16103.
77. Itoh Y, Takamura A, Ito N, et al. Homophilic complex formation of MT1-MMP facilitates proMMP-2 activation on the cell surface and promotes tumor cell invasion. *EMBO J.* 2001;20(17):4782–4793.
78. Tsukioka K, Suzuki J, Fujimori M, et al. Expression of matrix metalloproteinases in cardiac allograft vasculopathy and its attenuation by anti MMP-2 ribozyme gene transfection. *Cardiovasc Res.* 2002;56(3):472–478.
79. Johnson C, Galis ZS. Matrix metalloproteinase-2 and -9 differentially regulate smooth muscle cell migration and cell-mediated collagen organization. *Arterioscler Thromb Vasc Biol.* 2004;24(1):54–60.
80. Zaltsman AB, Newby AC. Increased secretion of gelatinases A and B from the aortas of cholesterol fed rabbits: relationship to lesion severity. *Atherosclerosis.* 1997;130(1-2): 61–70.
81. Kuge Y, Takai N, Ishino S, Temma T, Shiomi M, Saji H. Distribution profiles of membrane Type-1 matrix metalloproteinase (MT1-MMP), matrix metalloproteinase-2 (MMP-2) and cyclooxygenase-2 (COX-2) in rabbit atherosclerosis: comparison with plaque instability analysis. *Biol Pharm Bull.* 2007;30(9):1634–1640.
82. Wagsater D, Zhu C, Bjorkegren J, Skogsberg J, Eriksson P. MMP-2 and MMP-9 are prominent matrix metalloproteinases during atherosclerosis development in the Ldlr (-/-)Apob(100/100) mouse. *Int J Mol Med.* 2011;28(2):247–253.
83. Liu CC, Tsai CW, Deak F, et al. Deficiency in LRP6-mediated Wnt signaling contributes to synaptic abnormalities and amyloid pathology in Alzheimer's disease. *Neuron.* 2014;84:63–77.
84. Kuzuya M, Nakamura K, Sasaki T, Cheng XW, Itohara S, Iguchi A. Effect of MMP-2 deficiency on atherosclerotic lesion formation in apoE-deficient mice. *Arterioscler Thromb Vasc Biol.* 2006;26(5):1120–1125.
85. Sasaki T, Nakamura K, Sasada K, et al. Matrix metalloproteinase-2 deficiency impairs aortic atherosclerotic calcification in ApoE-deficient mice. *Atherosclerosis.* 2013;227(1):43–50.
86. Engelberger RP, Limacher A, Kucher N, et al. Biological variation of established and novel biomarkers for atherosclerosis: results from a prospective, parallel-group cohort study. *Clin Chim Acta.* 2015;447:16–22.

87. Lien LM, Hsieh YC, Bai CH, et al. Association of blood active matrix metalloproteinase-3 with carotid plaque score from a community population in Taiwan. *Atherosclerosis*. 2010;212(2):595–600.
88. Huang XY, Han LY, Huang XD, Guan CH, Mao XL, Ye ZS. Impact of 5A/6A polymorphism of matrix metalloproteinase-3 on recurrent atherosclerotic ischemic stroke in Chinese. *Int J Neurosci*. 2016;126(10):936–941.
89. Biscetti F, Straface G, Bertoletti G, et al. Identification of a potential proinflammatory genetic profile influencing carotid plaque vulnerability. *J Vasc Surg*. 2015;61(2): 374–381.
90. Koch W, de Waha A, Hoppmann P, Schömig A, Kastrati A. Haplotypes and 5A/6A polymorphism of the matrix metalloproteinase-3 gene in coronary disease: case-control study and a meta-analysis. *Atherosclerosis*. 2010;208(1):171–176.
91. Clee SM. A role for MMP-3 genetic variation in atherosclerosis susceptibility? *Atherosclerosis*. 2010;208(1):30–31.
92. Johnson JL, Dwivedi A, Somerville M, George SJ, Newby AC. Matrix metalloproteinase (MMP)-3 activates MMP-9 mediated vascular smooth muscle cell migration and neointima formation in mice. *Arterioscler Thromb Vasc Biol*. 2011;31(9):e35–e44.
93. Johnson JL, George SJ, Newby AC, Jackson CL. Divergent effects of matrix metalloproteinases 3, 7, 9, and 12 on atherosclerotic plaque stability in mouse brachiocephalic arteries. *Proc Natl Acad Sci USA*. 2005;102(43):15575–15580.
94. Silence J, Lupu F, Collen D, Lijnen HR. Persistence of atherosclerotic plaque but reduced aneurysm formation in mice with stromelysin-1 (MMP-3) gene inactivation. *Arterioscler Thromb Vasc Biol*. 2001;21(9):1440–1445.
95. Liu XQ, Mao Y, Wang B, et al. Specific matrix metalloproteinases play different roles in intraplaque angiogenesis and plaque instability in rabbits. *PLoS One*. 2014;9(9): e107851.
96. Abbas A, Aukrust P, Russell D, et al. Matrix metalloproteinase 7 is associated with symptomatic lesions and adverse events in patients with carotid atherosclerosis. *PLoS One*. 2014;9(1): e84935.
97. Goncalves I, Bengtsson E, Colhoun HM, et al. Elevated plasma levels of MMP-12 are associated with atherosclerotic burden and symptomatic cardiovascular disease in subjects with type 2 diabetes. *Arterioscler Thromb Vasc Biol*. 2015;35(7):1723–1731.
98. Imai K, Yokohama Y, Nakanishi I, et al. Matrix metalloproteinase 7 (matrilysin) from human rectal carcinoma cells. Activation of the precursor, interaction with other matrix metalloproteinases and enzymic properties. *J Biol Chem*. 1995;270(12):6691–6697.
99. Crabbe T, Smith B, O'Connell J, Docherty A. Human progelatinase A can be activated by matrilysin. *FEBS Lett*. 1994;345(1):14–16.
100. Henney AM, Wakeley PR, Davies MJ, et al. Localization of stromelysin gene expression in atherosclerotic plaques by *in situ* hybridization. *Proc Natl Acad Sci USA*. 1991;88:8154–8158.
101. Williams H, Johnson JL, Jackson CL, White SJ, George SJ. MMP-7 mediates cleavage of N-cadherin and promotes smooth muscle cell apoptosis. *Cardiovasc Res*. 2010;87(1):137–146.
102. Panayiotou AG, Griffin MB, Tyllis T, et al. Association of genotypes at the matrix metalloproteinase (MMP) loci with carotid IMT and presence of carotid and femoral atherosclerotic plaques. *Vasc Med*. 2013;18(5):298–306.
103. Park JY, Park JH, Jang W, et al. Apolipoprotein A-IV is a novel substrate for matrix metalloproteinases. *J Biochem*. 2012;151(3):291–298.
104. Tuomainen AM, Nyyssönen K, Laukkanen JA, et al. Serum matrix metalloproteinase-8 concentrations are associated with cardiovascular outcome in men. *Arterioscler Thromb Vasc Biol*. 2007;27(12):2722–2728.

105. Kato R, Momiyama Y, Ohmori R, Taniguchi H, Nakamura H, Ohsuzu F. Plasma matrix metalloproteinase-8 concentrations are associated with the presence and severity of coronary artery disease. *Circ J*. 2005;69(9):1035–1040.
106. Djuric T, Zivkovic M, Stankovic A, et al. Plasma levels of matrix metalloproteinase-8 in patients with carotid atherosclerosis. *J Clin Lab Anal*. 2010;24(4):246–251.
107. Herman MP, Sukhova GK, Libby P, et al. Expression of neutrophil collagenase (matrix metalloproteinase-8) in human atheroma: a novel collagenolytic pathway suggested by transcriptional profiling. *Circulation*. 2001;104(16):1899–1904.
108. Turu MM, Krupinski J, Catena E, et al. Intraplaque MMP-8 levels are increased in asymptomatic patients with carotid plaque progression on ultrasound. *Atherosclerosis*. 2006;187(1):161–169.
109. Molloy KJ, Thompson MM, Jones JL, et al. Unstable carotid plaques exhibit raised matrix metalloproteinase-8 activity. *Circulation*. 2004;110(3):337–343.
110. Verhoeven B, Hellings WE, Moll FL, et al. Carotid atherosclerotic plaques in patients with transient ischemic attacks and stroke have unstable characteristics compared with plaques in asymptomatic and amaurosis fugax patients. *J Vasc Surg*. 2005;42(6):1075–1081.
111. Laxton RC, Hu Y, Duchene J, et al. A role of matrix metalloproteinase-8 in atherosclerosis. *Circ Res*. 2009;105(9):921–929.
112. Ye S. Putative targeting of matrix metalloproteinase-8 in atherosclerosis. *Pharmacol Ther*. 2015;147:111–122.
113. Salminen A, Åström P, Metso J, et al. Matrix metalloproteinase 8 degrades apolipoprotein A-I and reduces its cholesterol efflux capacity. *FASEB J*. 2015;29(4):1435–1445.
114. Li Y. Correlation analysis of levels of adiponectin and matrix metalloproteinase-9 with stability of coronary heart disease. *Technol Health Care*. 2015;23(suppl 1):S95–S98.
115. Inokubo Y, Hanada H, Ishizaka H, Fukushi T, Kamada T, Okumura K. Plasma levels of matrix metalloproteinase-9 and tissue inhibitor of metalloproteinase-1 are increased in the coronary circulation in patients with acute coronary syndrome. *Am Heart J*. 2001;141(2):211–217.
116. Ju C, Ye M, Li F. Plasma brain natriuretic peptide, endothelin-1, and matrix metalloproteinase 9 expression and significance in type 2 diabetes mellitus patients with ischemic heart disease. *Med Sci Monit*. 2015;21:2094–2099.
117. Silvello D, Narvaes LB, Albuquerque LC, et al. Serum levels and polymorphisms of matrix metalloproteinases (MMPs) in carotid artery atherosclerosis: higher MMP-9 levels are associated with plaque vulnerability. *Biomarkers*. 2014;19(1):49–55.
118. Jefferis BJ, Whincup P, Welsh P, et al. Prospective study of matrix metalloproteinase-9 and risk of myocardial infarction and stroke in older men and women. *Atherosclerosis*. 2010;208(2):557–563.
119. Tan C, Liu Y, Li W, et al. Associations of matrix metalloproteinase-9 and monocyte chemoattractant protein-1 concentrations with carotid atherosclerosis, based on measurements of plaque and intima-media thickness. *Atherosclerosis*. 2014;232(1):199–203.
120. Blankenberg S, Rupprecht HJ, Poirier O, et al. Plasma concentrations and genetic variation of matrix metalloproteinase 9 and prognosis of patients with cardiovascular disease. *Circulation*. 2003;107(12):1579–1585.
121. Konstantino Y, Nguyen TT, Wolk R, Aiello RJ, Terra SG, Fryburg DA. Potential implications of matrix metalloproteinase-9 in assessment and treatment of coronary artery disease. *Biomarkers*. 2009;14(2):118–129.
122. Zhang B, Ye S, Herrmann SM, et al. Functional polymorphism in the regulatory region of gelatinase B gene in relation to severity of coronary atherosclerosis. *Circulation*. 1999;99(14):1788–1794.
123. Morgan AR, Zhang B, Tapper W, Collins A, Ye S. Haplotypic analysis of the MMP-9 gene in relation to coronary artery disease. *J Mol Med*. 2003;81(5):321–326.

124. Mason DP, Kenagy RD, Hasenstab D, et al. Matrix metalloproteinase-9 overexpression enhances vascular smooth muscle cell migration and alters remodeling in the injured rat carotid artery. *Circ Res.* 1999;85(12):1179–1185.
125. Uglow EB, Slater S, Sala-Newby GB, et al. Dismantling of cadherin-mediated cell-cell contacts modulates smooth muscle cell proliferation. *Circ Res.* 2003;92(12): 1314–1321.
126. Dwivedi A, Slater SC, George SJ. MMP-9 and -12 cause N-cadherin shedding and thereby beta-catenin signalling and vascular smooth muscle cell proliferation. *Cardiovasc Res.* 2009;81(1):178–186.
127. Sini S, Deepa D, Harikrishnan S, Jayakumari N. Evidence for an exclusive association of matrix metalloproteinase-9 with dysfunctional high-density lipoprotein: a novel finding. *Atherosclerosis.* 2014;236(1):162–168.
128. Godin D, Ivan E, Johnson C, Magid R, Galis ZS. Remodeling of carotid artery is associated with increased expression of matrix metalloproteinases in mouse blood flow cessation model. *Circulation.* 2000;102(23):2861–2866.
129. Galis ZS, Johnson C, Godin D, et al. Targeted disruption of the matrix metalloproteinase-9 gene impairs smooth muscle cell migration and geometrical arterial remodeling. *Circ Res.* 2002;91(9):852–859.
130. Luttun A, Lutgens E, Manderveld A, et al. Loss of matrix metalloproteinase-9 or matrix metalloproteinase-12 protects apolipoprotein E-deficient mice against atherosclerotic media destruction but differentially affects plaque growth. *Circulation.* 2004;109(11): 1408–1414.
131. Choi ET, Collins ET, Marine LA, et al. Matrix metalloproteinase-9 modulation by resident arterial cells is responsible for injury-induced accelerated atherosclerotic plaque development in apolipoprotein E-deficient mice. *Arterioscler Thromb Vasc Biol.* 2005;25(5):1020–1025.
132. Jin ZX, Xiong Q, Jia F, Sun CL, Zhu HT, Ke FS. Investigation of RNA interference suppression of matrix metalloproteinase-9 in mouse model of atherosclerosis. *Int J Clin Exp Med.* 2015;8(4):5272–5278.
133. Lemaitre V, Kim HE, Forney-Prescott M, Okada Y, D'Armiento J. Transgenic expression of matrix metalloproteinase-9 modulates collagen deposition in a mouse model of atherosclerosis. *Atherosclerosis.* 2009;205(1):107–112.
134. Gough PJ, Gomez IG, Wille PT, Raines EW. Macrophage expression of active MMP-9 induces acute plaque disruption in apoE-deficient mice. *J Clin Invest.* 2006;116(1):59–69.
135. de Nooijer R, Verkleij CJ, von der Thusen JH, et al. Lesional overexpression of matrix metalloproteinase-9 promotes intraplaque hemorrhage in advanced lesions but not at earlier stages of atherogenesis. *Arterioscler Thromb Vasc Biol.* 2006;26(2):340–346.
136. Glagov S, Weisenberg E, Zarins CK, Stankunavicius R, Kolettis GJ. Compensatory enlargement of human atherosclerotic coronary arteries. *N Engl J Med.* 1987;316(22): 1371–1375.
137. Lessner SM, Martinson DE, Galis ZS. Compensatory vascular remodeling during atherosclerotic lesion growth depends on matrix metalloproteinase-9 activity. *Arterioscler Thromb Vasc Biol.* 2004;24(11):2123–2129.
138. Martinez-Aguilar E, Gomez-Rodriguez V, Orbe J, et al. Matrix metalloproteinase 10 is associated with disease severity and mortality in patients with peripheral arterial disease. *J Vasc Surg.* 2015;61(2):428–435.
139. Coll B, Rodríguez JA, Craver L, et al. Serum levels of matrix metalloproteinase-10 are associated with the severity of atherosclerosis in patients with chronic kidney disease. *Kidney Int.* 2010;78(12):1275–1280.
140. Orbe J, Montero I, Rodríguez JA, Beloqui O, Roncal C, Páramo JA. Independent association of matrix metalloproteinase-10, cardiovascular risk factors and subclinical atherosclerosis. *J Thromb Haemost.* 2007;5(1):91–97.

141. Huang WC, Sala-Newby GB, Susana A, Johnson JL, Newby AC. Classical macrophage activation up-regulates several matrix metalloproteinases through mitogen activated protein kinases and nuclear factor-κB. *PLoS One*. 2012;7(8): e42507.
142. Reel B, Sala-Newby GB, Huang WC, Newby AC. Diverse patterns of cyclooxygenase-independent metalloproteinase gene regulation in human monocytes. *Br J Pharmacol*. 2011;163(8):1679–1690.
143. Nakamura H, Fujii Y, Ohuchi E, Yamamoto E, Okada Y. Activation of the precursor of human stromelysin 2 and its interactions with other matrix metalloproteinases. *Eur J Biochem*. 1998;253(1):67–75.
144. Vacek TP, Rehman S, Neamtu D, Yu S, Givimani S, Tyagi SC. Matrix metalloproteinases in atherosclerosis: role of nitric oxide, hydrogen sulfide, homocysteine, and polymorphisms. *Vasc Health Risk Manag*. 2015;11:173–183.
145. Schonbeck U, Mach F, Sukhova GK, et al. Expression of stromelysin-3 in atherosclerotic lesions: regulation via CD40-CD40 ligand signaling in vitro and in vivo. *J Exp Med*. 1999;189(5):843–853.
146. Mach F, Schonbeck U, Sukhova GK, et al. Functional CD40 ligand is expressed on human vascular endothelial cells, smooth muscle cells, and macrophages: implications for CD40-CD40 ligand signaling in atherosclerosis. *Proc Natl Acad Sci USA*. 1997;94(5):1931–1936.
147. Lijnen HR, Van Hoef B, Vanlinthout I, Verstreken M, Rio MC, Collen D. Accelerated neointima formation after vascular injury in mice with stromelysin-3 (MMP-11) gene inactivation. *Arterioscler Thromb Vasc Biol*. 1999;19(12):2863–2870.
148. Potempa J, Korzus E, Travis J. The serpin superfamily of proteinase inhibitors: structure, function, and regulation. *J Biol Chem*. 1994;269(23):15957–15960.
149. Manes S, Mira E, Barbacid MM, et al. Identification of insulin-like growth factor-binding protein-1 as a potential physiological substrate for human stromelysin-3. *J Biol Chem*. 1997;272(41):25706–25712.
150. Lind L, Ärnlöv J, Lindahl B, Siegbahn A, Sundström J, Ingelsson E. Use of a proximity extension assay proteomics chip to discover new biomarkers for human atherosclerosis. *Atherosclerosis*. 2015;242(1):205–210.
151. Johnson JL, Devel L, Czarny B, et al. A selective matrix metalloproteinase-12 inhibitor retards atherosclerotic plaque development in apolipoprotein E-knockout mice. *Arterioscler Thromb Vasc Biol*. 2011;31(3):528–535.
152. Shipley JM, Wesselschmidt RL, Kobayashi DK, Ley TJ, Shapiro SD. Metalloelastase is required for macrophage-mediated proteolysis and matrix invasion in mice. *Proc Natl Acad Sci USA*. 1996;93(9):3942–3946.
153. Liang J, Liu E, Yu Y, et al. Macrophage metalloelastase accelerates the progression of atherosclerosis in transgenic rabbits. *Circulation*. 2006;113(16):1993–2001.
154. Yamada S, Wang KY, Tanimoto A, et al. Matrix metalloproteinase 12 accelerates the initiation of atherosclerosis and stimulates the progression of fatty streaks to fibrous plaques in transgenic rabbits. *Am J Pathol*. 2008;172(5):1419–1429.
155. Matsumoto S, Kobayashi T, Katoh M, et al. Expression and localization of matrix metalloproteinase-12 in the aorta of cholesterol-fed rabbits: relationship to lesion development. *Am J Pathol*. 1998;153(1):109–119.
156. Newby AC. Metalloproteinase expression in monocytes and macrophages and its relationship to atherosclerotic plaque instability. *Arterioscler Thromb Vasc Biol*. 2008;28(12):2108–2114.
157. Quillard T, Tesmenitsky Y, Croce K, et al. Selective inhibition of matrix metalloproteinase-13 increases collagen content of established mouse atherosclerosis. *Arterioscler Thromb Vasc Biol*. 2011;31(11):2464–2472.
158. Quillard T, Araújo HA, Franck G, Tesmenitsky Y, Libby P. Matrix metalloproteinase-13 predominates over matrix metalloproteinase-8 as the functional interstitial collagenase in mouse atheromata. *Arterioscler Thromb Vasc Biol*. 2014;34(6):1179–1186.

159. Deguchi JO, Aikawa E, Libby P, et al. Matrix metalloproteinase-13/collagenase-3 deletion promotes collagen accumulation and organization in mouse atherosclerotic plaques. *Circulation*. 2005;112(17):2708–2715.
160. Hu JH, Touch P, Zhang J, et al. Reduction of mouse atherosclerosis by urokinase inhibition or with a limited-spectrum matrix metalloproteinase inhibitor. *Cardiovasc Res*. 2015;105(3):372–382.
161. Tarín C, Gomez M, Calvo E, López JA, Zaragoza C. Endothelial nitric oxide deficiency reduces MMP-13-mediated cleavage of ICAM-1 in vascular endothelium: a role in atherosclerosis. *Arterioscler Thromb Vasc Biol*. 2009;29(1):27–32.
162. Yang SW, Lim L, Ju S, Choi DH, Song H. Effects of matrix metalloproteinase 13 on vascular smooth muscle cells migration via Akt-ERK dependent pathway. *Tissue Cell*. 2015;47(1):115–121.
163. Marco M, Fortin C, Fulop T. Membrane-type matrix metalloproteinases: key mediators of leukocyte function. *J Leukoc Biol*. 2013;94(2):237–246.
164. Kuge Y, Takai N, Ogawa Y, et al. Imaging with radiolabelled anti-membrane type 1 matrix metalloproteinase (MT1-MMP) antibody: potentials for characterizing atherosclerotic plaques. *Eur J Nucl Med Mol Imaging*. 2010;37(11):2093–2104.
165. Rajavashisth TB, Xu XP, Jovinge S, et al. Membrane type 1 matrix metalloproteinase expression in human atherosclerotic plaques: evidence for activation by proinflammatory mediators. *Circulation*. 1999;99(24):3103–3109.
166. Johnson JL, Jenkins NP, Huang WC, et al. Relationship of MMP-14 and TIMP-3 expression with macrophage activation and human atherosclerotic plaque vulnerability. *Mediat Inflamm*. 2014;2014:276457.
167. Di Gregoli K, George SJ, Jackson CL, Newby AC, Johnson JL. Differential effects of tissue inhibitor of metalloproteinase (TIMP)-1 and TIMP-2 on atherosclerosis and monocyte/macrophage invasion. *Cardiovasc Res*. 2016;109(2):318–330.
168. Johnson JL, Sala-Newby GB, Ismail Y, Aguilera CM, Newby AC. Low tissue inhibitor of metalloproteinases 3 and high matrix metalloproteinase 14 levels defines a subpopulation of highly invasive foam-cell macrophages. *Arterioscler Thromb Vasc Biol*. 2008;28(9):1647–1653.
169. Li C, Jin XP, Zhu M, et al. Positive association of MMP 14 gene polymorphism with vulnerable carotid plaque formation in a Han Chinese population. *Scand J Clin Lab Invest*. 2014;74(3):248–253.
170. Schneider F, Sukhova GK, Aikawa M, et al. Matrix-metalloproteinase-14 deficiency in bone-marrow-derived cells promotes collagen accumulation in mouse atherosclerotic plaques. *Circulation*. 2008;117(7):931–939.
171. Di Gregoli K, Jenkins N, Salter R, White S, Newby AC, Johnson JL. MicroRNA-24 regulates macrophage behavior and retards atherosclerosis. *Arterioscler Thromb Vasc Biol*. 2014;34(9):1990–2000.
172. Ohkawara H, Ikeda K, Ogawa K, Takeishi Y. Membrane type 1-matrix metalloproteinase (MT1-MMP) identified as a multifunctional regulator of vascular responses. *Fukushima J Med Sci*. 2015;61(2):91–100.
173. Su W, Gao F, Lu J, Wu W, Zhou G, Lu S. Levels of matrix metalloproteinase-9 and tissue inhibitor of metalloproteinase-1 mRNAs in patients with primary hypertension or hypertension-induced atherosclerosis. *J Int Med Res*. 2012;40(3):986–994.
174. Lemaitre V, Soloway PD, D'Armiento J. Increased medial degradation with pseudo-aneurysm formation in apolipoprotein E-knockout mice deficient in tissue inhibitor of metalloproteinases-1. *Circulation*. 2003;107(2):333–338.
175. Silence J, Collen D, Lijnen HR. Reduced atherosclerotic plaque but enhanced aneurysm formation in mice with inactivation of the tissue inhibitor of metalloproteinase-1 (TIMP-1) gene. *Circ Res*. 2002;90(8):897–903.

176. Johnson JL, Baker AH, Oka K, et al. Suppression of atherosclerotic plaque progression and instability by tissue inhibitor of metalloproteinase-2: involvement of macrophage migration and apoptosis. *Circulation*. 2006;113(20):2435–2444.
177. Rouis M, Adamy C, Duverger N, et al. Adenovirus-mediated overexpression of tissue inhibitor of metalloproteinase-1 reduces atherosclerotic lesions in apolipoprotein E-deficient mice. *Circulation*. 1999;100(5):533–540.
178. Casagrande V, Menghini R, Menini S, et al. Overexpression of tissue inhibitor of metalloproteinase 3 in macrophages reduces atherosclerosis in low-density lipoprotein receptor knockout mice. *Arterioscler Thromb Vasc Biol*. 2012;32(1):74–81.
179. Stohr R, Cavalera M, Menini S, et al. Loss of TIMP3 exacerbates atherosclerosis in ApoE null mice. *Atherosclerosis*. 2014;235(2):438–443.
180. Qin H, Zhao Y, Zhang J, Pan X, Yang S, Xing D. Inflammation-targeted gold nanorods for intravascular photoacoustic imaging detection of matrix metalloproteinase-2 (MMP2) in atherosclerotic plaques. *Nanomedicine*. 2016;12(7):1765–1774.
181. Newby AC. Matrix metalloproteinase inhibition therapy for vascular diseases. *Vasc Pharmacol*. 2012;56(5–6):232–244.
182. Prescott MF, Sawyer WK, Von Linden-Reed J, et al. Effect of matrix metalloproteinase inhibition on progression of atherosclerosis and aneurysm in LDL receptor-deficient mice overexpressing MMP-3, MMP-12, and MMP-13 and on restenosis in rats after balloon injury. *Ann N Y Acad Sci*. 1999;878:179–190.
183. Cherr GS, Motew SJ, Travis JA, et al. Metalloproteinase inhibition and the response to angioplasty and stenting in atherosclerotic primates. *Arterioscler Thromb Vasc Biol*. 2002;22(1):161–166.
184. Johnson JL, Fritsche-Danielson R, Behrendt M, et al. Effect of broad-spectrum matrix metalloproteinase inhibition on atherosclerotic plaque stability. *Cardiovasc Res*. 2006;71(3):586–595.
185. Peterson JT. The importance of estimating the therapeutic index in the development of matrix metalloproteinase inhibitors. *Cardiovasc Res*. 2006;69(3):677–687.
186. Kim D, Lee D, Jang YL, et al. Facial amphipathic deoxycholic acid-modified polyethyleneimine for efficient MMP-2 siRNA delivery in vascular smooth muscle cells. *Eur J Pharm Biopharm*. 2012;81(1):14–23.
187. Lee D, Kim D, Mok H, Jeong JH, Choi D, Kim SH. Bioreducible crosslinked polyelectrolyte complexes for MMP-2 siRNA delivery into human vascular smooth muscle cells. *Pharm Res*. 2012;29(8):2213–2224.
188. Shon SM, Park JH, Nahrendorf M, et al. Exercise attenuates matrix metalloproteinase activity in preexisting atherosclerotic plaque. *Atherosclerosis*. 2011;216(1):67–73.
189. Gliesche DG, Hussner J, Witzigmann D, et al. Secreted matrix metalloproteinase-9 of proliferating smooth muscle cells as a trigger for drug release from stent surface polymers in coronary arteries. *Mol Pharm*. 2016;13(7):2290–2300.

CHAPTER SEVEN

The Role Matrix Metalloproteinases in the Production of Aortic Aneurysm

Simon W. Rabkin[1]
University of British Columbia, Vancouver, BC, Canada
[1]Corresponding author: e-mail address: rabkin@mail.ubc.ca

Contents

Abstract

Matrix metalloproteinases (MMPs) have been implicated in the pathogenesis of aortic aneurysm because the histology of thoracic aortic aneurysm (TAA) and abdominal aortic aneurysm (AAA) is characterized by the loss of smooth muscle cells in the aortic media and the destruction of extracellular matrix (ECM). Furthermore, AAA have evidence of inflammation and the cellular elements involved in inflammation such as

Progress in Molecular Biology and Translational Science, Volume 147
ISSN 1877-1173
http://dx.doi.org/10.1016/bs.pmbts.2017.02.002

macrophages can produce and/or activate MMPs This chapter focuses on human aortic aneurysm that are not due to specific known genetic causes because this type of aneurysm is the more common type. This chapter will also focus on MMP protein expression rather than on genetic data which may not necessarily translate to increased MMP protein expression. There are supporting data that certain MMPs are increased in the aortic wall. For TAA, it is most notably MMP-1, -9, -12, and -14 and MMP-2 when a bicuspid aortic valve is present. For AAA, it is MMP-1, -2, -3, -9, -12, and -13. The data are weaker or insufficient for the other MMPs. Several studies of gene polymorphisms support MMP-9 for TAA and MMP-3 for AAA as potentially important factors. The signaling pathways in the aorta that can lead to MMP activation include JNK, JAK/stat, osteopontin, and AMP-activated protein kinase alpha2. Substrates in the human vasculature for MMP-3, MMP-9, or MMP-14 include collagen, elastin, ECM glycoprotein, and proteoglycans. Confirmed and potential substrates for MMPs, maintain aortic size and function so that a reduction in their content relative to other components of the aortic wall may produce a failure to maintain aortic size leading to dilatation and aneurysm formation.

1. INTRODUCTION

Aortic aneurysm is a localized enlargement or dilation of the aorta and originates from the Greek word "aneurysma" or "a widening." The condition warrants considerable attention because it can lead to the potentially highly lethal conditions of aortic dissection or rupture and has been increasing in prevalence in many countries.[1–3] In the United States in 2014, approximately 10,000 people died because of aortic aneurysm/dissection.[4] The histology of thoracic aortic aneurysm (TAA) and abdominal aortic aneurysm (AAA) is characterized by the loss of smooth muscle cells (SMCs) in the aortic media and the destruction of extracellular matrix (ECM).[5–7] Although there is some controversy around the extent of loss of SMCs in the aortic media in TAA, compared to AAA, the destruction of ECM is evident in TAA.[8] Matrix metalloproteinases (MMPs) have been implicated in the pathogenesis of aortic aneurysm, because of the important role they play in connective tissue homeostasis.[9] Disruption of the balance between the production of active MMP enzymes and their inhibitors can lead to the action of active MMPs to produce degradation of ECM.[10–14] MMP cleavage of elastin suggests that elastin cleavage sites are readily accessible to enzymatic attack by MMPs.[15] The purpose of this chapter is to examine the data on the putative role of MMP in the production of aortic aneurysm and its complications, which for TAA is usually dissection (TAD) and for AAA is aortic rupture. Because of the differences between the thoracic and abdominal aorta,[16] these two entities will be considered separately. This chapter will focus on human aortic aneurysm that are not due to specific known genetic causes, because this type of aneurysm is the more common type and differences in aortic structure between humans and animals suggests that the animal data need to be relied on with caution. Animal models will be discussed when there is limited human data or the animal data supplements and/or explains the clinical data. Another focus of the chapter will be on MMPs protein expression rather than exclusive on genetic data which may not be necessarily translate to increased MMP protein expression in the aorta.

2. TAA AND DISSECTION (TAD)

2.1 MMP-1

There is limited data on MMP-1 in TAA and/or dissection. The data are based on immunohistochemical evaluation of the aorta. Koullias *et al.* evaluated the amount of MMP immunohistochemical staining of various MMPs and graded the amount of staining from 0 (none) to 4 (intense presence in greater than 75% of cells).[17] There was significantly more, about twofold, greater MMP-1 in TAA compared to controls who did not have an aortic aneurysm.[17] The magnitude of the increase in MMP-1 in TAD was also increased and was over twofold greater than controls and slightly greater than the amount in TAA.[17] In TAD, MMP-1 protein expression was significantly increased in the cytoplasm of the aortic SMCs in both the intima and media at the entry site of the TAD.[18]

Specific genotypes of the MMP-1 gene were associated with TAA which may accompany bicuspid aortic valve (BAV).[19] Patients with larger aortic aneurysms are more likely to have an abnormal MMP-2/TIMP-1 genotypes.[20]

2.2 MMP-2

MMP-2 has been assessed in the aorta of patients with TAA with a normal tricuspid aortic valve as well as a BAV which can be associated with an ascending aortic aneurysm. There have not been a many studies in persons with trileaflet aortic valves (TAVs).[21–23] A metaanalysis did not suggest an abnormality of MMP-2,[24] but subsequent investigations have indicated that MMP-2 protein levels in the aorta of TAA patients are significantly greater than persons with CAD.[25,26] In addition, active MMP-2 was identified in the aorta of patients with TAA[27] and immunohistochemistry showed areas of MMP-2 were much more intense in the aortic wall of TAA than in controls.[28] Inclusion of these new data would indicate that MMP-2 is increased in TAA although the data were not always in a format that would permit their inclusion into a metaanalysis.[26] In TAD, there is a significant expression of MMP-2 in the SMCs of the aortic intima.[18] Interestingly, AAA tissue has a greater capacity to activate exogenous pro-MMP-2 compared with atherosclerotic or normal aortic tissue.[29]

MMP-2 haplotypes are associated with genetic susceptibility to thoracic aortic dissection in Chinese Han population.[30]

Plasma MMP-2 concentrations in TAA patients and their significance are controversial. Plasma MMP-2 concentrations have been reported to be elevated in patients compared with controls.[26,28] MMP-2 aortic tissue levels were significantly higher in TAD compared to control.[25] However, no correlation was found between serum MMP-2 and aortic tissue total MMP-2 or tissue pro-MMP-2 or tissue active MMP-2.[27] These data suggest that circulating MMP-2 is not a valid indicator of tissue MMP-2.[27]

Regulation of the MMP-2 protein expression is complex.[13]

To the extent that MMP-2 maybe increased in the aorta of the aneurysm, it is noteworthy that it may in part occur by Ang II, the Ang II type 1 receptor (AT1R) through activation of the major mitogen-activated protein kinases, JNK, ERK1/2, and p38 MAPK.[26] Brahma-related gene 1 (Brg-1), the ATPase subunit of the SWI/SNF complex, enhances MMP-2 transcription.[31] MMP-2, in TAD, is upregulated by Brg-1 in human TAD.[32] There is a negative correlation between the percentage of contractile aortic SMCs in TAD and SMC line, suggesting that BRG1 through MMP-2 increases SMC apoptosis.[32] The upstream factors leading to the increase in MMP-2 may involve specific miR expression. MMP-2 is a miR-29a target and there is a significant inverse relationship between miR-29a and MMP-2.[33]

BAV disease, the most common congenital heart defect, is associated with TAA which is ascribed to structural abnormalities occurring at the cellular level independent of the hemodynamic effects of BAV.[34] The pathogenesis of TAA in BAV is of considerable interest because of the recommendation for earlier surgical intervention if TAA develops in the presence of BAV.[35] Patients with a BAV are more likely to have TAA than patients with a normal TAV. Metaanalysis showed a highly significant increase in aortic tissue MMP-2 in BAV compared to TAV[24] and the increased aortic MMP-2 was more strongly associated with BAV rather than TAV.[24] Subsequent data support this conclusion. Wang *et al.* found that MMP-2 in plasma was elevated in patients with BAV and mild or severe aortic stenosis when an ascending aortic dilatation was present.[36] There is also a significant relationship between aortic diameter and MMP-2 activity in TAA.[37] Thus the data support a potential role for MMP-2 in patients with BAV and TAA.

2.3 MMP-3 (Stromelysin) to MMP-8

There is very little data on these MMPs in TAA and TAD. For MMP-3, there are not enough data to implicate or dismiss a role for MMP-3 in TAA or TAD. Plasma stromelysin (MMP-3) level was significantly higher in patients with hypertension-induced aortic root dilatation compared to individuals without aortic dilatation.[38] The frequency of MMP-3 promoter 5A/6A genotypes, which is associated with higher tissue MMP-3 concentrations, was not differ between patients with TAA and a random sample of the population.[39] There is, however, an interesting association. In TAA there is a very low prevalence of the combined genotype of MMP-3 6A/6A and angiotensin-converting enzyme I/I suggesting that the combination produces a lower expression of MMP-3 and of ACE resulting in less angiotensin II in the aortic wall.[40]

MMP-8 levels are increased in different regions of a TAA. Higher MMP-8 levels were present in the convex aortic sites than in the concave aortic sites of the TAA in patients with BAV.[41] Plasma MMP-8, a neutrophil collagenase that targets type 1 collagen, is present in higher concentrations in the plasma of patients with TAD compared to controls.[42,43] There are suggestive data that 799C/T polymorphism in the promoter region of MMP-8 is associated with the development of TAD and that the T allele may increase the risk of development of TAD.[44]

2.4 MMP-9

A recent metaanalysis showed that there was a significant increase in MMP-9 in the aorta from persons with TAA compared with persons without TAA[24] (Fig. 1). There was also a reduction in TIMP-1 and TIMP-2.[24] Because TIMP-1 and TIMP-2 inhibit the activities of all MMPs and play a role in regulating ECM in different physiological processes,[45] the reduction in TIMP-1, TIMP-2, heightens the impact of MMP-9. In addition, there were other studies presenting data on increased protein expression of MMP-9 in

MMP-9 in TAA compared to control

Study name	Std. diff. in means	Standard error	Z-value	p-value
Schmoker *et al.*[23]	0.223	0.266	0.841	0.401
Koullias *et al.*[17]	1.550	0.457	3.393	0.001
Mi *et al.*[22]	1.322	0.638	2.071	0.038
	0.646	0.216	2.990	0.003

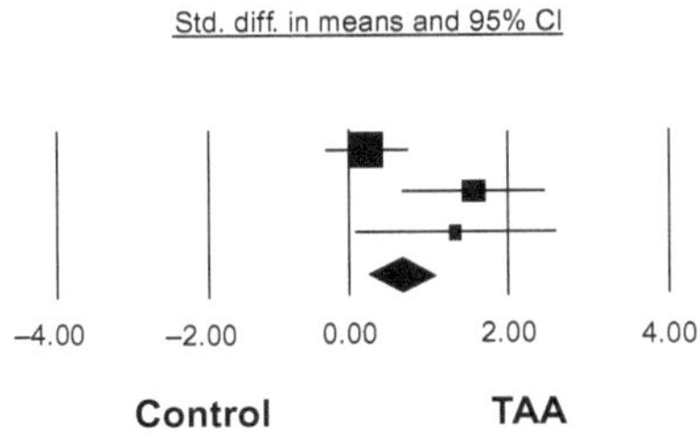

Fig. 1 The changes in MMP-9 in thoracic aortic aneurysm. *Reproduced from Rabkin SW. Differential expression of MMP-2, MMP-9 and TIMP proteins in ascending thoracic aortic aneurysm—comparison with and without bicuspid aortic valve: a meta-analysis.* VASA. *2014;43:433–442.*

TAA but without the quantitative type of data to be included in metaanalysis.[48] Two studies used immunohistochemistry and a semiquantitative method to assess MMP-9 in aorta removed at surgery and control aorta obtained at autopsy or at time of surgery for another procedure.[21,22] Both studies showed an increase in MMP-9 staining. Schmoker *et al.*[23] evaluated aortic tissue at the time of surgery for TAA compared to aorta at time of CABG which was the control. The total activity of MMP-2 and MMP-9 was quantified in the supernatants by activity assays which were based on an antibody capture technique.[23] They reported a lower MMP-2 concentration. In contrast, Hu *et al.* examined the aorta removes at surgery from 16 patients with TAD without a genetic cause compared to aorta from 9 patients who had aortic samples removed at the time of aortic valve replacement and used an ELISA-based system.[25] They found an over fourfold increase in MMP-9 in TAD.[25] Thus the majority of studies favor an increase in MMP-9 in the aorta or TAA or TAD patients. Supportive data are that studies of mRNA found increases in MMP-2, and MMP-9 mRNA levels in ascending aortic aneurysms.[28]

Examining almost 1200 genes in TAA found significant changes in 112 genes with MMP-9 showing a statistically significant and 8.6-fold increase compared to normal thoracic aorta.[46] Genetic studies suggest an association between MMP-9 gene polymorphism and TAD. In 206 Chinese patients with TAD, the rs2274756 polymorphism was significantly associated with TAD compared to controls.[47] Furthermore, the association was significant after adjusting for traditional cardiovascular risk factors (sex, age, hypertension, dyslipidemia, diabetes, and smoking habit).[47]

MMP-9 was found mainly in the media and occasionally in the adventitia and the neointima of TAA walls; and was rarely detected in control walls.[48] The cellular location of the MMP-9 is preferentially in the vascular SMCs, rather than inflammatory cells of the TAA.[49] MMP-9 was localized mainly in areas associated with severe tissue destruction.[48] In TAD, a significant expression of MMP-9, as well as MMP-2 was found in SMCs of the intima.[18]

MMP-9 protein expression is regulated by a large number of factors that operate at the level of transcription as well as signal transduction pathways.[14] MMP-9 production can be increased by hormones, cytokines, proteases, signaling molecules as well as specific mRNAs.[13,14] TNF-α promotes elastin breakdown through enhanced release of MMP-9 as well as MMP-2 by vascular SMCs.[50] Tissue IFN-γ expression correlates with the amount of MMP-9 in TAA.[51] IFN-γ may operate through a JNK signaling pathway to activating MMP-9, producing apoptosis and aneurysm formation.[51] AngII/ERK pathway can mediate the production of MMP-9 in human TAA walls, independent of TGF-β signaling.[48] Angiotensin II can produces Smad2 activation leading to MMP-9 production through a pathway involving intracellular signal regulated kinase (ERK).[48]

MMP-9 is subject to regulatory control through different signal transduction pathways. MAP kinases regulate MMP-9 expression[14] as well as AKT2 (RAC-beta serine/threonine-protein kinase) or protein kinase B (PKB).[52]

Thrombus formation in TAA or in TAD maybe operative in MMP activation. Plasmin-induced activation of MMP can degrade the important constituents of the aortic wall including elastin, collagen, fibronectin, and laminin.[53] Plasmin-induced activation of MMP-3, -9, -12, and -13 produces collagen and elastin degradation.[54]

MMP-9 expression is upregulated by Brg-1 in aortic SMCs which is associated with an increase in apoptosis.[32] The resultant decrease in the percentage of contractile phenotype of cells[32] would be anticipated to facilitate aortic expansion and TAA.

Decreased expression of miRNAs-1, -21, -29a, -133a, and -486 is present in TAA and there is a significant inverse relationship between expression of these miRNAs and aortic diameter.[33] MMP-2 and MMP-9 are potential targets for miRNA-29a and miRNA-133a.[33] These data suggest that these miRNAs inhibit MMP production so that their reduction may lead to TAA formation.[33]

In TAD, several miRNAs are considerably different compared to normal aorta.[55] These findings suggested several pathways especially those involved in the focal adhesion and the mitogen-activated protein kinase (MAPK) signaling pathways affecting vascular smooth muscle leading to TAA/TAD.[55]

2.5 MMP-12 to MMP-19

MMP-12 activity in the aorta wall of TAD is increased.[56]

Serum MMP-12 proteolysis was present and was greater in TAD cases compared to a healthy control.[56,57] MMP-12 activity in serum was higher than in the aorta wall.[56]

Immunohistochemically staining of TAA demonstrated increased expression of MMP-14, mt1-MMP, a potent collagenase, as well as MMP-19 in dilated aorta.[58] Messenger RNA expression for MMP-14 and all membrane bound MMPs (MMP-14, MMP-15, MMP-16, MMP-17, MMP-24, and MMP-25) is present in the aorta.[58] MMP-14 as well as MMP-19 showed a higher expression in dilated aortas.[58]

MMP-17 (also called membrane-type 4-MMP) plays a role in anchoring several components of the arterial wall. A missense mutation in MMP-17 was found to be associated with TAD.[59]

In summary, the strongest evidence linking MMPs to TAA is for MMP-9. There are a number of factors that regulate MMP-9 and the availability of MMP-9 to produce ECM degradation (Fig. 2).

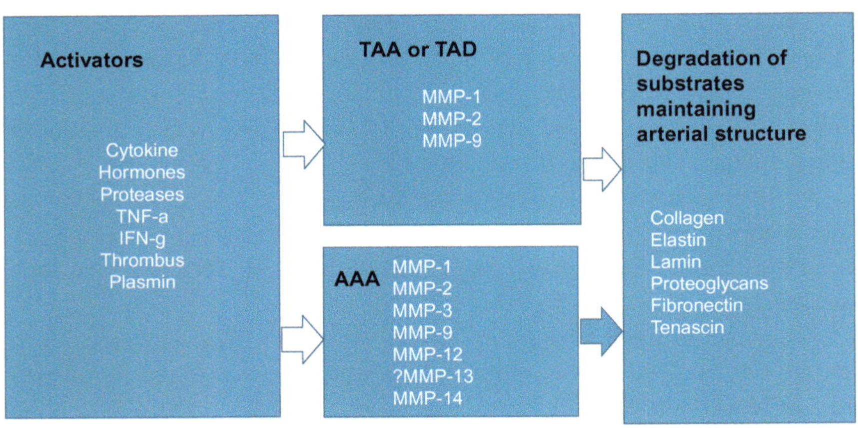

Fig. 2 A schematic which shows putative factors that increase the amount or activate MMPs in thoracic aortic aneurysm or dissection or abdominal aortic aneurysm.

3. ABDOMINAL AORTIC ANEURYSM (AAA)

The development of AAA is a complex process that involves different factors and likely includes as well inflammation in the arterial wall.[60] A role for MMPs in AAA is based on several lines of evidence. First, aortic elastin, collagen, and their associated proteins, such as glycosaminoglycans, maintain aortic size, and function. Second, aortic elastin and glycosaminoglycan content are reduced in AAA relative to its proportion to other components of the aortic wall in the normal aorta[61] implicating factors in the aorta that could reduce the content of elastin and glycosaminoglycans. Third, disruption of the balance between the production of active enzymes and their inhibitors, favor MMP activation which leads to accelerated turnover of ECM.[10,12–14] Fourth, MMPs are synthesized by a number of the cellular components of the aorta[62] so that local production would act locally on aortic constituents. Fifth, MMPs in the aorta are increased in AAA[63–68] especially the active form of the MMPs.[69,70] Sixth, AAA have evidence of inflammation and the cellular elements involved in inflammation such as macrophages can produce and/or activate MMPs.[66]

A discussion of some of the MMPs in AAA is useful because it provides additional specific data.

3.1 MMP-1

In human AAA, MMP-1 is present in an increased amounts compared to controls.[69] The increased MMP-1 includes not only zymogen levels but also MMP-1 proteolytic activity.[71] The increase in MMP-1, as well as MMP-9, and MMP-12, was accompanied by a decrease of their inhibitors.[72] Macrophage inhibitory factor (MIF) is upregulated in stable AAA and even higher levels are present in ruptured AAA.[72] MIF was localized to endothelial cells, SMCs, macrophages, and cytotoxic T cells.[72] These data led to the suggestion that inflammation in the abdominal aorta involves macrophage infiltration which in turn liberates MMP-1 that destroys elastin in the media and downregulates and/or destroys vascular SMCs.[71] ^{18}F-FDG uptake, a marker of inflammatory activity is present in AAA, and correlates with aortic biopsy evidence of a marked increased number of adventitial inflammatory cells, along with a marked increase of several MMPs, notably MMP-1 and MMP-13.[73]

3.2 MMP-2

MMP-2 and TIMP-2 are present in the arterial wall of AAA as demonstrated by immunohistochemistry, in situ hybridization and in situ zymography.[74,75] Aortic SMCs cultured from aneurysmal tissue express MMP-2 protein and messenger RNA at a significantly higher level than controls; a finding that was not present in other mesenchymal tissue.[74] The colocalization of MMP-2-and TIMP-2 with medial SMCs and elastin fibers supports the postulate that MMP-2 can be involved in the pathogenesis of AAA.[75]

The thrombus in the aortic wall can be a trigger for MMP-2 expression and activation. Both the total amount of MMP-2 as well as active MMP-2 correlate directly with the amount of intraluminal thrombus.[76] The luminal and parietal parts of the thrombus contain, respectively, 20- and 10-fold more gelantinolytic activity than serum.[77] The amount of active MMP-2 correlates with the amount of inflammation.[77]

3.3 MMP-3 (Stromelysin-1)

MMP-3 is overexpressed in AAA.[78] More activated forms of MMP-3 are present in AAA compared to controls.[66] MMP-3 protein is also detected in the macrophage-like mononuclear cells infiltrating the AAA.[66] This is consistent with a role for MMP-3 in production of AAA both directly as well as indirectly through MMP-3 activation by inflammation.

A role for MMP-3 in AAA development is supported by another line of evidence. A familial occurrence of AAA is well recognized and several efforts have been undertaken to identify responsible genetic defects. These studies suggest that an abnormality in the MMP-3 gene is part of the genetic profile that predisposes to AAA but the nature of the defect varies between studies. Several studies have implicated polymorphism in MMP-3. One polymorphism of MMP-3 specifically MMP-3 rs3025058 was significantly more common in patients with AAA.[79] MMP-3 nt-1612 polymorphisms also had a high odds or are at an increased risk of developing AAA.[80] An increased frequency of the 5A allele in the promoter region of the MMP-3 gene was associated with AAA.[81] Other investigators found that the genotype distribution was significantly different between patients with AAA compared to controls. In a multivariable logistic regression analysis adjusted for traditional cardiovascular factors and chronic obstructive pulmonary disease, the presence of three or four genetic risk conditions was a strong and independent determinant of AAA disease specifically -1306C/T MMP-2, 5A/6A MMP-3, -77A/G MMP-13, and G1355A ELN polymorphisms.[82] These data were confirmed in a metaanalysis including other data.[82]

The expression of mRNA for 14 MMPs and 4 tissue inhibitors of metalloproteinases (TIMPs) was estimated in samples of aortic wall from eight patients with AAA and eight with atherosclerotic obstructive arterial disease.[78] The greatest change and difference between the two conditions was for MMP-3.[78]

3.4 MMP-9

MMP-9 zymogen levels and proteolytic activities were increased in human AAAs when compared with healthy aorta.[22,66,71,77] MMP-9 protein and activity were markedly increased in mesenchymal stromal cells.[83] MMP-9 was present in macrophages.[66] Increased MMP-9 expression, associated with disruption of elastic lamellae in human TAA compared to control aorta.[48] Macrophage-derived MMP-9 and mesenchymal cell MMP-2 may work in concert as both appear required to produce AAA in experimental murine models of AAA.[84]

MMP-9 gene expression is increased in AAA.[46,85] MMP-9 gene expression showed a significant and over 85-fold increase in AAA compared to normal abdominal aorta.[46] The change in MMP-9 was the largest increase compared to over 100 other genes.[46] Recognizing the caveat that increases in gene expression is not equivalent to increases in protein expression, the data on increased MMP-9 protein expression is important confirmation that the gene expression is translated into increases in MMP-9 protein.[66,71]

The stimulus for increased protein expression and activation can originate from the thrombus if/when present in the AAA. Clot formation and clot lysis induce the release of promatrix-metalloproteinase (pro-MMP)-9.[86] Thrombus and wall extracts generated plasmin in the presence of a fibrin matrix and activate MMPs.[86] MMP-9 concentration in AAA correlates with the amount of intraluminal thrombus suggesting that thrombus activates MMP-9.[76] The luminal and parietal parts of the thrombus contained, respectively, 20- and 10-fold more gelantinolytic activity than the serum.[77] The proportion of MMP-9 to MMP-2 increases markedly in AAA compared to the normal aorta.[77] Importantly a significant proportion of MMP-9 is in its processed active form, which is not a finding in normal aorta[77] as it was never observed in normal samples. These data are consistent with the proposal that mural thrombus, by trapping polymorphonuclear leukocytes and adsorbing plasma components acts as a source of proteases in aneurysms that may play a critical role in enlargement and rupture[86] or that thrombus-induced MMP activation accelerates MMP activation in the aortic wall. Other investigators contend that intramural thrombus correlates more strongly with MMP-2 but not MMP-9 and MMP-2 correlates more strongly with lumen thrombus thickness, vascular smooth muscle apoptosis, and elastin degradation.[87] This supports the contention that the amount of luminal thrombus in AAA influences AAA growth, AAA wall stability, and perhaps rupture.[87,88]

An important treatment for AAA is the insertion of a stent—a procedure labeled endovascular repair (EVAR). Plasma MMP-9 concentrations measured 3 months after EVAR are higher in patients that have EVAR failure or an endoleak.[89] This finding raises the question whether MMP-9 activation plays a role in TEVAR failure.

3.5 MMP-10

There is very little data on MMP-10 and AAA. One study is intriguing. DNA samples from 812 unrelated white subject of whom 387 had AAA and the rest without AAA, were genotyped for 14 polymorphisms in 13 different candidate genes.[90] There was an association of AAA with two TIMP1 gene polymorphisms (nt + 434 and rs2070584) in men without a family history of AAA. In addition, there was a significant interaction between this polymorphism and MMP-10.[90]

3.6 MMP-12

MMP-12 (macrophage elastase) zymogen levels and proteolytic activities were increase in human AAAs when compared with healthy human aorta.[71,91] Importantly, immunoreactive MMP-12 was localized to residual elastin fragments within the media of AAA.[91] Enhanced expression of MMP-12 paralleled the increased expression of aneurysmal macrophage migration inhibitory factor.[72] The associated presence of macrophage infiltration and destruction of elastin suggests that chronic aortic wall inflammation, mediated by macrophage infiltration, increases levels of active MMP-12, and accounts for the destruction of medial elastin.[71] In vivo evidence supports a causal relationship. Incubation of control aortic tissue with recombinant MMP-12 produced extensive fragmentation of glycoproteins[92] confirming certain glycoproteins as substrates of MMP-12 and the potential of MMP-12 to damage aortic wall composition.

Gene expression of MMP-12 as well as -1, -7, and -9, were upregulated in the thrombus-free AAA wall compared with the thrombus-covered wall.[93]

The data from proteomic approaches to study AAA have been controversial as some studies did not find clear cut evidence of MMPs in AAA.[94] Other studies, however, present more compelling data. Didangelos *et al.* found accumulation of MMP-12 in AAA along with degradation of collagen XII, thrombospondin 2, aortic carboxypeptidase-like protein, periostin, fibronectin, and tenascin.[92]

3.7 MMP-13

MMP-13 is expressed in the aortic wall[95] and is localized especially to endothelial cells and SMCs in AAA.[96] The protein expression of MMP-13 was 1.8-fold higher in AAA compared to atherosclerotic aorta.[97] There are some conflicting data specifically that analysis of gene products in AAA compared to normal human aorta, using a membrane-based complementary DNA expression array, did not report an increase in MMP-13 expression.[98] This may, however, be explained by technical factors in the gene expression studies.[97]

Polymorphisms in MMP-13 are associated with AAA and are proposed to contribute to the pathogenesis of AAA.[82]

^{18}F-FDG uptake in patients with AAA is associated with a marked increased number of adventitial inflammatory cells, along with a marked increase of several MMPs, notably the MMP-1 and MMP-13.[73]

Cluster of differentiation 147 (CD147) bore a number of different names, including EMMPRIN in human tissue, which are now incorporated into one name.[99] It was initially called ECM metalloproteinase inducer (EMMPRIN) because of its capacity to stimulated collagenase (MMP-1) production.[99] Its action in the cardiovascular system was later identified.[100,101] CD147 and MMP-13 are both expressed in endothelial cells and SMCs in AAA.[96] Experimentally nitric oxide appears to regulate AAA development, NO can regulate the development of AAA in part by inducing the CD147 expression and in turn modulating the activity of MMP-13 activity.[96]

3.8 MMP-14 or Membrane Type-1 MMP (MT1-MMP)

MMP-14 or membrane type-1 matrix metalloproteinase (MT1-MMP) has been demonstrated in the normal aorta wall as well as in AAA based on data from immunohistochemistry, in situ hybridization, and in situ zymography.[29,75] Colocalization with medial SMCs provides a reasonable explanation for the damage to this cell type following MMP-14 activation, which in turn would weaken the ability of the aortic wall to withstand the distending force of intraluminal blood pressure. MMP-14 is localized to aortic SMCs and macrophages in aneurysmal tissue.[29]

Altered MMP-14 proteolytic turnover and differential regulation of TIMP expression in AAAs suggest that tight regulatory mechanisms are involved in the molecular regulation of MMP activation processes in the pathogenesis of AAAs. In the mouse, macrophage-derived membrane-anchored MMP-14 acts on elastin to promote AAA formation and MMP-1 is also a direct-acting regulator of macrophage proteolytic activity.[71] MMP-14 can play a dominant role in macrophage-mediated elastin destruction.[102] The result is that MMP-14 is operative to produce progressive enlargement of AAA.[102]

4. PUTATIVE SIGNALING PATHWAYS INVOLVED IN AORTIC ANEURYSM DEVELOPMENT: RELATIONSHIP TO MMPs

JNK has been implicated in the pathogenesis of AAA because of the high level of phosphorylated JNK in AAA.[103,104] JNK programs a gene expression pattern that leads to ECM degradation. In two experimental murine models of AAA, inhibition of JNK prevented AAA development.[104] Several agents that reduce JNK phosphorylation, namely the nitrogen-containing bisphosphonate zoledronate,[105] the thiazolidinedione rosiglitazone, the antioxidant flavonoid quercetin, and the natural phenolic compound, curcumin inhibited experimentally induced AAA.[106–108] However, there are conflicting data as inhibition of JNK with SP600125, did not prevent cigarette-smoke extract-induced MMP-1 expression and cigarette-smoke extract produced AAA.[109]

A related factor is of interest in aortic aneurysm is osteopontin (OPN) which is both increased in the wall of aortic aneurysms and correlates with MMP expression. OPN expression is increased in TAA TAVs as well as BAVs.[110,111] Aortic medial SMCs from patients with TAD had 10-fold more OPN than control aorta.[112] OPN expression in SMCs is especially elevated in inflammatory cells.[49] In TAA, OPN protein levels in the aortic wall correlate directly with aortic diameter.[113] Elevated expression of osteopontin is also found in human AAA.[114] In both AAA as well as

TAA, there is a significant positive correlation between MMP-2 expression and OPN expression suggesting OPN can upregulate MMP.[22] Indeed, data suggest that OPN provides a paracrine signal augmenting vascular pro-MMP-9 activity, mediated in part via superoxide generation and oxylipid formation.[115]

5. CIGARETTES AND AAA DEVELOPMENT

The role of cigarettes in AAA development has been recently reviewed.[116] Smoking is not only strongly associated with AAA, in both men and women,[117–119] but is also associated with an accelerated AAA[120] and TAA expansion.[121,122] In addition, cigarette smoking is a critical risk factor for fatal AAA rupture.[123] Cigar smoking has a similar association with aortic aneurysm, depending on the level of cigar exposure.[124]

Proof of the ability of cigarette smoke and its constituents to produce aortic aneurysm comes from several different animal models in which animal are exposed to inhaled cigarette smoke or several of its components—nicotine[125,126] or 3,4-benzopyrene[127] often in conjunction with an agent the increases aortic wall stress such as angiotensin II or an agent that weakens aortic wall structure such as elastase.[125,127–130]

Cigarette smoke or its constituents usually increase MMPs gene expression, quantity, and/or activity.[116] For MMP-1, studies found an increase in MMP-1 assessed either by mRNA, protein, or immunofluorescent microscopy.[109,125,131] For MMP-2, most studies reported an increase,[126,127,132] following exposure to cigarette smoke or its components.[109,133] For MMP-8, several studies indicated an increase[130,131] although there is some contradictory data.[109] For MMP-9, most studies found an increase[126,127,130–132,134] with few exceptions.[109,128] For MMP-12, two studies showed an increase[127,130] while one study did not find a change following exposure to cigarette smoke or its components.[128] On balance, MMPs are increased with the strongest evidence for MMP-1, MMP-2, MMP-9 followed by MMP-8 and MMP-12.[116]

Several signaling pathways have been implicated to mediate the effect of cigarettes or tobacco products on aortic MMPs.[116] It is helpful to review this briefly as it identifies the signally pathways that act on MMPs. Cigarette smoke phosphorylates JNK and nicotine, at concentrations equivalent to plasma levels of cigarette smokers, augment MMP-2, and MMP-9 expressions through a JNK pathway.[134] JNK inhibition suppresses MMP-2 and MMP-9 expression.[134] AMP-activated protein kinase alpha2 (AMPK-α2) is also a mediator of cigarette-induced AAA and uses the AP-2 family of transcription factors.[126,135] Genetic deletion of AMPK-alpha2 (Ape(−/−); Prkaa2(−/−) mice) do not develop nicotine- or AngII-induced AAA.[126] In vascular SMCs, nicotine or AngII-activated AMPK-alpha2 with resultant phosphorylation of (AP-2alpha) and MMP-2 gene expression.[126] This pathway may be cell specific because in other cell types, nicotine decreases AP-2.[136] Janus kinase (JAK) and signal transducer and activator of transcription (STAT) pathway are also involved as mediators of the effect of cigarettes on the vasculature. In rat aortic vascular SMCs, aqueous extract of cigarette smoke significantly increased pro-MMP-9 and modestly increased pro-MMP-2.[132] Increased phosphorylation of Jak2 and Stat3 (pStat3 Tyr 705) occurs in vascular SMCs after exposure to aqueous cigarette smoke extract which also translocates Jak2 and Stat3 to the nucleus.[132] Small interfering RNAs for Jak2 and/or Stat3 significantly reduce pro-MMP-9 and pro-MMP-2.[132]

6. MMP SUBSTRATES AND PATHOGENESIS OF AORTIC ANEURYSM

Stegemann *et al.* used a proteomics approach to identify vascular substrates for three MMPs, by incubation of human radial arteries with MMP-3, MMP-9, or MMP-14.[137] Using mass spectrometry, they identified a number of compounds released from the arterial tissue providing evidence for arterial wall substrates for these MMPs.[137] The likely substrates, from this experiment based on the magnitude of the change or the nature of the protein, include, in addition to collagen which is expected, the ECM glycoprotein Emilin-1, fibronectin, laminin subunit α-5, latent-TGF β-binding protein 2, Periostin, Tenascin-C, Tenascin-X, and the proteoglycan perlecan (Table 1).[137] The majority of these molecules functions to maintaining arterial structure and regulates arterial function. Emilin-1 (elastin microfibril interface located protein) is a glycoprotein localized at the interface between elastin and microfibrils in the artery and undoubtedly operates to facilitate the function of elastin.[139] Fibronectins are glycoproteins that bind to a number of ECM components such as collagen, fibrin as well as integrins, and play a role in cell adhesion, cell growth, inflammatory, and fibrotic processes. Perlecan A is a proteoglycans which inhibit SMC adhesion to fibronectin, influencing SMC activation, migration, and proliferation.[138]

Comparing this proteomic analysis with analysis of aortic tissues from patients with TAD reveal several interesting matches which delve deeper into the mechanisms by which MMP activation might produce TAA,

Table 1 The Roles of Extracellular Proteins Degraded and Released From Human Arteries on Incubation With MMP-3, -9, Or -14

Substrates	**Function in Arteries or Aorta**
Vascular MMP Protein	
Perlecan	A proteoglycans which inhibit smooth muscle cell (SMC) adhesion to fibronectin, influencing SMC activation, migration, and proliferation[138]
Emilin-1	Localized at the interface between elastin and microfibrils and likely operates to facilitate the function of elastin[139]
Fibronectin	Fibronectin binds to collagen, fibrin, and integrins
Laminin subunit a-5	Glycoproteins play a structural scaffolding role
Periostin	Enhance cell migration and fibrillogenesis[140] associated with inflammatory cell infiltrations[141]
Tenascin	Cell growth and adhesion
Collagen	Stiffness of aorta

AAA, dissection, or rupture. Emilin-1 is downregulated by approximately 2.3-fold in the aorta of patients with TAD.[142] The potential linkage of Emilin-1 and hypertension with TAD has been reviewed.[143] It is reasonable to contend that loss of Emilin-1, which appears to operate to facilitate the function of elastin,[139] would produce aortic aneurysm because failure of the aortic elastin function predisposes to aortic aneurysmal dilatation and TAD.[144]

Fibronectin is distorted in TAD.[145] Whether this is a cause or an effect of TAD is not certain, however, considering that fibronectin binds to collagen, fibrin, and integrins, the loss of these components would be anticipated to lead to aortic dilatation. Impaired splicing of fibronectin is associated with TAA formation in patients with BAV.[146] TAA are characterized by a loss of the normal arrangement of lamin in the aortic wall.[147] Collagen I, Laminin alpha2 chain, and fibronectin are all decreased in TAA.[147]

Periostin functions to enhance cell migration and fibrillogenesis in association with ECM molecules.[140] In addition, periostin is associated with inflammatory cell infiltration and destruction of elastic fibers.[141] Mechanical strain increases periostin expression in cultured rat vascular SMCs as well as increasing MMP-2.[141] Increased Periostin expression can produce FAK activation and MCP-1 upregulation that can in turn produce cellular infiltration.[141]

There are different types of Tenascins that subserve many different functions. The role of Tenascins in aneurysm development is more complicated. Tenascin-X exerts a structural function as it regulates both the structure and stability of elastic fibers and organizes collagen fibrils in ECM, influencing tissue elasticity or rigidity.[148] Tenascin-X expression is markedly decreased in AAA tissue.[149]

The potential role of Perlecan in aneurysm formation is unclear and requires further investigation. Perlecan is degraded by MMP-3 in human endothelial cells.[150] However, perlecan levels are increased in AAA and in SMC culture from AAA.[151] Biglycan-deficient mice exhibited significantly increased vascular perlecan content, a deficiency of dense collagen fibers, elastin breaks, and aneurysms.[152]

7. SUMMARY

The increase in MMPs in TAA and AAA support the proteolytic theory of aneurysm development which contends that increased aortic concentrations of active MMPs lead to ECM degradation which weakens the ability of the aorta to withstand the distending intraarterial pressure (Fig. 2). This theory is supported by evidence of increased MMP activity in TAA and AAA along with evidence of destruction of collagen and elastin in the aortic wall which experimentally can be suppressed with MMP inhibition.[153–155] An alternate theory relies on the ability of MMPs, to cleave molecules involved in signal transduction which in turn alters signal transduction pathways[12] that constrain aortic dilatation.

A role for MMP in aneurysm development is supported by the ability of cigarette smoke, which is a major factor producing AAA, to activate a number of MMPs.

MMP substrates have been identified in the arterial wall. Some of these have been implicated in aneurysm development. The evidence presented in this chapter focused on human studies and protein content of MMPs. Based on these data, there is sufficient evidence to implicate MMPs in the pathogenesis of TAA and AAA.

REFERENCES

1. Davies RR, Goldstein LJ, Coady MA, et al. Yearly rupture or dissection rates for thoracic aortic aneurysms: simple prediction based on size. *Ann Thorac Surg*. 2002;73:17–27.
2. Olsson C, Thelin S, Stahle E, Ekbom A, Granath F. Thoracic aortic aneurysm and dissection: increasing prevalence and improved outcomes reported in a nationwide

population-based study of more than 14,000 cases from 1987 to 2002. *Circulation.* 2006;114:2611–2618.
3. Sampson UKA, Norman PE, Fowkes FGR, et al. Estimation of global and regional incidence and prevalence of abdominal aortic aneurysms 1990 to 2010. *Glob Heart.* 2014;9:159–170.
4. Centers for Disease Control and Prevention. National Center for Health Statistics. Underlying Cause of Death 1999–2014 on CDC WONDER Online Database, released 2015. Data are from the Multiple Cause of Death Files, 1999–2014, 2015. [WWW Document]. URL: http://wonder.cdc.gov/ucd-icd10.html. Centers for Disease Control and Prevention. *National Center for Health Statistics. Underlying Cause of Death 1999–2014 on CDC WONDER Online Database, released 2015. Data are from the Multiple Cause of Death Files, 1999–2014.* [WWW Document]. URL:2015. http://wonder.cdc.gov/ucd-icd10.html
5. Nordon IM, Hinchliffe RJ, Loftus IM, Thompson MM. Pathophysiology and epidemiology of abdominal aortic aneurysms. *Nat Rev Cardiol.* 2011;8:92–102.
6. Norman P, Curci J. Understanding the effects of tobacco smoke on the pathogenesis of aortic aneurysm. *Arterioscler Thromb Vasc Biol.* 2013;33:1473–1477.
7. Schlatmann TJ, Becker AE. Pathogenesis of dissecting aneurysm of aorta. Comparative histopathologic study of significance of medial changes. *Am J Cardiol.* 1977;39:21–26.
8. Tang PC, Coady MA, Lovoulos C, et al. Hyperplastic cellular remodeling of the media in ascending thoracic aortic aneurysms. *Circulation.* 2005;112:1098–1105.
9. Nagase H, Visse R, Murphy G. Structure and function of matrix metalloproteinases and TIMPs. *Cardiovasc Res.* 2006;69:562–573.
10. Bode W, Fernandez-Catalan C, Tschesche H, Grams F, Nagase H, Maskos K. Structural properties of matrix metalloproteinases. *Cell Mol Life Sci.* 1999;55:639–652.
11. Bonnans C, Chou J, Werb Z. Remodelling the extracellular matrix in development and disease. *Nat Rev Mol Cell Biol.* 2014;15:786–801.
12. Brew K, Nagase H. The tissue inhibitors of metalloproteinases (TIMPs): an ancient family with structural and functional diversity. *Biochim Biophys Acta.* 2010;1803:55–71.
13. Clark IM, Swingler TE, Sampieri CL, Edwards DR. The regulation of matrix metalloproteinases and their inhibitors. *Int J Biochem Cell Biol.* 2008;40:1362–1378.
14. Vandooren J, Van den Steen PE, Opdenakker G. Biochemistry and molecular biology of gelatinase B or matrix metalloproteinase-9 (MMP-9): the next decade. *Crit Rev Biochem Mol Biol.* 2013;48:222–272.
15. Mecham RP, Broekelmann TJ, Fliszar CJ, Shapiro SD, Welgus HG, Senior RM. Elastin degradation by matrix metalloproteinases. Cleavage site specificity and mechanisms of elastolysis. *J Biol Chem.* 1997;272:18071–18076.
16. Ruddy JM, Jones JA, Ikonomidis JS. Pathophysiology of thoracic aortic aneurysm (TAA): is it not one uniform aorta? Role of embryologic origin. *Prog Cardiovasc Dis.* 2013;56:68–73.
17. Koullias GJ, Ravichandran P, Korkolis DP, Rimm DL, Elefteriades JA. Increased tissue microarray matrix metalloproteinase expression favors proteolysis in thoracic aortic aneurysms and dissections. *Ann Thorac Surg.* 2004;78:2106–2110.
18. Ishii T, Asuwa N. Collagen and elastin degradation by matrix metalloproteinases and tissue inhibitors of matrix metalloproteinase in aortic dissection. *Hum Pathol.* 2000;31:640–646.
19. Martin M, Pichel IA, Florez Munoz JP, et al. Low transcriptional activity haplotype of matrix metalloproteinase 1 is less frequent in bicuspid aortic valve patients. *Gene.* 2013;524:304–308.
20. Wilton E, Bland M, Thompson M, Jahangiri M. Matrix metalloproteinase expression in the ascending aorta and aortic valve. *Interact Cardiovasc Thorac Surg.* 2008;7:37–40.

21. Koullias GJ, Korkolis DP, Ravichandran P, Psyrri A, Hatzaras I, Elefteriades JA. Tissue microarray detection of matrix metalloproteinases, in diseased tricuspid and bicuspid aortic valves with or without pathology of the ascending aorta. *Eur J Cardiothorac Surg*. 2004;26:1098–1103.
22. Mi T, Nie B, Zhang C, Zhou H. The elevated expression of osteopontin and NF-kappaB in human aortic aneurysms and its implication. *J Huazhong Univ Sci Technolog Med Sci*. 2011;31:602–607.
23. Schmoker JD, McPartland KJ, Fellinger EK, et al. Matrix metalloproteinase and tissue inhibitor expression in atherosclerotic and nonatherosclerotic thoracic aortic aneurysms. *J Thorac Cardiovasc Surg*. 2007;133:155–161.
24. Rabkin SW. Differential expression of MMP-2, MMP-9 and TIMP proteins in ascending thoracic aortic aneurysm—comparison with and without bicuspid aortic valve: a meta-analysis. *Vasa*. 2014;43:433–442.
25. Hu Z, Wang Z, Wu H, et al. Ang II enhances noradrenaline release from sympathetic nerve endings thus contributing to the up-regulation of metalloprotease-2 in aortic dissection patients' aorta wall. *PLoS One*. 2013;8: e76922.
26. Wang C, Chang Q, Sun X, et al. Angiotensin II induces an increase in matrix metalloproteinase 2 expression in aortic smooth muscle cells of ascending thoracic aortic aneurysms through JNK, ERK1/2, and p38 MAPK activation. *J Cardiovasc Pharmacol*. 2015;66:285–293.
27. Tscheuschler A, Meffert P, Beyersdorf F, et al. MMP-2 isoforms in aortic tissue and serum of patients with ascending aortic aneurysms and aortic root aneurysms. *PLoS One*. 2016;11: e0164308.
28. Huusko T, Salonurmi T, Taskinen P, et al. Elevated messenger RNA expression and plasma protein levels of osteopontin and matrix metalloproteinase types 2 and 9 in patients with ascending aortic aneurysms. *J Thorac Cardiovasc Surg*. 2013;145:1117–1123.
29. Nollendorfs A, Greiner TC, Nagase H, Baxter BT. The expression and localization of membrane type-1 matrix metalloproteinase in human abdominal aortic aneurysms. *J Vasc Surg*. 2001;34:316–322.
30. Liu O, Li J, Xin Y, et al. Association of MMP-2 gene haplotypes with thoracic aortic dissection in Chinese han population. *BMC Cardiovasc Disord*. 2016;16:11.
31. Ma Z, Chang MJ, Shah R, Adamski J, Zhao X, Benveniste EN. Brg-1 is required for maximal transcription of the human matrix metalloproteinase-2 gene. *J Biol Chem*. 2004;279:46326–46334.
32. Yuan Y, Wang C, Xu J, Tao J, Xu Z, Huang S. BRG1 overexpression in smooth muscle cells promotes the development of thoracic aortic dissection. *BMC Cardiovasc Disord*. 2014;14:144.
33. Jones JA, Stroud RE, O'Quinn EC, et al. Selective microRNA suppression in human thoracic aneurysms: relationship of miR-29a to aortic size and proteolytic induction. *Circ Cardiovasc Genet*. 2011;4:605–613.
34. Siu SC, Silversides CK. Bicuspid aortic valve disease. *J Am Coll Cardiol*. 2010;55:2789–2800.
35. Hiratzka LF, Bakris GL, Beckman JA, et al. 2010 ACCF/AHA/AATS/ACR/ASA/SCA/SCAI/SIR/STS/SVM guidelines for the diagnosis and management of patients with Thoracic Aortic Disease: a report of the American COLLEGE of cardiology Foundation/American Heart Association Task Force on Practice Guidelines, American Association for Thoracic Surgery, American College of Radiology, American Stroke Association, Society of Cardiovascular Anesthesiologists, Society for Cardiovascular Angiography and Interventions, Society of Interventional Radiology, Society of Thoracic Surgeons, and Society for Vascular Medicine. *Circulation*. 2010;121: e266–e369.

36. Wang Y, Wu B, Dong L, Wang C, Wang X, Shu X. Circulating matrix metalloproteinase patterns in association with aortic dilatation in bicuspid aortic valve patients with isolated severe aortic stenosis. *Heart Vessels*. 2016;31:189–197.
37. Fedak PW, de Sa MP, Verma S, et al. Vascular matrix remodeling in patients with bicuspid aortic valve malformations: implications for aortic dilatation. *J Thorac Cardiovasc Surg*. 2003;126:797–806.
38. Barutcu I, Karakaya O, Esen AM, et al. Circulating stromelysin concentration is elevated in hypertensive aortic root dilatation. *Heart Vessels*. 2009;24:138–141.
39. Lesauskaite V, Sinkunaite G, Benetis R, et al. Matrix metalloproteinase-3 gene polymorphism and dilatative pathology of ascending thoracic aorta. *Medicina (Kaunas)*. 2008;44:386–391.
40. Lesauskaite V, Sinkunaite-Marsalkiene G, Tamosiunas A, Benetis R. Protective effects of angiotensin-converting enzyme I/I and matrix metalloproteinase-3 6A/6A polymorphisms on dilatative pathology within the ascending thoracic aorta. *Eur J Cardiothorac Surg*. 2011;40:23–27.
41. Mohamed SA, Noack F, Schoellermann K, et al. Elevation of matrix metalloproteinases in different areas of ascending aortic aneurysms in patients with bicuspid and tricuspid aortic valves. *ScientificWorldJournal*. 2012;806261.
42. Li Y, Shao A-Z, Jiang H-T, et al. The prominent expression of plasma matrix metalloproteinase-8 in acute thoracic aortic dissection. *J Surg Res*. 2010;163:e99–e104.
43. Wang X-Z, Du X-M, Jing Q-M, et al. Impact of matrix metalloproteinase-8 gene variations on the risk of thoracic aortic dissection in a Chinese Han population. *Mol Biol Rep*. 2013;40:5953–5958.
44. Li Y, Li N, Ma W, Jiang Y, Ma G, Zhang L. Association between -799C/T single nucleotide polymorphism of the MMP-8 promoter region and thoracic aortic dissection. *Mol Med Rep*. 2014;10:1857–1862.
45. Gomez DE, Alonso DF, Yoshiji H, Thorgeirsson UP. Tissue inhibitors of metalloproteinases: structure, regulation and biological functions. *Eur J Cell Biol*. 1997;74:111–122.
46. Absi TS, Sundt 3rd TM, Tung WS, et al. Altered patterns of gene expression distinguishing ascending aortic aneurysms from abdominal aortic aneurysms: complementary DNA expression profiling in the molecular characterization of aortic disease. *J Thorac Cardiovasc Surg*. 2003;126:344–357.
47. Wang X-L, Liu O, Yan-wen Q, Zhang HJ, Lv Y. Association of the polymorphisms of MMP-9 and TIMP-3 genes with thoracic aortic dissection in Chinese Han population. *Acta Pharmacol Sin*. 2014;35:351–355.
48. Nagasawa A, Yoshimura K, Suzuki R, et al. Important role of the angiotensin II pathway in producing matrix metalloproteinase-9 in human thoracic aortic aneurysms. *J Surg Res*. 2013;183:472–477.
49. Lesauskaite V, Epistolato MC, Castagnini M, Urbonavicius S, Tanganelli P. Expression of matrix metalloproteinases, their tissue inhibitors, and osteopontin in the wall of thoracic and abdominal aortas with dilatative pathology. *Hum Pathol*. 2006;37:1076–1084.
50. Sproul EP, Argraves WS. A cytokine axis regulates elastin formation and degradation. *Matrix Biol*. 2013;32:86–94.
51. Zhang L, Liao MF, Tian L, et al. Overexpression of interleukin-1beta and interferon-in type I thoracic aortic dissections and ascending thoracic aortic aneurysms: possible correlation with matrix metalloproteinase-9 expression and apoptosis of aortic media cells. *Eur J Cardiothorac Surg*. 2011;40:17–22.
52. Shen YH, Zhang L, Ren P, et al. AKT2 confers protection against aortic aneurysms and dissections. *Circ Res*. 2013;112:618–632.
53. Rossignol P, Ho-Tin-Noe B, Vranckx R, et al. Protease nexin-1 inhibits plasminogen activation-induced apoptosis of adherent cells. *J Biol Chem*. 2004;279:10346–10356.

54. Carmeliet P, Moons L, Lijnen R, et al. Urokinase-generated plasmin activates matrix metalloproteinases during aneurysm formation. *Nat Genet.* 1997;17:439–444.
55. Liao M, Zou S, Weng J, et al. A microRNA profile comparison between thoracic aortic dissection and normal thoracic aorta indicates the potential role of microRNAs in contributing to thoracic aortic dissection pathogenesis. *J Vasc Surg.* 2011;53: 1341–1349:e3.
56. Song Y, Xie Y, Liu F, et al. Expression of matrix metalloproteinase-12 in aortic dissection. *BMC Cardiovasc Disord.* 2013;13:34.
57. Del Porto F, di Gioia C, Tritapepe L, et al. The multitasking role of macrophages in Stanford type A acute aortic dissection. *Cardiology.* 2014;127:123–129.
58. Jackson V, Olsson T, Kurtovic S, et al. Matrix metalloproteinase 14 and 19 expression is associated with thoracic aortic aneurysms. *J Thorac Cardiovasc Surg.* 2012;144: 459–466.
59. Martin-Alonso M, Garcia-Redondo AB, Guo D, et al. Deficiency of MMP17/MT4-MMP proteolytic activity predisposes to aortic aneurysm in mice. *Circ Res.* 2015;117: e13–e26.
60. Shimizu K, Mitchell RN, Libby P. Inflammation and cellular immune responses in abdominal aortic aneurysms. *Arterioscler Thromb Vasc Biol.* 2006;26:987–994.
61. Gandhi R, Irizarry E, Cantor J, et al. Analysis of elastin cross-linking and the connective tissue matrix of abdominal aortic aneurysms. *Surgery.* 1994;115:617–620.
62. Galis ZS, Khatri JJ. Matrix metalloproteinases in vascular remodeling and atherogenesis: the good, the bad, and the ugly. *Circ Res.* 2002;90:251–262.
63. Davis V, Persidskaia R, Baca-Regen L, et al. Matrix metalloproteinase-2 production and its binding to the matrix are increased in abdominal aortic aneurysms. *Arterioscler Thromb Vasc Biol.* 1998;18:1625–1633.
64. Freestone T, Turner R, Coady A, Higman D, Greenhalgh R, Powell J. Inflammation and matrix metalloproteinases in the enlarging abdominal aortic aneurysm. *Arterioscler Thromb Vasc Biol.* 1995;15:1145–1151.
65. Keeling W, Armstrong P, Stone P, Bandyk D, Shames M. An overview of matrix metalloproteinases in the pathogenesis and treatment of abdominal aortic aneurysms. *Vasc Endovascular Surg.* 2005;39:457–464.
66. Newman K, Jean-Claude J, Li H, et al. Cellular localization of matrix metalloproteinases in the abdominal aortic aneurysm wall. *J Vasc Surg.* 1994;20:814–820.
67. Newman K, Malon A, Shin R, Scholes J, Ramey W, Tilson M. Matrix metalloproteinases in abdominal aortic aneurysm: characterization, purification, and their possible sources. *Connect Tissue Res.* 1994;30:265–276.
68. Reeps C, Pelisek J, Seidl S, et al. Inflammatory infiltrates and neovessels are relevant sources of MMPs in abdominal aortic aneurysm wall. *Pathobiology.* 2009;76:243–252.
69. Irizarry E, Newman K, Gandhi R, et al. Demonstration of interstitial collagenase in abdominal aortic aneurysm disease. *J Surg Res.* 1993;54:571–574.
70. Newman K, Ogata Y, Malon A, et al. Identification of matrix metalloproteinases 3 (stromelysin-1) and 9 (gelatinase B) in abdominal aortic aneurysm. *Arterioscler Thromb.* 1994;14:1315–1320.
71. Annabi B, Shedid D, Ghosn P, et al. Differential regulation of matrix metalloproteinase activities in abdominal aortic aneurysms. *J Vasc Surg.* 2002;35:539–546.
72. Verschuren L, Lindeman JHN, van Bockel JH, Abdul-Hussien H, Kooistra T, Kleemann R. Up-regulation and coexpression of MIF and matrix metalloproteinases in human abdominal aortic aneurysms. *Antioxid Redox Signal.* 2005;7:1195–1202.
73. Courtois A, Nusgens BV, Hustinx R, et al. 18 F-FDG uptake assessed by PET/CT in abdominal aortic aneurysms is associated with cellular and molecular alterations prefacing wall deterioration and rupture. *J Nucl Med.* 2013;54:1740–1747.

74. Crowther M, Goodall S, Jones JL, Bell PR, Thompson MM. Increased matrix metalloproteinase 2 expression in vascular smooth muscle cells cultured from abdominal aortic aneurysms. *J Vasc Surg*. 2000;32:575–583.
75. Crowther M, Goodall S, Jones JL, Bell PR, Thompson MM. Localization of matrix metalloproteinase 2 within the aneurysmal and normal aortic wall. *Br J Surg*. 2000;87:1391–1400.
76. Khan JA, Abdul Rahman MNA, Mazari FAK, et al. Intraluminal thrombus has a selective influence on matrix metalloproteinases and their inhibitors (tissue inhibitors of matrix metalloproteinases) in the wall of abdominal aortic aneurysms. *Ann Vasc Surg*. 2012;26:322–329.
77. Sakalihasan N, Delvenne P, Nusgens BV, Limet R, Lapiere CM. Activated forms of MMP2 and MMP9 in abdominal aortic aneurysms. *J Vasc Surg*. 1996;24:127–133.
78. Carrell TWG, Burnand KG, Wells GMA, Clements JM, Smith A. Stromelysin-1 (matrix metalloproteinase-3) and tissue inhibitor of metalloproteinase-3 are overexpressed in the wall of abdominal aortic aneurysms. *Circulation*. 2002;105:477–482.
79. Saratzis A, Bown MJ, Wild B, et al. Association between seven single nucleotide polymorphisms involved in inflammation and proteolysis and abdominal aortic aneurysm. *J Vasc Surg*. 2015;61:1120–1128:e1.
80. McColgan P, Peck GE, Greenhalgh RM, Sharma P. The genetics of abdominal aortic aneurysms: a comprehensive meta-analysis involving eight candidate genes in over 16,700 patients. *Int Surg*. 2009;94:350–358.
81. Deguara J, Burnand KG, Berg J, et al. An increased frequency of the 5A allele in the promoter region of the MMP3 gene is associated with abdominal aortic aneurysms. *Hum Mol Genet*. 2007;16:3002–3007(Erratum appears in Hum Mol Genet. 2009 Dec 1;18(23):4688. Note: Stern, Rachel F [corrected to Stern, Rowena F]).
82. Saracini C, Bolli P, Sticchi E, et al. Polymorphisms of genes involved in extracellular matrix remodeling and abdominal aortic aneurysm. *J Vasc Surg*. 2012;55:171–179:e2.
83. Ciavarella C, Alviano F, Gallitto E, et al. Human vascular wall mesenchymal stromal cells contribute to abdominal aortic aneurysm pathogenesis through an impaired immunomodulatory activity and increased levels of matrix metalloproteinase-9. *Circ J*. 2015;79:1460–1469.
84. Longo GM, Xiong W, Greiner TC, Zhao Y, Fiotti N, Baxter BT. Matrix metalloproteinases 2 and 9 work in concert to produce aortic aneurysms. *J Clin Invest*. 2002;110:625–632.
85. Elmore JR, Keister BF, Franklin DP, Youkey JR, Carey DJ. Expression of matrix metalloproteinases and TIMPs in human abdominal aortic aneurysms. *Ann Vasc Surg*. 1998;12:221–228.
86. Fontaine V, Jacob M-P, Houard X, et al. Involvement of the mural thrombus as a site of protease release and activation in human aortic aneurysms. *Am J Pathol*. 2002;161:1701–1710.
87. Koole D, Zandvoort HJA, Schoneveld A, et al. Intraluminal abdominal aortic aneurysm thrombus is associated with disruption of wall integrity. *J Vasc Surg*. 2013;57:77–83.
88. Petersen E, Gineitis A, Wagberg F, Angquist KA. Activity of matrix metalloproteinase-2 and -9 in abdominal aortic aneurysms. Relation to size and rupture. *Eur J Vasc Endovasc Surg*. 2000;20:457–461.
89. Ng E, Morris DR, Golledge J. The association between plasma matrix metalloproteinase-9 concentration and endoleak after endovascular aortic aneurysm repair: a meta-analysis. *Atherosclerosis*. 2015;242:535–542.
90. Ogata T, Shibamura H, Tromp G, et al. Genetic analysis of polymorphisms in biologically relevant candidate genes in patients with abdominal aortic aneurysms. *J Vasc Surg*. 2005;41:1036–1042.

91. Curci JA, Liao S, Huffman MD, Shapiro SD, Thompson RW. Expression and localization of macrophage elastase (matrix metalloproteinase-12) in abdominal aortic aneurysms. *J Clin Invest.* 1998;102:1900–1910.
92. Didangelos A, Yin X, Mandal K, et al. Extracellular matrix composition and remodeling in human abdominal aortic aneurysms: a proteomics approach. *Mol Cell Proteomics.* 2011;10M111.008128.
93. Kazi M, Zhu C, Roy J, et al. Difference in matrix-degrading protease expression and activity between thrombus-free and thrombus-covered wall of abdominal aortic aneurysm. *Arterioscler Thromb Vasc Biol.* 2005;25:1341–1346.
94. Liao M, Liu Z, Bao J, et al. A proteomic study of the aortic media in human thoracic aortic dissection: implication for oxidative stress. *J Thorac Cardiovasc Surg.* 2008;136:63–65.
95. Tromp G, Gatalica Z, Skunca M, et al. Elevated expression of matrix metalloproteinase-13 in abdominal aortic aneurysms. *Ann Vasc Surg.* 2004;18:414–420.
96. Lizarbe TR, Tarin C, Gomez M, et al. Nitric oxide induces the progression of abdominal aortic aneurysms through the matrix metalloproteinase inducer EMMPRIN. *Am J Pathol.* 2009;175:1421–1430.
97. Mao D, Lee JK, VanVickle SJ, Thompson RW. Expression of collagenase-3 (MMP-13) in human abdominal aortic aneurysms and vascular smooth muscle cells in culture. *Biochem Biophys Res Commun.* 1999;261:904–910.
98. Tung WS, Lee JK, Thompson RW. Simultaneous analysis of 1176 gene products in normal human aorta and abdominal aortic aneurysms using a membrane-based complementary DNA expression array. *J Vasc Surg.* 2001;34:143–150.
99. Grass GD, Toole BP. How, with whom and when: an overview of CD147-mediated regulatory networks influencing matrix metalloproteinase activity. *Biosci Rep.* 2015;36: e00283.
100. Seizer P, Gawaz M, May AE. Cyclophilin A and EMMPRIN (CD147) in cardiovascular diseases. *Cardiovasc Res.* 2014;102:17–23.
101. Xiong L, Edwards 3rd CK, Zhou L. The biological function and clinical utilization of CD147 in human diseases: a review of the current scientific literature. *Int J Mol Sci.* 2014;15:17411–17441.
102. Xiong W, Knispel R, MacTaggart J, Greiner TC, Weiss SJ, Baxter BT. Membrane-type 1 matrix metalloproteinase regulates macrophage-dependent elastolytic activity and aneurysm formation in vivo. *J Biol Chem.* 2009;284:1765–1771.
103. DiMusto P, Lu G, Ghosh A, et al. Increased JNK in males compared with females in a rodent model of abdominal aortic aneurysm. *J Surg Res.* 2012;176:687–695.
104. Yoshimura K, Aoki H, Ikeda Y, et al. Regression of abdominal aortic aneurysm by inhibition of c-Jun N-terminal kinase. *Nat Med.* 2005;11:1330–1338.
105. Tsai S, Huang P, Peng Y, et al. Zoledronate attenuates angiotensin II-induced abdominal aortic aneurysm through inactivation of Rho/ROCK-dependent JNK and NF-kappaB pathway. *Cardiovasc Res.* 2013;100:501–510.
106. Fan J, Li X, Yan YW, et al. Curcumin attenuates rat thoracic aortic aneurysm formation by inhibition of the c-Jun N-terminal kinase pathway and apoptosis. *Nutrition.* 2012;28:1068–1074.
107. Pirianov G, Torsney E, Howe F, Cockerill GW. Rosiglitazone negatively regulates c-Jun N-terminal kinase and toll-like receptor 4 proinflammatory signalling during initiation of experimental aortic aneurysms. *Atherosclerosis.* 2012;225:69–75.
108. Wang L, Cheng X, Li H, et al. Quercetin reduces oxidative stress and inhibits activation of c-Jun N-terminal kinase/activator protein-1 signaling in an experimental mouse model of abdominal aortic aneurysm. *Mol Med Rep.* 2014;9:435–442.
109. Lemaitre V, Dabo AJ, D'Armiento J. Cigarette smoke components induce matrix metalloproteinase-1 in aortic endothelial cells through inhibition of mTOR signaling. *Toxicol Sci.* 2011;123:542–549.

110. Blunder S, Messner B, Aschacher T, et al. Characteristics of TAV- and BAV-associated thoracic aortic aneurysms—smooth muscle cell biology, expression profiling, and histological analyses. *Atherosclerosis.* 2012;220:355–361.
111. Majumdar R, Miller DV, Ballman KV, et al. Elevated expressions of osteopontin and tenascin C in ascending aortic aneurysms are associated with trileaflet aortic valves as compared with bicuspid aortic valves. *Cardiovasc Pathol.* 2007;16:144–150.
112. Zhang J, Wang L, Fu W, et al. Smooth muscle cell phenotypic diversity between dissected and unaffected thoracic aortic media. *J Cardiovasc Surg (Torino).* 2013;54:511–521.
113. Meng YH, Tian C, Liu L, Wang L, Chang Q. Elevated expression of connective tissue growth factor, osteopontin and increased collagen content in human ascending thoracic aortic aneurysms. *Vascular.* 2014;22:20–27.
114. Golledge J, Muller J, Shephard N, et al. Association between osteopontin and human abdominal aortic aneurysm. *Arterioscler Thromb Vasc Biol.* 2007;27:655–660.
115. Lai C-F, Seshadri V, Huang K, et al. An osteopontin-NADPH oxidase signaling cascade promotes pro-matrix metalloproteinase 9 activation in aortic mesenchymal cells. *Circ Res.* 2006;98:1479–1489.
116. Rabkin SW. The effect of nicotine and tobacco on aortic matrix metalloproteinases in the production of aortic aneurysm. *Curr Vasc Pharmacol.* 2016;14:514–522.
117. Kent K, Zwolak R, Egorova N, et al. Analysis of risk factors for abdominal aortic aneurysm in a cohort of more than 3 million individuals. *J Vasc Surg.* 2010;52:539–548.
118. Lederle F, Nelson D, Joseph A. Smokers' relative risk for aortic aneurysm compared with other smoking-related diseases: a systematic review. *J Vasc Surg.* 2003;38:329–334.
119. Wilmink T, Quick C, Day N. The association between cigarette smoking and abdominal aortic aneurysms. *J Vasc Surg.* 1999;30:1099–1105.
120. MacSweeney S, Ellis M, Worrell PC, Greenhalgh R, Powell J. Smoking and growth rate of small abdominal aortic aneurysms. *Lancet.* 1994;344:651–652.
121. Bonser RS, Pagano D, Lewis ME, et al. Clinical and patho-anatomical factors affecting expansion of thoracic aortic aneurysms. *Heart.* 2000;84:277–283. http://dx.doi.org/10.1136/heart.84.3.277.
122. Dapunt OE, Galla JD, Sadeghi AM, et al. The natural history of thoracic aortic aneurysms. *J Thorac Cardiovasc Surg.* 1994;107:1323–1333. http://dx.doi.org/10.1016/S0022-5223(94)70054-0.
123. Strachan DP. Predictors of death from aortic aneurysm among middle-aged men: the Whitehall study. *Br J Surg.* 1991;78:401–404.
124. Chang CM, Corey CG, Rostron BL, Apelberg BJ. Systematic review of cigar smoking and all cause and smoking related mortality. *BMC Public Health.* 2015;15:390.
125. Maegdefessel L, Azuma J, Toh R, et al. MicroRNA-21 blocks abdominal aortic aneurysm development and nicotine-augmented expansion. *Sci Transl Med.* 2012;4:122ra22.
126. Wang S, Zhang C, Zhang M, et al. Activation of AMP-activated protein kinase alpha2 by nicotine instigates formation of abdominal aortic aneurysms in mice in vivo. *Nat Med.* 2012;18:902–910.
127. Ji K, Zhang Y, Jiang F, et al. Exploration of the mechanisms by which 3,4-benzopyrene promotes angiotensin II-induced abdominal aortic aneurysm formation in mice. *J Vasc Surg.* 2014;59:492–499.
128. Bergoeing M, Arif B, Hackmann A, Ennis T, Thompson R, Curci J. Cigarette smoking increases aortic dilatation without affecting matrix metalloproteinase-9 and -12 expression in a modified mouse model of aneurysm formation. *J Vasc Surg.* 2007;45:1217–1227.
129. Buckley C, Wyble C, Borhani M, et al. Accelerated enlargement of experimental abdominal aortic aneurysms in a mouse model of chronic cigarette smoke exposure. *J Am Coll Surg.* 2004;199:896–903.

130. Stolle K, Berges A, Lietz M, Lebrun S, Wallerath T. Cigarette smoke enhances abdominal aortic aneurysm formation in angiotensin II-treated apolipoprotein E-deficient mice. *Toxicol Lett.* 2010;199:403–409.
131. Nordskog B, Blixt A, Morgan W, Fields W, Hellmann G. Matrix-degrading and pro-inflammatory changes in human vascular endothelial cells exposed to cigarette smoke condensate. *Cardiovasc Toxicol.* 2003;3:101–117.
132. Ghosh A, Pechota A, Coleman D, Upchurch Jr GR, Eliason JL. Cigarette smoke-induced MMP2 and MMP9 secretion from aortic vascular smooth cells is mediated via the Jak/Stat pathway. *Hum Pathol.* 2015;46:284–294.
133. Jacob-Ferreira A, Palei A, Cau S, et al. Evidence for the involvement of matrix metalloproteinases in the cardiovascular effects produced by nicotine. *Eur J Pharmacol.* 2010;627:216–222.
134. Li Z, Guo Z, Zhang Z, et al. Nicotine-induced upregulation of VCAM-1, MMP-2, and MMP-9 through the alpha7-nAChR-JNK pathway in RAW264.7 and MOVAS cells. *Mol Cell Biochem.* 2015;399:49–58.
135. Eckert D, Buhl S, Weber S, Jager R, Schorle H. The AP-2 family of transcription factors. *Genome Biol.* 2005;6:246.
136. Sun X, Ritzenthaler J, Zhong X, Zheng Y, Roman J, Han S. Nicotine stimulates PPARbeta/delta expression in human lung carcinoma cells through activation of PI3K/mTOR and suppression of AP-2alpha. *Cancer Res.* 2009;69:6445–6453.
137. Stegemann C, Didangelos A, Barallobre-Barreiro J, et al. Proteomic identification of matrix metalloproteinase substrates in the human vasculature. *Circ Cardiovasc Genet.* 2013;6:106–117.
138. Lundmark K, Tran PK, Kinsella MG, Clowes AW, Wight TN, Hedin U. Perlecan inhibits smooth muscle cell adhesion to fibronectin: role of heparan sulfate. *J Cell Physiol.* 2001;188:67–74.
139. Colombatti A, Doliana R, Bot S, et al. The EMILIN protein family. *Matrix Biol.* 2000;19:289–301.
140. Halper J, Kjaer M. Basic components of connective tissues and extracellular matrix: elastin, fibrillin, fibulins, fibrinogen, fibronectin, laminin, tenascins and thrombospondins. *Adv Exp Med Biol.* 2014;802:31–47.
141. Yamashita O, Yoshimura K, Nagasawa A, et al. Periostin links mechanical strain to inflammation in abdominal aortic aneurysm. *PLoS One.* 2013;8: e79753.
142. Zhang K, Pan X, Zheng J, Xu D, Zhang J, Sun L. Comparative tissue proteomics analysis of thoracic aortic dissection with hypertension using the iTRAQ technique. *Eur J Cardiothorac Surg.* 2015;47:431–438.
143. Rabkin SW. Is Emilin-1 a molecular link contributing to the extension of thoracic aortic aneurysm dissection and increasing the magnitude of the associated hypertension? *IMM.* 2016. http://dx.doi.org/10.15761/IMM.1000254.
144. Pratt B, Curci J. Arterial elastic fiber structure function and potential roles in acute aortic dissection. *J Cardiovasc Surg (Torino).* 2010;51:647–656.
145. Sariola H, Viljanen T, Luosto R. Histological pattern and changes in extracellular matrix in aortic dissections. *J Clin Pathol.* 1986;39:1074–1081.
146. Paloschi V, Kurtovic S, Folkersen L, et al. Impaired splicing of fibronectin is associated with thoracic aortic aneurysm formation in patients with bicuspid aortic valve. *Arterioscler Thromb Vasc Biol.* 2011;31:691–697.
147. Cotrufo M, De Santo L, Della Corte A, et al. Basal lamina structural alterations in human asymmetric aneurismatic aorta. *Eur J Histochem.* 2005;49:363–370.
148. Valcourt U, Alcaraz LB, Exposito J-Y, Lethias C, Bartholin L. Tenascin-X: beyond the architectural function. *Cell Adh Migr.* 2015;9:154–165.

149. Zweers MC, Peeters ACTM, Graafsma S, et al. Abdominal aortic aneurysm is associated with high serum levels of tenascin-X and decreased aneurysmal tissue tenascin-X. *Circulation*. 2006;113:1702–1707.
150. Whitelock JM, Murdoch AD, Iozzo RV, Underwood PA. The degradation of human endothelial cell-derived perlecan and release of bound basic fibroblast growth factor by stromelysin, collagenase, plasmin, and heparanases. *J Biol Chem*. 1996;271: 10079–10086.
151. Melrose J, Whitelock J, Xu Q, Ghosh P. Pathogenesis of abdominal aortic aneurysms: possible role of differential production of proteoglycans by smooth muscle cells. *J Vasc Surg*. 1998;28:676–686.
152. Tang T, Thompson JC, Wilson PG, et al. Biglycan deficiency: increased aortic aneurysm formation and lack of atheroprotection. *J Mol Cell Cardiol*. 2014;75:174–180.
153. Ejiri J, Inoue N, Tsukube T, et al. Oxidative stress in the pathogenesis of thoracic aortic aneurysm: protective role of statin and angiotensin II type 1 receptor blocker. *Cardiovasc Res*. 2003;59:988–996.
154. Nosoudi N, Nahar-Gohad P, Sinha A, et al. Prevention of abdominal aortic aneurysm progression by targeted inhibition of matrix metalloproteinase activity with batimastat-loaded nanoparticles. *Circ Res*. 2015;117:e80–e89.
155. Zhang T, Xu J, Li D, et al. Salvianolic acid A, a matrix metalloproteinase-9 inhibitor of Salvia miltiorrhiza, attenuates aortic aneurysm formation in apolipoprotein E-deficient mice. *Phytomedicine*. 2014;21(10):1137–1145.

CHAPTER EIGHT

Matrix Metalloproteinases in Remodeling of Lower Extremity Veins and Chronic Venous Disease

Yunfei Chen, Wei Peng, Joseph D. Raffetto, Raouf A. Khalil[1]
Vascular Surgery Research Laboratories, Brigham and Women's Hospital, Harvard Medical School, Boston, MA, United States
[1]Corresponding author: e-mail address: raouf_khalil@hms.harvard.edu

Contents

Abstract

The veins of the lower extremity are equipped with efficient wall, contractile vascular smooth muscle (VSM), and competent valves in order to withstand the high venous hydrostatic pressure in the lower limb and allow unidirectional movement of deoxygenated blood toward the heart. The vein wall structure and function are in part regulated by matrix metalloproteinases (MMPs). MMPs are zinc-dependent endopeptidases that are secreted as inactive pro-MMPs by different cells in the venous wall including fibroblasts, VSM, and leukocytes. Pro-MMPs are activated by other MMPs, proteinases, and

Progress in Molecular Biology and Translational Science, Volume 147
ISSN 1877-1173
http://dx.doi.org/10.1016/bs.pmbts.2017.02.003

other endogenous and exogenous activators. MMPs degrade various extracellular matrix (ECM) proteins including collagen and elastin, and could affect other cellular processes including endothelium-mediated dilation, VSM cell migration, and proliferation as well as modulation of Ca^{2+} signaling and contraction in VSM. It is thought that increased lower limb venous hydrostatic pressure increases hypoxia-inducible factors and other MMP inducers such as extracellular matrix metalloproteinase inducer, leading to increased MMP expression/activity, ECM protein degradation, vein wall relaxation, and venous dilation. Vein wall inflammation and leukocyte infiltration cause additional increases in MMPs, and further vein wall dilation and valve degradation, that could lead to chronic venous disease and varicose veins (VVs). VVs are often presented as vein wall dilation and tortuosity, incompetent venous valves, and venous reflux. Different regions of VVs show different MMP levels and ECM proteins with atrophic regions showing high MMP levels/activity and little ECM compared to hypertrophic regions with little or inactive MMPs and abundant ECM. Treatment of VVs includes compression stockings, venotonics, sclerotherapy, or surgical removal. However, these approaches do not treat the cause of VVs, and other lines of treatment may be needed. Modulation of endogenous tissue inhibitors of metalloproteinases (TIMPs), and exogenous synthetic MMP inhibitors may provide new approaches in the management of VVs.

ABBREVIATIONS

ADAM a disintegrin and metalloproteinase
ADAMTS a disintegrin and metalloproteinase with thrombospondin motif
AP-1 activator protein-1
BK_{Ca} large conductance Ca^{2+}-activated K^+ channel
CEAP clinical–etiology–anatomy–pathophysiology
CVD chronic venous disease
CVI chronic venous insufficiency
ECM extracellular matrix
EDHF endothelium-derived hyperpolarizing factor
EMMPRIN extracellular matrix metalloproteinase inducer
GPCR G protein-coupled receptor
GSV great saphenous vein
HIF hypoxia-inducible factor
ICAM-1 intercellular adhesion molecule-1
IL interleukin
IVC inferior vena cava
MAPK mitogen-activated protein kinase
MMP matrix metalloproteinase
MT-MMP membrane-type MMP
NF-κB nuclear factor kappa-light-chain-enhancer of activated B cells
NGAL neutrophil gelatinase-associated lipocalin
NO nitric oxide
PAR protease-activated receptor
PDGF platelet-derived growth factor
PGE_2 prostglandin E2

PI_3K phosphoinositide 3-kinase
RGD Arg-Gly-Asp
ROS reactive oxygen species
SDX sulodexide
siRNA small interfering RNA
TGF-β transforming growth factor-β
TIMP tissue inhibitors of metalloproteinases
TNF-α tumor necrosis factor-α
TRP transient receptor potential
VCAM-1 vascular cell adhesion molecule-1
VEGF vascular endothelial growth factor
VSM vascular smooth muscle
VSMC VSM cell
VVs varicose veins
ZBG Zn^{2+} binding globulin
Zn^{2+} zinc

1. INTRODUCTION

Veins are a complex network of blood vessels that help transfer deoxygenated blood from various tissues and organs toward the heart. In the lower extremities, an intricate network of superficial, perforator, and deep veins accounts for the transfer of blood toward the heart against venous hydrostatic pressure (Fig. 1). Superficial veins carry blood from the skin and subcutaneous tissue and include the great saphenous vein (GSV) and small saphenous vein. The GSV is located in the medial side of the lower limb and runs from the ankle upward until it joins the common femoral vein at the saphenofemoral junction. The small saphenous vein is located in the back of the lower limb and runs from the ankle upward until it joins the popliteal vein at the saphenopopliteal junction. In addition, the anterior and posterior accessory saphenous veins run in the thigh and leg. The deep veins are embedded in the muscles and carry blood from all other parts of the lower extremity. The deep veins include the common femoral, deep femoral, femoral, popliteal, and tibial veins.[1] In all parts of the lower extremity, blood flows from the superficial veins to the deep veins.[1,2] An exception is the foot, where the blood flow is bidirectional. Connecting the superficial and deep venous system are the perforator veins, with inward direction of blood flow to the deep veins. The lower limb veins are equipped with bicuspid valves that protrude from the inner wall. The vein valves ensure the flow of blood in one direction from the superficial to the

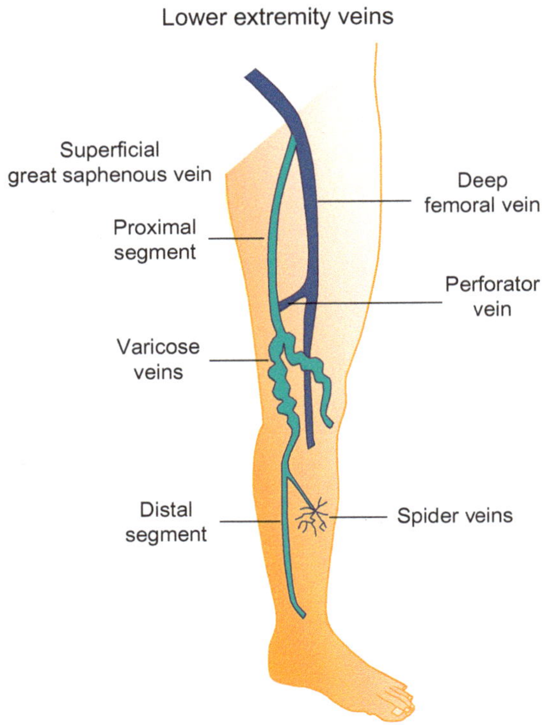

Fig. 1 The lower extremity venous system, and changes in VVs. The lower extremity has an intricate system of superficial and deep veins connected by perforator veins. Excessive vein wall dilation and incompetent venous valves could lead to superficial dilated spider vein or engorged and tortuous varicose veins.

deep veins and toward the heart. In addition, contraction of skeletal muscle in the calf, foot, and thigh helps to drive blood flow toward the heart, and against gravity and the venous hydrostatic pressure. In the standing position, the venous hydrostatic pressure could reach as high as 90–100 mm Hg at the ankle.[1,3]

When compared to the arteries, the veins are relatively thin. However, the structural integrity of the veins is still maintained and presented in three histological layers. The innermost layer or the tunica intima comprises mainly endothelial cells which line the venous wall and therefore are in direct contact with the changes in venous blood flow. The tunica media is separated from the tunica intima by the internal elastic lamina and contains several layers of vascular smooth muscle (VSM). The tunica adventitia is the outermost layer in the venous wall and mainly contains fibroblasts that are embedded in an extracellular matrix (ECM) of several proteins including collagen, elastin, and other proteins.[4]

The vein wall structure and function are regulated by a host of ions, signaling molecules, and enzymes. Matrix metalloproteinases (MMPs) are Zn^{2+}-dependent endopeptidases that are largely known for their ability to degrade various ECM proteins. MMPs could play a major role in venous tissue remodeling by degrading various components of ECM. In addition to their effects on ECM, MMPs may interact with bioactive molecules on the cell membrane and could regulate G protein-coupled receptors (GPCRs) and cell signaling. MMPs could play a role in various physiological processes and could affect cell proliferation, migration, and differentiation. MMPs could also be involved in cell apoptosis, immune response, tissue repair, and angiogenesis. MMPs are regulated at the mRNA expression and enzymatic activity levels. MMPs expression and activity could be altered in uteroplacental and vascular tissues and could play a role in the uteroplacental and vascular remodeling during normal pregnancy. MMPs are also regulated by endogenous tissue inhibitors of metalloproteinases (TIMPs). MMP/TIMP imbalance has been implicated in various vascular diseases including atherosclerosis, hypertension, and aortic aneurysm. MMPs may also play an important role in the regulation of venous structure and function, MMP imbalance has been implicated in venous dysfunction, and the pathogenesis of chronic venous disease (CVD).[3] One of the manifestations of CVD is varicose veins (VVs). VVs are a common health problem manifested as large, unsightly dilated, and tortuous veins of the lower extremities. If untreated, VVs may lead to chronic venous insufficiency (CVI) with skin changes and venous leg ulcers. VVs may also lead to other venous complications including thrombophlebitis and deep venous thrombosis. Therefore, it is imperative to carefully examine the mechanisms involved in CVD in order to develop better treatment approaches.

In this chapter, we will review data reported in PubMed and other scientific databases as well as data from our laboratory to provide insights on the role of MMPs in the regulation of vein structure and function, the remodeling of lower extremity veins, and the pathogenesis of CVD. We will describe the structural and functional abnormalities observed in VVs, and the changes in MMP expression/activity associated with VVs. We will discuss the potential factors that could drive the changes in venous tissue MMPs including increases in the lower limb venous hydrostatic pressure, the inflammatory response, hypoxia, and the various endogenous and exogenous MMP activators and inducers. We will also discuss the mechanisms of action of MMPs and how they could cause increases in ECM turnover as well as endothelial cell and VSM dysfunction leading to progressive venous dilation and VVs formation. We will conclude the chapter by summarizing some of the current medical and surgical approaches used for treatment of VVs and explore the potential benefits of overexpression of endogenous TIMPs or exogenous application of synthetic MMP inhibitors as novel tools in the management of VVs.

2. CHRONIC VENOUS DISEASE (CVD)

CVD is a common disorder of the lower extremity venous system with major social and economic implications. According to the clinical–etiology–anatomy–pathophysiology (CEAP) classification, CVD could have several clinical stages, C0–6. The C0 stage shows no visible signs of CVD. C1 is manifested as telangiectasies or spider veins. C2 is presented as VVs. C3 is associated with edema. C4a shows skin pigmentation or eczema and C4b shows lipodermatosclerosis or atrophie blanche. C5 stage shows healed ulcer and C6 stage shows active ulcer. The advanced stages C4–6 of CVD are often described as CVI.[5]

VVs are common venous disorder affecting approximately 25 million adults in the United States.[6] VVs are presented as abnormally distended and tortuous superficial veins of the lower extremity. In addition to vein wall dilation and tortuosity, VVs often show incompetent venous valves and measurable venous reflux (Fig. 2). While VVs are often detected in the lower extremity, the vein pathology may not be confined to the lower limb veins. It is possible that VVs is a sign of a generalized pathology in the venous system that is mainly manifested in the lower limb veins because of the high venous hydrostatic pressure. In support of this paradigm, patients with VVs also show increased distensibility in their arm veins, suggesting a generalized disorder in the venous system.[7] In addition to their socioeconomical impact and unsightly cosmetic appearance, VVs can lead to major complications such as thrombophlebitis, deep venous thrombosis, and venous leg ulcers.[5]

Several risk factors may lead to VVs, e.g., advanced age, female gender, contraceptive pills and estrogen therapy, pregnancy, overweight and obesity, prior leg injury, vein inflammation, and phlebitis. Estrogen may activate estrogen receptors in the vein wall leading to venous dilation, and females show enhanced estrogen receptor-mediated venous dilation and more distended veins when compared with males.[8] Estrogen is markedly increased

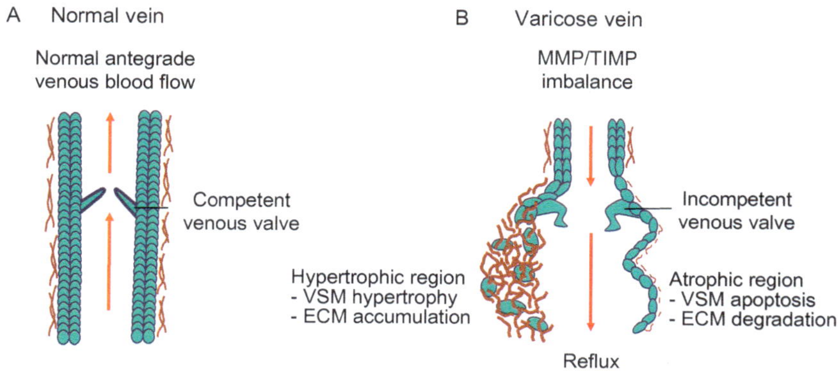

Fig. 2 Vein valves and blood flow in normal veins and VVs. Competent venous valves allow blood flow in the antegrade direction toward the heart (A). Vein dysfunction could progress to large dilated VVs with incompetent valves (B). VVs mainly show atrophic regions where an increase in MMPs increases ECM degradation, but could also show hypertrophic regions in which MMP/TIMP imbalance would promote VSMC hypertrophy and ECM accumulation, leading to tortuosity, dilation, defective valves, and venous reflux (B).

and may contribute to the increased venous dilation during pregnancy. Also, during pregnancy the progressive increases in uterine size and maternal body weight along with the changes in hemodynamics and cardiac output could lead to increased hydrostatic pressure in the lower extremity veins and the development of VVs. Behavioral factors such as sedentary lifestyle and prolonged standing could also increase the risk for CVD.[9–11]

Family history and hereditary and genetic factors may represent potential risk factors for VVs.[12] Primary lymphedema–distichiasis is a rare syndrome involving a mutation in the FOXC2 region of chromosome 16 and is associated with VVs in early age.[13] A study on nine families have shown a link between VVs and the D16S520 marker on chromosome 16q24 near the FOXC2 region, providing evidence that VVs could be linked to FOXC2, and that CVD could be inherited in an autosomal dominant mode with incomplete penetrance.[14] Patients with Klippel–Trenaunay syndrome also have VVs, supporting heritability of VVs.[15] The lower limb venous dynamic and vein wall elasticity may also be reduced in children of VVs patients.[16] Evidence also suggests a genetic component of VVs. A heterozygous mutation in the Notch3 gene has been identified in the cerebral autosomal dominant arteriopathy with subcortical infarcts and leukoencephalopathy (CADASIL) pedigree with VVs.[17] Microarray analysis of 3063 human cDNAs from VVs showed upregulation of 82 genes, particularly those associated with the regulation of ECM, cytoskeletal proteins, and myofibroblasts.[12] Subjects with Ehlers–Danlos syndrome type-IV are prone to developing vascular pathology and VVs.[18] A single nucleotide polymorphism in the promoter region of MMP-9 gene has been identified in Chinese individuals with VVs, and a significant correlation has been found between patients with VVs and controls at -1562C/T in the gene promoter of MMP-9.[19]

Desmuslin is an intermediate filament protein important in smooth muscle function, and mutations or nucleotide polymorphic variant may be associated with VVs. In support, human saphenous vein smooth muscle cells (SMCs) treated with desmuslin siRNA showed increased collagen synthesis and MMP-2 expression and decreased expression of the phenotype and differentiation markers SM α-actin, SM-myosin heavy chain, and smoothelin and exhibited disassembly of actin stress fibers when compared with the control cells. These observations have suggested that desmuslin is required for maintaining the VSMC phenotype, and that decreased desmuslin expression may affect differentiation of VSMCs and ultimately contribute to the development of VVs. Other genetic defects have been implicated in advanced stages of CVD especially the venous leg ulcer risk and delayed healing, and include the genes for hemochromotosis, ferroportin, factor XIII, fibroblast growth factor receptor-2, and MMP-12.[20]

3. STRUCTURAL AND FUNCTIONAL ABNORMALITIES IN VVs

VVs often appear as dilated, engorged, and tortuous veins, giving the impression that the lower limb veins may be undergoing marked hypertrophic remodeling. However, structural and histological evidence suggest that VVs may have not only hypertrophic but also atrophic regions (Fig. 2).[21] The hypertrophic regions of VVs often show abnormal VSMC shape and orientation and ECM accumulation. In contrast, the atrophic regions show ECM degradation and an increase in inflammatory cell infiltration.[22] Histological examination of tissue sections of VVs does not show distinct vascular layers, and there are no clear boundaries between the tunica intima, media, and adventitia. Tissue sections of VVs show focal intimal thickening, increased medial thickening, and fragmentation of elastin fibers.[23] In tissue sections of VVs, VSMCs appear disorganized in the tunica media and the adjacent intima, with abundant ill-defined unstructured material. Also, collagen fibers appear disorganized making it difficult to delineate the tunica media from the adventitia, and the elastic fibers appear thick and fragmented in the tunica intima and adventitia.[4]

VVs show imbalance in the protein components of ECM mainly due to changes in collagen and/or elastin content. Collagen measurements in the walls of VVs show marked variability ranging from an increase,[24] to a decrease,[25] or no change.[26] Experiments on cultured VSMCs from VVs and cultured dermal fibroblasts from patients with CVD have shown an increase in the synthesis of type-I collagen and a decreased in the synthesis of type-III collagen, with no apparent change in gene transcription. These observations suggest that patients with VVs have posttranslational inhibition of type-III collagen synthesis and may have a systemic abnormality in collagen production in various tissues. Type-III collagen is a critical factor in determining the elasticity and distensibility of blood vessels, and alterations in collagen synthesis and the collagen type-I/type-III ratio could cause marked changes in the vein wall integrity, leading to structural weakness in the vein wall, venous dilation, and formation of VVs.[4] Some studies have suggested that a decrease in the elastin content could play a role in the pathogenesis of VVs, as it may cause a decrease in the vein wall elasticity and lead to vein wall dilation.[27] However, other studies have suggested that the elastin network may be increased in VVs.[4]

In addition to the changes in the vein wall, VVs also show incompetent venous valves. However, whether valve dysfunction is a primary event that leads to vein wall dilation, or vice versa, the changes in VVs wall lead to valve dysfunction is debatable. It has been suggested that valve dysfunction could lead to venous reflux and high venous hydrostatic pressure, and the excessive and prolonged pressure could lead to progressive damage and dilation of the vein wall. The dilation of the vein segments in close proximity to the vein valves would then cause further distortion and disruption of the valves leading to further increases in venous reflux, venous hydrostatic pressure, and progressive vein wall dilation. This view has been supported by the observations that VVs show hypertrophy of the vein valves, increased width of the valvular annulus,[28] decreased collagen content and viscoelasticity,[29] and increased inflammation and monocyte and macrophage infiltration in the valvular sinuses compared to distal VVs walls.[30] However, this view has been challenged by the observation that vein wall dilation and VVs may sometimes be seen below competent venous valves.[21] Also, increased collagen and decreased elastin have been observed not only in VVs segments but also in competent saphenous vein segments in close proximity to the varices, suggesting that imbalance in ECM proteins may occur in the vein wall prior to the vein valve insufficiency.[24] Also, VVs do not always develop in a descending direction, i.e., from the thigh, to the calf, and ankle, and antegrade progression in VVs in the normal direction of venous flow from the ankle, to the calf and the thigh may be caused by primary changes in the vein wall, which could lead to dysfunction of the venous valves.[21] These observations lend support to the suggestion that vein wall dilation could be a primary pathological event that could cause distortion and dysfunction of the venous valves, and lead to venous reflux and higher venous hydrostatic pressure, and ultimately causing progressive venous dilation and VVs.[31] Regardless of which one is the primary pathological event, both vein wall dilation and venous valve dysfunction appear to contribute to the pathogenesis of VVs and abnormal venous blood flow. This is typically manifested as a venous reflux or backflow of blood away from the heart that lasts longer than 0.5 s in the superficial VVs.[31]

4. MMP LEVELS IN VVS

VVs may show significant changes in MMP expression/activity.[3] Studies have shown an increase in the levels of MMP-1, -2, -3, and -7 with a prominent increase in MMP-2 activity in VVs.[4] Increased plasma levels of MMP-10 and the hemostatic markers D-dimers, prothrombin fragments 1 and 2, von Willebrand factor, and activity of plasminogen activator inhibitor (PAI-1) have also been observed in patients with primary VVs, suggesting a prothrombotic and proinflammatory states.[32] Other studies have shown an increase in MMP-1 protein level in the GSV, and an increase in the levels of MMP-1 and -13 in the proximal vs distal segments of VVs, with no change in MMPs mRNA expression, suggesting that the increased MMPs levels are related to changes in MMP posttranscriptional modification or protein degradation.[33] The levels of MMPs may also vary within the different tissue layers and cellular components of the VVs wall. Immunohistochemical analysis in the tissue sections of VVs showed prominent localization of MMP-1 in fibroblasts, VSMCs, and endothelial cells, MMP-9 in endothelial cells, medial VSMCs and adventitial microvessels, and MMP-12 in VSMCs and fibroblasts.[34] Other studies have shown increased MMP-1 expression in all layers of VVs and MMP-9 expression in the intimal and adventitial layers of VVs.[23] The localization of MMPs in the tunica adventitia and fibroblasts is consistent with the role of MMPs degradation of ECM proteins. Interestingly, studies also showed increased levels of MMP-2 levels in all layers of the vein wall, and of MMP-1, -3, and -7 in the tunica intima and media of VVs,[4] suggesting additional effects of MMPs on the endothelium and VSMCs.

Although several studies have shown increases in the levels of certain MMPs in VVs, some studies have shown no change or even a decrease in the levels of MMPs. One study has shown that the levels of active MMP-1 and both pro- and active forms of MMP-2 are decreased in VVs.[35] The variability in the levels of MMPs may explain the variability in the measurements of collagen content in VVs wall which ranged from a decrease[25] to no change,[26] or even an increase.[24] The variability in the levels of MMPs may be due to examining different vein segments from different regions of VVs, i.e., hypertrophic vs atrophic regions, or examining vein specimens from patients at different stages of CVD, or inability to distinguish between pro- and active forms of MMPs.

Changes in MMP expression/activity have also been associated with the progression of CVD and advanced stages of CVI. Studies have shown elevated serum levels of MMP-2, ADAMTS-1, and ADAMTS-7 in the initial stages of CVD, whereas the serum levels of MMP-1, -8, -9, neutrophil gelatinase-associated lipocalin (NGAL), ADAM-10 and -17, and ADAMTS-4 were particularly elevated during CVD complications and skin changes.[36] The collagenases MMP-1 and -8 are overexpressed in the fluids and tissues of long-lasting nonhealing chronic venous ulcers.[37] The levels of MMP-1 and -8 were even higher in patients with infected ulcers than those with uninfected ulcers.[38]

5. POTENTIAL MMP INDUCERS/ACTIVATORS IN VVs

Multiple factors can induce or activate MMPs in vitro, ex vivo, and in vivo. Many factors could modulate the expression/activity of MMPs in VVs including increases in lower extremity venous hydrostatic pressure, inflammation of the vein wall, hypoxia, and other factors.

5.1 Venous Hydrostatic Pressure and MMPs in VVs

Increased lower extremity venous hydrostatic pressure is a major factor that could lead to increased expression/activity of MMPs in VVs (Fig. 3). Studies have suggested that mechanical stretch may lead to increases in the expression of MMPs in endothelial cells, VSMCs, and fibroblasts.[39] We have also shown that prolonged increases in mechanical tension or wall stretch of isolated rings of rat inferior vena cava (IVC) are associated with increased expression of MMP-2 and -9 in the tunica intima and increased MMP-9

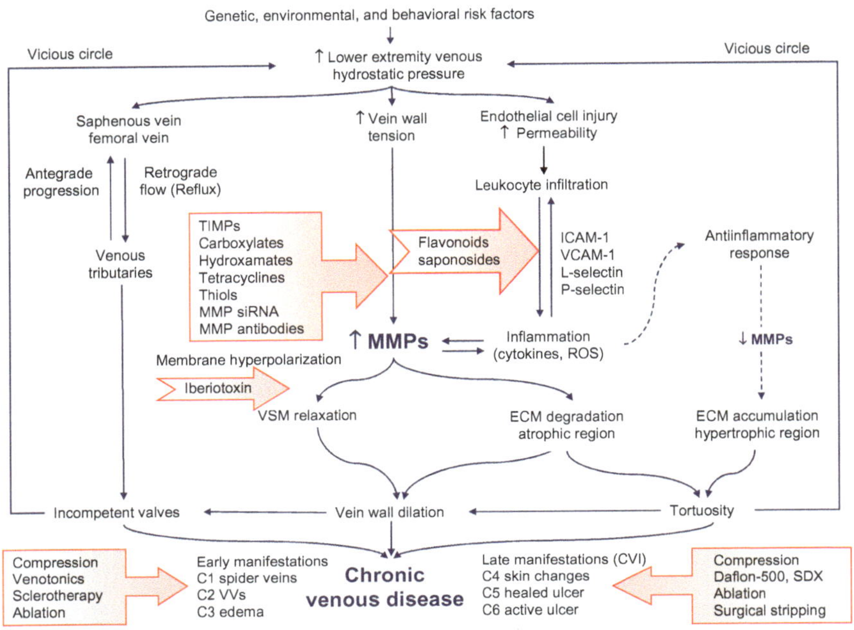

Fig. 3 Pathophysiology and management of CVD. Certain genetic, environmental, and behavioral risk factors cause an increase in venous hydrostatic pressure in the lower extremity saphenous and femoral veins leading to valve dysfunction and venous reflux. Increased venous hydrostatic pressure also increases vein wall tension leading to increases in MMPs. Increased venous hydrostatic pressure could also cause endothelial cell injury, increased permeability, leukocyte infiltration, and increased adhesion molecules, inflammatory cytokines, and reactive oxygen species (ROS) leading to further increases in MMPs. Increased MMPs may cause VSM hyperpolarization and relaxation as well as ECM degradation leading to vein wall dilation, valve dysfunction, and progressive increases in venous hydrostatic pressure (vicious cycle). Increased MMPs generally promote ECM degradation particularly in atrophic regions. Other theories (indicated by *interrupted arrows*) suggest a compensatory antiinflammatory pathway involving prostaglandins and their receptors that lead to decreased MMPs and thereby ECM accumulation, particularly in hypertrophic regions of VVs. Persistent valve dysfunction, progressive vein wall dilation, and tortuosity lead to different stages of CVD and CVI. Current treatment of CVD and CVI (presented in *shaded arrows*) includes physical, pharmacological, and surgical approaches. Inhibitors of the activity or action of MMPs (also presented in *shaded arrows*) may provide potential tools for the management of CVD/CVI.

in the tunica media of the vein wall. Prolonged IVC stretch was also associated with decreased vein contraction to the α-adrenergic agonist phenylephrine. Importantly, in IVC pretreated with specific MMP inhibitors, prolonged mechanical stretch did not cause decreases in IVC contraction.

These observations suggested that prolonged increases in venous pressure/wall tension may cause changes in MMP expression/activity, which in turn cause decreases in vein contraction, and thereby increase venous dilation.[40] The factors linking the increased venous pressure to increased MMP expression in the vein wall are not clearly understood but may involve intermediary factors such as inflammation or hypoxia-inducible factors (HIFs).[41]

5.2 Inflammation and MMPs in VVs

Endothelial cells are exposed to marked fluctuations in blood flow, and increases in venous pressure could cause endothelial cell injury, increased permeability, activation of adhesion molecules, leukocyte infiltration of the vein wall, and collectively these factors could contribute to inflammation of the vein wall.[42] Rat models of increased lower extremity venous pressure have been produced by induction of femoral arteriovenous fistula. These rat models show increased venous pressure in the saphenous vein, and the prolonged increases in venous pressure are associated with leukocyte infiltration, increased expression of intercellular adhesion molecule-1 (ICAM-1) and P-selectin, and inflammation of the vein wall.[43] Leukocytes are a major source of MMPs.[44] Accumulation of adhesion molecules facilitates leukocyte adhesion and infiltration of the vein wall, and leads to further inflammation and increased expression/activity of MMPs. MMPs in turn cause degradation of ECM proteins, weakening of the vein wall, vein wall dilation, vein valve dysfunction, further increases in lower limb venous hydrostatic pressure, and advanced stages of CVD (Fig. 3).[45] The link between increased lower extremity venous hydrostatic pressure, vein wall inflammation, and increased MMP expression/activity may be well presented in the atrophic regions of VVs, where increased degradation of ECM proteins often occurs.

Studies have shown increased monocyte/macrophage infiltration in the walls and valves of saphenous vein specimens from patients with CVD.[30,46] In addition to the increased inflammatory cell infiltration, VVs specimens show increased endothelial cell expression of ICAM-1 and vascular cell adhesion molecule-1 (VCAM-1).[47] Also, the plasma levels of endothelial cell and leukocyte activation markers such as ICAM-1, VCAM-1, angiotensin converting enzyme, and L-selectin are increased in patients with VVs, and the increases in inflammatory markers are associated with increases in the plasma levels of proMMP-9, lending support to a potential relation between postural blood stasis and increase in lower extremity venous hydrostatic pressure, polymorphonuclear leukocyte infiltration and activation and increased release of MMPs in VVs.[48]

Inflammatory cytokines could play a role linking pressure-induced leukocyte infiltration, vein wall inflammation, and increases in MMP expression/activity. Urokinase could contribute to the inflammatory response by increasing the expression of tumor necrosis factor-α (TNF-α) in damaged vessels. TNF-α could in turn increase the activity of MMP-9 gene promoter partly through activation of activator protein-1 (AP-1), specificity protein-1 (Sp-1), or nuclear factor κ-light-chain-enhancer of activated B cells (NF-κB).[49] Other cytokines including interleukins IL-17 and -18 may induce MMP-9 expression via activation of AP-1 and NF-κB-dependent pathways.[50] Interestingly, studies have shown that higher levels of MMP-1 and -8 are associated with higher levels of IL-1, -6, -8, vascular endothelial growth factor (VEGF), and TNF-α in patients with infected venous leg ulcers compared to those with uninfected ulcers, documenting a possible association between infection, MMP activation, cytokine secretions, and CVD symptoms.[38]

Cytokines are known to increase reactive oxygen species (ROS), which could in turn affect MMP expression/activity. Studies in fibroblasts have suggested that the levels of MMP expression may be influenced by the levels of NADPH oxidase-1 (Nox-1).[51] Urokinase may affect MMP-9 expression partly through increasing the generation of ROS.[52] While leukocytes are a major source of MMPs, they also generate ROS that can influence MMP activity. For instance, ROS may activate MMPs via oxidation of the MMP prodomain thiol followed by autolytic cleavage. On the other hand, ROS may inactivate MMPs by modifying the amino acids critical for catalytic activity, thus providing a feedback mechanism that could control any bursts in MMP proteolytic activity.[53]

5.3 Hypoxia and MMPs in VVs

HIFs could provide a potential mechanism linking increases in lower extremity venous hydrostatic pressure, to the increases in MMP expression/activity and reduced venous contraction (Fig. 4). HIFs are major nuclear transcriptional factors that typically regulate most of the genes involved in oxygen homeostasis. Interestingly, mechanical stretch may influence the expression/activity of HIFs. Studies have shown that exposure of rat skeletal muscle fibers to prolonged mechanical stretch is associated with increased mRNA

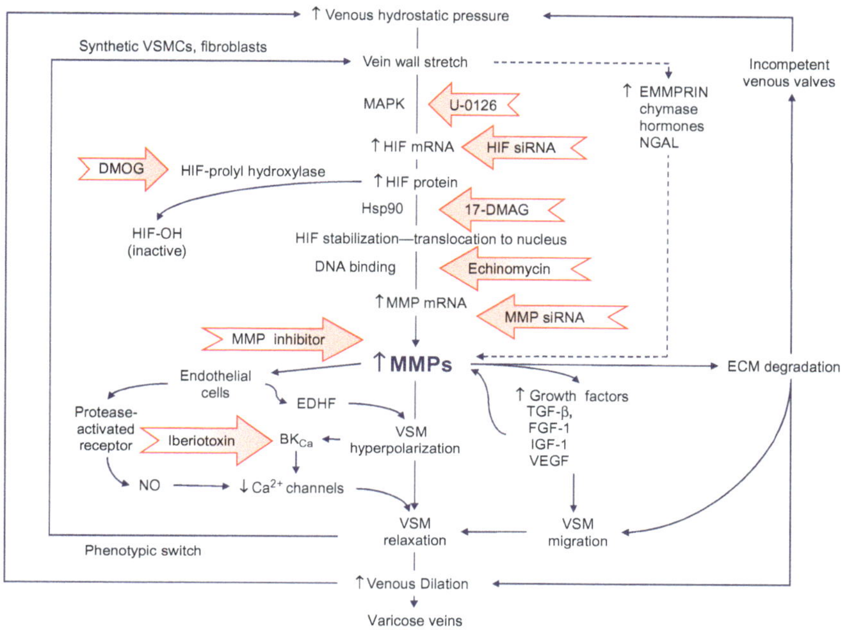

Fig. 4 Mechanisms linking increased venous hydrostatic pressure to increased MMP expression and VVs. Increased venous hydrostatic pressure causes vein wall stretch, which increases HIF mRNA expression and protein levels, and in turn MMP levels. Increased wall stretch may also increase other MMP inducers such as EMMPRIN, chymase, hormones, and NGAL. Increased MMPs may activate protease-activated receptors in endothelial cells leading to NO production and venous dilation. MMPs may also stimulate endothelial cells to produce EDHF which in turn opens BK_{Ca} channels in VSM, leading to hyperpolarization, decreased Ca^{2+} entry through Ca^{2+} channels, and VSM relaxation. Loss of contractile function in VSM could cause a phenotypic switch to synthetic VSMCs. MMPs may also increase the release of growth factors, leading to VSMC hypertrophy. MMPs also cause ECM degradation leading to VSMC migration, further decreases in vein contraction and increases in venous dilation, and VVs. MMP-induced ECM degradation may also cause valve degeneration leading to further increases in venous hydrostatic pressure. As indicated in *shaded arrows*, inhibitors of MMP synthesis (U-0126, HIF siRNA, 17-DMAG, Echinomycin, MMP siRNA), activity (MMP Inhibitor), or actions (Iberiotoxin) may provide new tools for management of VVs. BK_{Ca}, large conductance Ca^{2+}-activated K^+ channels; DMOG, dimethyloxaloylglycine, is an experimental inhibitor of HIF-prolyl hydroxylase; FGF, fibroblast growth factor; HIF, hypoxia-inducible factor; Hsp90, heat-shock protein 90; IGF, insulin-like growth factor; MAPK, mitogen-activated protein kinase; NGAL, neutrophil gelatinase-associated lipocalin; TGF-β, transforming growth factor β; VEGF, vascular endothelial growth factor.

expression and protein levels of HIF-1α and -2α in the skeletal muscle capillary endothelial cells.[54] Also, experiments on the rat ventricle have suggested that mechanical stretch of the ventricular wall is associated with upregulation of HIF-1α.[55] We have tested the role of HIFs in mechanical stretch-induced reduction of vein contraction in rat IVC. We have found that prolonged increases in wall tension in isolated segments of rat IVC are associated with increases in the mRNA expression and protein levels of not only MMP-2 and -9, but also HIF-1α and -2α. The increases in MMP and HIF expression and protein levels were associated with a decrease in the magnitude of IVC contraction to phenylephrine. Interestingly, vein contraction was further reduced in IVC pretreated with the HIF stabilizer dimethyloxaloylglycine (DMOG), which prevents inactivation of HIF by HIF-prolyl hydroxylase. On the other hand, the reduction in IVC contraction was reversed in tissues pretreated with the HIF inhibitors U0126 and echinomycin, supporting a role of HIF as a linking mechanism between increased venous hydrostatic pressure and reduced venous contraction (Fig. 4).[41] An important question is how mechanical stretch could affect HIFs. One possibility is that mechanical stretch could activate Ca^{2+} influx through transient receptor potential (TRP) channels such as TRPV4, and the increases in Ca^{2+} could activate phosphoinositide 3-kinase (PI_3K), which could in turn affect HIF.[56] Another possibility is that mechanical stretch could interact with the cell membrane integrins which could activate a cascade of intracellular signaling pathways that ultimately activate mitogen-activated protein kinase (MAPK) and affect HIF expression. As a form of biomechanical stress, mechanical stretch could activate G protein-coupled receptors (GPCRs) or receptor tyrosine kinases or increase the formation of ROS, which could lead to activation of MAPK. We have found that the increase in HIF mRNA expression and the reduction in IVC contraction associated with prolonged vein wall stretch are reversed in IVC treated with MAPK inhibitors, supporting a role of MAPK in transducing the effects of mechanical stretch on HIF.[41]

Studies in patients with VVs support a role of HIF in the pathogenesis of CVD. The expression of HIF-1α and -2α and HIF target genes are upregulated in VVs.[57] Studies also suggest that HIF-1α may regulate the expression of MMP-2 and -9 in patients with hemodialysis polytetrafluoroethylene grafts or arteriovenous fistulas.[58] In addition to mechanical stretch, other factors such as low oxygen tension, low pH, cytokines, hormones, and heat may influence HIF expression and, in turn, affect venous MMP expression/activity.

5.4 Other MMP Inducers/Activators in VVs

Other MMP inducers/activators have been identified in the veins and may promote MMP expression/activity in VVs. Extracellular MMP inducer (EMMPRIN, CD147, Basigin) is a widely expressed membrane protein of the immunoglobulin superfamily. EMMPRIN has been suggested to play a role in tissue remodeling and has been implicated in pathological conditions such as atherosclerosis, aneurysm, heart failure, rheumatoid arthritis, and cancer. Studies have shown that high volume mechanical ventilation causes acute lung injury and is associated with upregulation of MMP-2, MMP-9, and MT1-MMP as well as EMMPRIN.[59] Also, EMMPRIN along with MMP-2, MT1-MMP and MT2-MMP are overexpressed in dermal structures of venous leg ulcers, which could lead to unrestrained activation of MMPs and enhanced ECM turnover.[60]

Prostanoids are bioactive lipids produced by many vascular cells and may interact with MMPs in the pathogenesis of VVs. Prostaglandin E2 (PGE2) through activation of EP1–4 receptor subtypes play a role in the regulation of vascular tone, inflammation, and vascular wall remodeling.[61] Activation of EP2/EP4 receptors by PGE2 is associated with increased MMP activity in human endometriotic epithelial and stromal cells.[62] PGE2 synthesis may be decreased in VVs due to a compensatory increase in antiinflammatory 15-deoxy-delta-12,14-PGJ2, a decrease in membrane-associated prostaglandin E synthase-1, and an increase in the degrading enzyme 15-hydroxyprostaglandin dehydrogenase. The overall decrease in PGE2 and its EP4 receptor activity may then cause a decrease in the activity of MMP-1 and -2 and lead to increased collagen deposition, which may explain the hypertrophic remodeling observed in some regions of VVs.[35]

Chymase is a chymotrypsin-like serine protease purified from mast cell granules and mammalian cardiovascular tissues. Chymase has been implicated in the increased MMP-9 activity and the accumulation of monocytes and macrophages observed in the aorta of stroke-prone spontaneously hypertensive rats.[63] Also, gonadal hormones such as estrogen and progesterone may increase the expression/activity of MMP-2 and -9 in uteroplacental and vascular tissues.[64,65] Some MMP modulators could prevent MMP degradation, and thereby increase MMP levels and activities. For instance, NGAL may form a complex with MMP-9, which would protect MMP-9 from proteolytic degradation and thereby increase its levels and activity.[66] Whether these MMP inducers/activators are increased in VVs need to be further examined.

6. MECHANISMS OF MMP ACTIONS IN VVs

MMPs are widely recognized for their proteolytic effects on ECM proteins, and the effects of MMPs on ECM protein degradation and tissue remodeling could play an important role in the pathogenesis of VVs. However, MMPs could also affect other components of the vein wall and influence other cellular and molecular pathways in VSMCs and the endothelium. These additional signaling effects of MMPs could affect vein function, and may play a role at least in the initial vasodilatory stages of VVs.

6.1 MMPs and ECM Abnormalities in VVs

Changes in MMP activity could alter the composition of ECM and, in turn, contribute to the structural and functional abnormalities associated with in VVs. While it is widely believed that venous tissue levels of MMPs increase in VVs, decreases in MMP levels have also been observed,[35] and this may partially explain the different venous pathologies in the atrophic vs hypertrophic regions of VVs. An increase in MMP activity is predicted to degrade and decrease ECM proteins in the atrophic regions of VVs.[22] On the other hand, a decrease in MMP activity may lead to ECM accumulation in the media of hypertrophic regions of VVs, which could interfere with the contractile function of VSMCs, thus hindering venous contraction and leading to venous dilation and VVs.[67]

The content of ECM could be determined by MMP-induced degradation of major ECM proteins such as collagen and elastin. Collagen content varies in VVs compared to normal veins, with an increase in collagen type-I and a decrease in collagen type-III.[68–70] In cultured VSMCs from VVs, collagen type-III and fibronectin are decreased likely due to posttranscriptional degradation by MMP-3.[70] Elastin levels may also show a decrease in VVs possibly due to increased elastolytic activity of MMPs or other elastases produced by macrophages, monocytes, platelets, and fibroblasts.[27] The changes in collagen and elastin content in VVs are dynamic processes that may depend on the stage of CVD. For instance, increased collagen content may compensate for the decreased elastin levels at early stages of VVs. On the other hand, collagen levels may decrease at later stages of CVD. This may partly explain the divergent reports regarding the collagen content in VVs, with some studies showing a decrease,[25] while other studies showing hardly any change[26,27] or even an increase.[24] Other ECM proteins may show changes in VVs, e.g., increases tenascin and decreases in laminin.[68–70]

6.2 MMPs and VSM Dysfunction in VVs

In addition to changes in venous tissue remodeling and ECM proteins, VVs may show alterations in VSMC growth, migration, and contractile function (Fig. 4). When compared with contractile VSMCs from normal veins, VSMCs from VVs are more dedifferentiated and show increased migration, and MMP-2 levels, which may contribute to vein wall remodeling and weakening against increased venous hydrostatic pressure.[71] MMP-1 and -9 have also been shown to increase human aortic SMC migration.[72,73] MMP-induced ECM proteolysis may modulate cell–matrix adhesion either by removal of sites of adhesion or by exposing a binding site and in turn facilitate VSMC migration. MMPs can also facilitate VSMC growth by promoting permissive interactions between VSMCs and components of ECM, possibly via integrin-mediated pathways.[74] MT1-MMP may stimulate the release of transforming growth factor-β (TGF-β) and promotes the maturation of osteoblasts.[75] Also, upregulation of MMP-2 increases the expression of VEGFa, while downregulation of MMP-2 decreases VEGFa expression in human gastric cancer cell line SNU-5.[76] MMPs may also promote the release of growth factors by cleaving the growth factor-binding proteins or matrix molecules, and this may partly explain the VSMC hypertrophy observed in some parts of the hypertrophic regions of VVs.[77]

While MMPs could stimulate growth factor release, MMPs can be regulated by growth factors.[78] For example, overexpression of VEGFa in SNU-5 cells increases MMP-2 expression, while downregulation of VEGFa decreases MMP-2 expression.[76] Also, platelet-derived growth factor-BB (PDGF-BB) increases MMP-2 expression in rat VSMCs, possibly via Rho-associated protein kinase, extracellular signal-regulated kinases, and p38 MAPK phosphorylation.[79] Also, in a study on carotid plaques, EGF upregulated MMP-9 activity and increased MMP-1, -9, and EGFR mRNA transcripts in VSMCs.[80] Other studies have shown that TGF-β1 could induce the expression of MMP-9 and -12 and TIMP-1 and -2 in the GSV, and suggested the involvement of TGF-β1 in the vein wall pathology.[23]

Because synthetic VSMCs do not contract, MMP and growth factor-mediated VSMC dedifferentiation and migration are likely to decrease venous contraction and promote dilation (Fig. 4). MMPs may also affect VSM contraction mechanisms. VSM contraction is triggered by increases in Ca^{2+} release from the intracellular stores and Ca^{2+} entry from the extracellular space. MMPs do not inhibit phenylephrine-induced aortic contraction in Ca^{2+}-free medium, suggesting that they do not inhibit the Ca^{2+} release mechanism in the sarcoplasmic reticulum.[81] On the other hand, MMP-2 and -9 cause aortic relaxation by inhibiting Ca^{2+} influx,[81] and MMP-2 inhibits Ca^{2+}-dependent contraction in rat IVC.[82] It has been proposed that during substrate degradation MMPs may produce Arg-Gly-Asp (RGD)-containing peptides, which could bind to $\alpha_v\beta_3$ integrin receptors and inhibit Ca^{2+} entry into VSM.[83] This is an unlikely mechanism as our data have shown that RGD peptides do not affect contraction of IVC segments.[82] The mechanism by which MMPs inhibit Ca^{2+} entry may likely involve direct effects on plasma membrane Ca^{2+} or K^+ channels. In rat IVC, MMP-2-induced relaxation is abolished in high KCl depolarizing solution, which prevents K^+ ion from moving out of the cell via K^+ channels. Importantly, blockade of large conductance Ca^{2+}-activated K^+ channels (BK_{Ca}) by iberiotoxin inhibited MMP-2-induced IVC relaxation, suggesting that MMP-2 actions may involve hyperpolarization and activation of BK_{Ca}, which in turn lead to decreased Ca^{2+} influx through voltage-gated Ca^{2+} channels (Fig. 4).[84] Sustained MMP-induced inhibition of venous tissue Ca^{2+} influx and contraction mechanisms may lead to progressive venous dilation and VVs.

6.3 MMPs and Endothelium-Dependent Relaxation

The endothelium controls vascular tone by releasing relaxing factors such as nitric oxide (NO) and prostacyclin (PGI_2) and through hyperpolarization of the underlying VSMCs by endothelium-derived hyperpolarizing factor (EDHF).[85] MMPs may stimulate protease-activated receptors (PARs), which are GPCRs that may play a role in venous dilation in VVs (Fig. 4). PARs 1–4 have been identified in humans, and MMP-1 has been shown to activate PAR-1.[86] PAR-1 is expressed in VSMCs,[87] endothelial cells, and platelets[88] and is coupled to increased NO production,[89] which could contribute to progressive venous dilation and the formation of VVs.

EDHF-mediated relaxation may involve the opening of small and intermediate conductance Ca^{2+}-activated K^+ channels and hyperpolarization of endothelial cells. Endothelial cell hyperpolarization may spread via myoendothelial gap junctions causing relaxation of VSMCs. Although the exact nature of EDHF is unclear, studies suggest that EDHF-mediated responses could involve epoxyeicosatrienoic acids, which are epoxides of arachidonic acid generated by cytochrome P450 epoxygenases. Other possible EDHFs include hydrogen peroxide or even the potassium ion.[84] EDHF could in turn cause vascular hyperpolarization through opening of BK_{Ca} in VSMCs.[85] MMP-2 may increase EDHF release and enhance K^+ efflux via BK_{Ca}, leading to venous tissue hyperpolarization and relaxation.[84] In contrast, MMP-3 may be associated with impaired endothelium-dependent vasodilation,[90] making it important to further examine the effects of MMPs on endothelium-derived relaxing factors.

7. MANAGEMENT OF VVs

Management of VVs comprises physical approaches such as graduated compression stockings. Graduated elastic compression stockings enhance venous emptying, reduce the pain and edema, and may slow the progression of VVs to the more advanced forms of CVI manifested with skin changes and venous leg ulcer.[91] Compression stockings could also decrease the incidence of venous thromboembolism following VVs surgery and improve hemodynamic performance in postthrombotic syndrome.[92] Pharmacological treatment of VVs includes venotonic drugs that improve venous tone and capillary permeability, and reduce leukocyte infiltration. Venotonic drugs include naturally occurring plant extracts and glycosides such as α-benzopyrones (coumarins), γ-benzopyrones (flavonoids), plant extracts (blueberry and grape seed, ergots, Ginkgo biloba), and saponosides (Centella asiatica, escin, horse chestnut seed extract, ruscus extract).[45] Benzopyrones include catechin (Green tea), dicoumarols, diosmin (Daflon-500), escletin, flavonoic acid, hesperitin, hesperidine, oxerutin, quercetin, rutosides, troxerutin, umbelliferone, and venoruton.[93] Diosmin is a flavonoid and an active ingredient in Daflon-500 that may improve venous tone, microvascular permeability, lymphatic activity, and microcirculatory flow.[94] Rutosides enhance endothelial function in patients with CVI.[95] Flavonoids may affect the endothelium and leukocytes and reduce edema and inflammation. Saponosides, such as escin (horse chestnut seed extract), may limit venous wall distensibility and morphologic changes. We have shown that escin promotes Ca^{2+}-dependent venous contraction.[96] Other compounds have been tested in advanced CVI and include pentoxifylline, a xanthine derivative with beneficial antiinflammatory and hemorheologic properties including inhibition of TNF-α and leukotriene synthesis and improved red blood cell deformability,[97] and red vine leaves (AS 195) and PGE1, which may increase microcirculatory blood flow and transcutaneous oxygen tension and alleviate edema.[45]

Other approaches have been developed to obliterate dilated VVs and improve venous hemodynamics. Sclerotherapy under Duplex ultrasound guidance entails injection of concentrated sclerosing agents such as hypertonic saline, sodium morrhuate, and ethanolamine oleate in the dilated veins. Some of the FDA-approved sclerosing agents for treatment of VVs include sodium tetradecyl sulfate (STS), which is a liquid detergent, sodium morrhuate, and polidocanol.[98] STS and polidocanol are also used in a foam state to displace blood and reduce thrombosis in VVs. Proprietary polidocanol endovenous microfoam has recently been approved by the FDA and has shown excellent results and improved quality of life in patients with VVs.[99]

Management of VVs may also include surgical approaches such as endovenous ablation with a radiofrequency or infrared laser typically at wavelengths 810–1320 nm, but as high as 1470 and 1550 nm. The high endoluminal thermal heat causes denaturation of endothelial proteins and leads to vein occlusion.[100] Ablation therapy shows relatively good outcome and vein occlusion rates with a 2% vein recanalization rate 4 years after radiofrequency therapy[101] and 3%–7% VVs recurrence rate 2–3 years after infrared laser therapy.[102] Surgical stripping of the saphenous vein with high ligation of the saphenofemoral junction is another surgical approach with low recurrence rates. "Stab phlebectomy" is another surgical approach that entails avulsion of large VVs clusters that communicate with the incompetent saphenous vein. Transilluminated power phlebectomy is an alternative to open phlebectomy that removes clusters of VVs using fewer incisions and shorter operation time.[103]

In addition to ultrasound guided foam sclerotherapy, innovative technologies for endovenous treatment of VVs are emerging and involve the use of nonthermal and nontumescent techniques such as cyanoacrylate glue and other mechanochemical methods.[104,105]

Initial results are encouraging, but additional studies are needed to further evaluate the benefits of these new techniques vs thermal ablation and surgery.

8. POTENTIAL BENEFITS OF MMP INHIBITORS IN VVs

Currently available treatment options for VVs focus mainly on the symptoms rather than the causes of CVD. The identification of the role of MMPs in the pathogenesis of CVD has prompted the search for inhibitors of MMP expression or activity. MMP inhibitors can be used to prevent

MMPs from binding with their substrates and degrading the ECM, and thereby prevent the development or recurrence of VVs. MMP inhibitors include endogenous inhibitors such as TIMPs and α2-Macroglobulin, as well as synthetic Zn^{2+}-dependent and Zn^{2+}-independent inhibitors.

8.1 TIMPs and MMP/TIMP Ratio

TIMPs are endogenous, naturally occurring MMP inhibitors that bind MMPs in a 1:1 stoichiometry.[106,107] TIMPs include four homologous, TIMP-1, -2, -3, and -4. TIMP-1 and -3 are glycoproteins, but neither TIMP-2 nor TIMP-4 protein contain carbohydrates. TIMPs do not show high specificity toward a specific MMP and inhibit different MMPs with different efficacies. For example, TIMP-1 is a poor inhibitor of MT1-MMP, MT3-MMP, MT5-MMP, and MMP-19, while TIMP-2 and -3 can inhibit MT1-MMP and MT2-MMP.[108] Also, while TIMP-1 and -2 can bind MMP-10 (stromelysin-2), their binding is 10-fold weaker than that to MMP-3 (stromelysin-1).[109] Importantly, TIMP-1 has a threonine-2 (Thr2) side chain that enters the MMP S1′ pocket in a manner similar to that of a substrate P1′ substituent, largely determining the affinity to MMP-3. Substitutions at Thr2 could affect the stability of the TIMP–MMP complex and the TIMP specificity for different MMPs. For example, a substitution of alanine for Thr2 is associated with a 17-fold decrease in binding of TIMP-1 to MMP-1 relative to MMP-3.[110]

TIMPs have been localized in different regions within the veins. Studies have tested whether histological changes in VVs wall may correlate with alterations in the expression of MMPs and TIMPs. VVs were compared with GSV segments from arterial bypass, and with arm and neck veins from fistula and carotid operations. There was a higher expression of TIMP-2 and increased connective tissue accumulation in the tunica media of VVs compared with control arm and neck veins. TIMP-2 and -3 expression was higher in hypertrophic than atrophic segments, and in the thicker proximal segments compared to the distal segments of VVs. It was suggested that a higher TIMP expression would suppress protease activity, reduce ECM turnover, and favor deposition of connective tissue and thicker vein wall.[111] Other studies showed TIMP-1, -2, and -3 in the intima and TIMP-1 and -2 in the media of control veins, as compared to TIMP-1 and -3 in the intima and TIMP-1, -2, and -3 in the media of VVs.[4]

An imbalance of MMP/TIMP ratio may contribute to the development of VVs. A change in either TIMP or MMP levels could alter the

MMP/TIMP ratio and cause a net change in specific MMP activity. In one study, MMP-7 and -9, and TIMP-1, -2, and -3 levels were only slightly modified, while MMP-1, -2, and -3 levels were increased, and these changes were accompanied by an increase in the elastic network and accumulation of collagen type-I, fibrillin-1, and laminin in both the veins and the skin of patients with VVs compared with control subjects undergoing coronary bypass surgery. These findings suggest that an imbalance MMP/TIMP ratio could lead to disruption of ECM production/degradation balance, and the observed remodeling in both the veins and the skin of patients with VVs suggests systemic alterations of the connective tissue.[4] Other studies showed a decrease in MMP-2/TIMP-1 ratio in avulsed VVs and suggested that the decrease in MMP-2 proteolytic activity could be the cause of the extensive accumulation of ECM observed in hypertrophic regions of VVs.[112] Also, marked increases in plasma levels of MMP-2 and -9, TIMP-1 and -2, and the MMP-2/TIMP-2 ratio were observed in patients with leg venous ulcers compared with normal controls. In subjects with healed venous ulcers, there was a decrease in MMP-9 and TIMP-1 levels and in the MMP-2/TIMP-2 ratio compared to the baseline values.[113] These observations highlight the importance of further examining the levels of TIMPs in comparison with MMPs in different regions of VVs at different stages of CVD.

8.2 Synthetic MMP Inhibitors

Deep sea water components such as Mg^{2+}, Cu^{2+}, and Mn^{2+} may inhibit MMP activity via a mechanism involving interference with Zn^{2+} binding at the MMP catalytic active site.[114] Utilizing the Zn^{2+} binding property, several MMP inhibitors have been developed.[45,115] MMP inhibitors often have a Zn^{2+} binding group, e.g., hydroxamic acid, carboxylic acid, and sulfhydryl group.[116] Zn^{2+} binding globulins (ZBGs) displace the Zn^{2+}-bound water molecule in an MMP and inactivate the enzyme. A ZBG is also an anchor that keeps the MMP inhibitor in the MMP active site and allows the backbone of the MMP inhibitor to enter the MMP substrate-binding pockets.[117] Hydroxamic acids include succinyl, sulfonamide, and phosphinamide hydroxamates.[116,118,119] Batimastat (BB-94), marimastat (BB-2516), and ilomastat (GM6001) are broad spectrum succinyl hydroxamates with a structure mimicking collagen, and inhibit MMPs by bidentate chelation of Zn^{2+}.[116,120] Other ZBGs include carboxylic acids, sulfonylhydrazides, thiols, aminomethyl benzimidazole-containing ZBGs, phosphorous- and nitrogen-based ZBGs, and heterocyclic bidentate chelators.[117,121,122] Tetracyclines and mechanism-based inhibitors also inhibit MMPs by chelating Zn^{2+} ion.[116] SB-3CT (compound **40**) is a mechanism-based MMP inhibitor that coordinates with MMP Zn^{2+}, thus allowing the conserved Glu202 in the MMP molecule to perform a nucleophilic attack and form a covalent bond with the compound.[117] When compared to the traditional competitive Zn^{2+} chelating MMP inhibitors, the strong covalent bond in SB-3CT prevents dissociation of the MMP inhibitor and decreases the rate of catalytic turnover, and therefore reduces the amount of MMP inhibitor needed to saturate the MMP active site.[123]

Some MMP inhibitors such as compound **37** do not have ZBGs and do not bind to the highly conserved Zn^{2+} binding group.[124] Instead, these MMP inhibitors undergo noncovalent interaction with the S1′, S2′, S3′, and S4′ pockets in the MMP molecule in a fashion similar to that of the substrate P1′, P2′, P3′, and P4′ substituents. The specificity and efficacy of the MMP inhibitor are determined by which pockets it blocks in the MMP molecule.[116]

Small interfering RNA (siRNA) may inhibit the transcriptional product of an MMP.[125] Also, sulodexide (SDX) is a glycosaminoglycan that decreases MMP-9 secretion from white blood cells without displacement of the MMP prodomain, and may inhibit MMPs with cysteine residues such as MMP-2 and -9.[22,66] Statins such as atorvastatin inhibit the expression of MMP-1, -2, and -9 in human retinal pigment epithelial cells,[126] and decrease the release of MMP-1, -2, -3, and -9 from rabbit macrophages and rabbit aortic and human saphenous vein VSMCs.[127] Also, pravastatin suppresses the increase in cardiac MMP-2 and -9 activity in a rat model of heart failure.[128] Despite the marked advances in the design of MMP inhibitors, doxycycline is the only FDA-approved MMP inhibitor.[129] In a recent study examining the effects of doxycycline in leg venous ulcer, patients were randomized into two groups, one group received the most appropriate basic treatment including compression therapy followed or not by vein surgery plus oral low dose of doxycycline 20 mg b.i.d. for 3 months, while the second group of patients received basic treatment only. Patients receiving basic treatment plus doxycycline showed a higher healing rate of venous ulcers compared with patients receiving basic treatment only. In patients receiving basic treatment only, the lower healing rate was associated with higher levels of MMP-9. NGAL and VEGF in plasma, wound fluid, and biopsies. It was suggested that doxycycline administration through its immunomodulatory and antiinflammatory actions, and inhibitory effects on MMP, could improve ECM function and speed venous leg ulcer and wound healing.[130] A major limitation of MMP inhibitors is that they cause musculoskeletal side effects manifested as joint inflammation, pain, stiffness, and tendonitis.[131] MMP inhibitors have the potential to be used clinically in CVD if their selectivity against specific MMPs is enhanced and their general side effects are minimized using targeted site-specific delivery.[132]

9. CONCLUDING REMARKS

MMPs play a major role in venous tissue remodeling and could be important biomarkers for the progression of CVD and potential targets for the management of VVs. However, understanding the role of MMPs in the pathogenesis of VVs has been challenging. More than one MMP are likely involved in VVs. Also, the changes in MMPs in the venous system may not be uniform. Vein remodeling is a dynamic process, and an increase in one MMP in one region may be paralleled by a decrease of other MMPs in other regions. Also, MMPs have different proteolytic activities toward different substrates, and their activities may vary during the course of CVD. MMP activity is also regulated by endogenous TIMPs and the MMP/TIMP ratio could vary in atrophic vs hypertrophic regions of VVs. Therefore, it is important to measure different MMPs and TIMPs in various regions of VVs and at different stages of CVD. Another challenge is that currently available MMP inhibitors have poor selectivity and many biologic actions, and may cause side effects.[116] As more selective MMP inhibitors are developed, their effectiveness in treatment of VVs should be tested. Targeted approaches to modify MMP expression/activity locally in the vicinity of VVs may also minimize any systemic side effects.

ACKNOWLEDGMENTS

This work was supported by grants from National Heart, Lung, and Blood Institute (HL-65998, HL-111775). Y.C. was a visiting scholar from the Department of Vascular Surgery, Wuhan Union Hospital, Tongji Medical College, Huazhong University of Science and Technology, Wuhan, PR China, and a recipient of scholarship from the China Scholarship Council. W.P. was a visiting scholar from the Department of Otorhinolaryngology—Head and Neck Surgery, Wuhan Union Hospital of Tongji Medical College, Huazhong University of Science and Technology, Wuhan, China.

Conflict of Interest: None.

REFERENCES

1. Recek C. Conception of the venous hemodynamics in the lower extremity. *Angiology*. 2006;57(5):556–563.
2. Lim CS, Davies AH. Pathogenesis of primary varicose veins. *Br J Surg*. 2009;96(11):1231–1242.
3. MacColl E, Khalil RA. Matrix metalloproteinases as regulators of vein structure and function: implications in chronic venous disease. *J Pharmacol Exp Ther*. 2015;355(3):410–428.
4. Sansilvestri-Morel P, Fioretti F, Rupin A, et al. Comparison of extracellular matrix in skin and saphenous veins from patients with varicose veins: does the skin reflect venous matrix changes? *Clin Sci (Lond)*. 2007;112(4):229–239.
5. Eklof B, Rutherford RB, Bergan JJ, et al. Revision of the CEAP classification for chronic venous disorders: consensus statement. *J Vasc Surg*. 2004;40(6):1248–1252.
6. Beebe-Dimmer JL, Pfeifer JR, Engle JS, Schottenfeld D. The epidemiology of chronic venous insufficiency and varicose veins. *Ann Epidemiol*. 2005;15(3):175–184.
7. Zsoter T, Cronin RF. Venous distensibility in patients with varicose veins. *Can Med Assoc J*. 1966;94(25):1293–1297.
8. Raffetto JD, Qiao X, Beauregard KG, Khalil RA. Estrogen receptor-mediated enhancement of venous relaxation in female rat: implications in sex-related differences in varicose veins. *J Vasc Surg*. 2010;51(4):972–981.
9. Mekky S, Schilling RS, Walford J. Varicose veins in women cotton workers. An epidemiological study in England and Egypt. *Br Med J*. 1969;2(5657):591–595.
10. Jawien A. The influence of environmental factors in chronic venous insufficiency. *Angiology*. 2003;54(suppl 1):S19–S31.
11. Lacroix P, Aboyans V, Preux PM, Houles MB, Laskar M. Epidemiology of venous insufficiency in an occupational population. *Int Angiol*. 2003;22(2):172–176.
12. Lee S, Lee W, Choe Y, et al. Gene expression profiles in varicose veins using complementary DNA microarray. *Dermatol Surg*. 2005;31(4):391–395.
13. Ng MY, Andrew T, Spector TD, Jeffery S. Linkage to the FOXC2 region of chromosome 16 for varicose veins in otherwise healthy, unselected sibling pairs. *J Med Genet*. 2005;42(3):235–239.
14. Serra R, Buffone G, de Franciscis A, et al. A genetic study of chronic venous insufficiency. *Ann Vasc Surg*. 2012;26(5):636–642.
15. Delis KT, Gloviczki P, Wennberg PW, Rooke TW, Driscoll DJ. Hemodynamic impairment, venous segmental disease, and clinical severity scoring in limbs with Klippel–Trenaunay syndrome. *J Vasc Surg*. 2007;45(3):561–567.
16. Reagan B, Folse R. Lower limb venous dynamics in normal persons and children of patients with varicose veins. *Surg Gynecol Obstet*. 1971;132(1):15–18.
17. Saiki S, Sakai K, Saiki M, et al. Varicose veins associated with CADASIL result from a novel mutation in the Notch3 gene. *Neurology*. 2006;67(2):337–339.
18. Badauy CM, Gomes SS, Sant'Ana Filho M, Chies JA. Ehlers-Danlos syndrome (EDS) type IV: review of the literature. *Clin Oral Investig*. 2007;11(3):183–187.
19. Xu HM, Zhao Y, Zhang XM, Zhu T, Fu WG. Polymorphisms in MMP-9 and TIMP-2 in Chinese patients with varicose veins. *J Surg Res*. 2011;168(1):e143–e148.
20. Anwar MA, Georgiadis KA, Shalhoub J, Lim CS, Gohel MS, Davies AH. A review of familial, genetic, and congenital aspects of primary varicose vein disease. *Circ Cardiovasc Genet*. 2012;5(4):460–466.
21. Naoum JJ, Hunter GC, Woodside KJ, Chen C. Current advances in the pathogenesis of varicose veins. *J Surg Res*. 2007;141(2):311–316.
22. Mannello F, Medda V, Ligi D, Raffetto JD. Glycosaminoglycan sulodexide inhibition of MMP-9 gelatinase secretion and activity: possible pharmacological role against

collagen degradation in vascular chronic diseases. *Curr Vasc Pharmacol.* 2013;11(3):354–365.
23. Naik B, Kumar M, Khanna AK, Suman PK. Clinico-histopathological study of varicose vein and role of matrix metalloproteinases-1, matrix metalloproteinases-9 and tissue inhibitor of matrix metalloproteinase-1 in varicose vein formation. *Indian J Pathol Microbiol.* 2016;59(1):25–30.
24. Gandhi RH, Irizarry E, Nackman GB, Halpern VJ, Mulcare RJ, Tilson MD. Analysis of the connective tissue matrix and proteolytic activity of primary varicose veins. *J Vasc Surg.* 1993;18(5):814–820.
25. Haviarova Z, Weismann P, Stvrtinova V, Benuska J. The determination of the collagen and elastin amount in the human varicose vein by the computer morphometric method. *Gen Physiol Biophys.* 1999;18(suppl 1):30–33.
26. Kockx MM, Knaapen MW, Bortier HE, Cromheeke KM, Boutherin-Falson O, Finet M. Vascular remodeling in varicose veins. *Angiology.* 1998;49(11):871–877.
27. Venturi M, Bonavina L, Annoni F, et al. Biochemical assay of collagen and elastin in the normal and varicose vein wall. *J Surg Res.* 1996;60(1):245–248.
28. Corcos L, De Anna D, Dini M, Macchi C, Ferrari PA, Dini S. Proximal long saphenous vein valves in primary venous insufficiency. *J Mal Vasc.* 2000;25(1):27–36.
29. Psaila JV, Melhuish J. Viscoelastic properties and collagen content of the long saphenous vein in normal and varicose veins. *Br J Surg.* 1989;76(1):37–40.
30. Ono T, Bergan JJ, Schmid-Schonbein GW, Takase S. Monocyte infiltration into venous valves. *J Vasc Surg.* 1998;27(1):158–166.
31. Raffetto JD, Khalil RA. Mechanisms of varicose vein formation: valve dysfunction and wall dilation. *Phlebology.* 2008;23(2):85–98.
32. Dzieciuchowicz L, Espinosa G, Paramo JA. Increased levels of metalloproteinase 10 and hemostatic markers in patients with noncomplicated primary varicose veins. *Clin Appl Thromb Hemost.* 2015;21(7):684–687.
33. Gillespie DL, Patel A, Fileta B, et al. Varicose veins possess greater quantities of MMP-1 than normal veins and demonstrate regional variation in MMP-1 and MMP-13. *J Surg Res.* 2002;106(2):233–238.
34. Woodside KJ, Hu M, Burke A, et al. Morphologic characteristics of varicose veins: possible role of metalloproteinases. *J Vasc Surg.* 2003;38(1):162–169.
35. Gomez I, Benyahia C, Louedec L, et al. Decreased PGE(2) content reduces MMP-1 activity and consequently increases collagen density in human varicose vein. *PLoS One.* 2014;9(2):e88021
36. Serra R, Gallelli L, Butrico L, et al. From varices to venous ulceration: the story of chronic venous disease described by metalloproteinases. *Int Wound J.* 2017;14(1):233–240.
37. Amato B, Coretti G, Compagna R, et al. Role of matrix metalloproteinases in non-healing venous ulcers. *Int Wound J.* 2015;12(6):641–645.
38. Serra R, Grande R, Buffone G, et al. Extracellular matrix assessment of infected chronic venous leg ulcers: role of metalloproteinases and inflammatory cytokines. *Int Wound J.* 2016;13(1):53–58.
39. Asanuma K, Magid R, Johnson C, Nerem RM, Galis ZS. Uniaxial strain upregulates matrix-degrading enzymes produced by human vascular smooth muscle cells. *Am J Physiol Heart Circ Physiol.* 2003;284(5):H1778–H1784.
40. Raffetto JD, Qiao X, Koledova VV, Khalil RA. Prolonged increases in vein wall tension increase matrix metalloproteinases and decrease constriction in rat vena cava: potential implications in varicose veins. *J Vasc Surg.* 2008;48(2):447–456.
41. Lim CS, Qiao X, Reslan OM, et al. Prolonged mechanical stretch is associated with upregulation of hypoxia-inducible factors and reduced contraction in rat inferior vena cava. *J Vasc Surg.* 2011;53(3):764–773.

42. Schmid-Schonbein GW, Takase S, Bergan JJ. New advances in the understanding of the pathophysiology of chronic venous insufficiency. *Angiology*. 2001;52(suppl 1): S27–S34.
43. Takase S, Pascarella L, Bergan JJ, Schmid-Schonbein GW. Hypertension-induced venous valve remodeling. *J Vasc Surg*. 2004;39(6):1329–1334.
44. Saito S, Trovato MJ, You R, et al. Role of matrix metalloproteinases 1, 2, and 9 and tissue inhibitor of matrix metalloproteinase-1 in chronic venous insufficiency. *J Vasc Surg*. 2001;34(5):930–938.
45. Kucukguven A, Khalil RA. Matrix metalloproteinases as potential targets in the venous dilation associated with varicose veins. *Curr Drug Targets*. 2013;14(3):287–324.
46. Sayer GL, Smith PD. Immunocytochemical characterisation of the inflammatory cell infiltrate of varicose veins. *Eur J Vasc Endovasc Surg*. 2004;28(5):479–483.
47. Aunapuu M, Arend A. Histopathological changes and expression of adhesion molecules and laminin in varicose veins. *Vasa*. 2005;34(3):170–175.
48. Jacob MP, Cazaubon M, Scemama A, et al. Plasma matrix metalloproteinase-9 as a marker of blood stasis in varicose veins. *Circulation*. 2002;106(5):535–538.
49. Sato H, Kita M, Seiki M. v-Src activates the expression of 92-kDa type IV collagenase gene through the AP-1 site and the GT box homologous to retinoblastoma control elements. A mechanism regulating gene expression independent of that by inflammatory cytokines. *J Biol Chem*. 1993;268(31):23460–23468.
50. Reddy VS, Prabhu SD, Mummidi S, et al. Interleukin-18 induces EMMPRIN expression in primary cardiomyocytes via JNK/Sp1 signaling and MMP-9 in part via EMMPRIN and through AP-1 and NF-kappaB activation. *Am J Physiol Heart Circ Physiol*. 2010;299(4):H1242–H1254.
51. Arbiser JL, Petros J, Klafter R, et al. Reactive oxygen generated by Nox1 triggers the angiogenic switch. *Proc Natl Acad Sci U S A*. 2002;99(2):715–720.
52. Zubkova ES, Men'shikov MY, Plekhanova OS, Beloglazova IB, Ratner EI, Parfenova EV. Urokinase stimulates production of matrix metalloproteinase-9 in fibroblasts with involvement of reactive oxygen species. *Bull Exp Biol Med*. 2014;157(1):18–21.
53. Fu X, Kao JL, Bergt C, et al. Oxidative cross-linking of tryptophan to glycine restrains matrix metalloproteinase activity: specific structural motifs control protein oxidation. *J Biol Chem*. 2004;279(8):6209–6212.
54. Milkiewicz M, Doyle JL, Fudalewski T, Ispanovic E, Aghasi M, Haas TL. HIF-1alpha and HIF-2alpha play a central role in stretch-induced but not shear-stress-induced angiogenesis in rat skeletal muscle. *J Physiol*. 2007;583(pt 2):753–766.
55. Kim CH, Cho YS, Chun YS, Park JW, Kim MS. Early expression of myocardial HIF-1alpha in response to mechanical stresses: regulation by stretch-activated channels and the phosphatidylinositol 3-kinase signaling pathway. *Circ Res*. 2002;90(2):E25–E33.
56. Thodeti CK, Matthews B, Ravi A, et al. TRPV4 channels mediate cyclic strain-induced endothelial cell reorientation through integrin-to-integrin signaling. *Circ Res*. 2009;104(9):1123–1130.
57. Lim CS, Kiriakidis S, Paleolog EM, Davies AH. Increased activation of the hypoxia-inducible factor pathway in varicose veins. *J Vasc Surg*. 2012;55(5):1427–1439.
58. Misra S, Fu AA, Rajan DK, et al. Expression of hypoxia inducible factor-1 alpha, macrophage migration inhibition factor, matrix metalloproteinase-2 and -9, and their inhibitors in hemodialysis grafts and arteriovenous fistulas. *J Vasc Interv Radiol*. 2008;19(2 pt 1):252–259.
59. Foda HD, Rollo EE, Drews M, et al. Ventilator-induced lung injury upregulates and activates gelatinases and EMMPRIN: attenuation by the synthetic matrix metalloproteinase inhibitor, Prinomastat (AG3340). *Am J Respir Cell Mol Biol*. 2001;25(6):717–724.

60. Norgauer J, Hildenbrand T, Idzko M, et al. Elevated expression of extracellular matrix metalloproteinase inducer (CD147) and membrane-type matrix metalloproteinases in venous leg ulcers. *Br J Dermatol.* 2002;147(6):1180–1186.
61. Majed BH, Khalil RA. Molecular mechanisms regulating the vascular prostacyclin pathways and their adaptation during pregnancy and in the newborn. *Pharmacol Rev.* 2012;64(3):540–582.
62. Lee J, Banu SK, Subbarao T, Starzinski-Powitz A, Arosh JA. Selective inhibition of prostaglandin E2 receptors EP2 and EP4 inhibits invasion of human immortalized endometriotic epithelial and stromal cells through suppression of metalloproteinases. *Mol Cell Endocrinol.* 2011;332(1–2):306–313.
63. Takai S, Jin D, Chen H, et al. Chymase inhibition improves vascular dysfunction and survival in stroke-prone spontaneously hypertensive rats. *J Hypertens.* 2014;32(8): 1637–1649.
64. Yin Z, Sada AA, Reslan OM, Narula N, Khalil RA. Increased MMPs expression and decreased contraction in the rat myometrium during pregnancy and in response to prolonged stretch and sex hormones. *Am J Physiol Endocrinol Metab.* 2012;303(1):E55–E70.
65. Dang Y, Li W, Tran V, Khalil RA. EMMPRIN-mediated induction of uterine and vascular matrix metalloproteinases during pregnancy and in response to estrogen and progesterone. *Biochem Pharmacol.* 2013;86(6):734–747.
66. Serra R, Gallelli L, Conti A, et al. The effects of sulodexide on both clinical and molecular parameters in patients with mixed arterial and venous ulcers of lower limbs. *Drug Des Devel Ther.* 2014;8:519–527.
67. Badier-Commander C, Verbeuren T, Lebard C, Michel JB, Jacob MP. Increased TIMP/MMP ratio in varicose veins: a possible explanation for extracellular matrix accumulation. *J Pathol.* 2000;192(1):105–112.
68. Kirsch D, Dienes HP, Kuchle R, et al. Changes in the extracellular matrix of the vein wall—the cause of primary varicosis? *Vasa.* 2000;29(3):173–177.
69. Sansilvestri-Morel P, Rupin A, Badier-Commander C, et al. Imbalance in the synthesis of collagen type I and collagen type III in smooth muscle cells derived from human varicose veins. *J Vasc Res.* 2001;38(6):560–568.
70. Sansilvestri-Morel P, Rupin A, Jullien ND, et al. Decreased production of collagen Type III in cultured smooth muscle cells from varicose vein patients is due to a degradation by MMPs: possible implication of MMP-3. *J Vasc Res.* 2005;42(5):388–398.
71. Xiao Y, Huang Z, Yin H, Lin Y, Wang S. In vitro differences between smooth muscle cells derived from varicose veins and normal veins. *J Vasc Surg.* 2009;50(5):1149–1154.
72. Shi ZD, Ji XY, Berardi DE, Qazi H, Tarbell JM. Interstitial flow induces MMP-1 expression and vascular SMC migration in collagen I gels via an ERK1/2-dependent and c-Jun-mediated mechanism. *Am J Physiol Heart Circ Physiol.* 2010;298(1): H127–H135.
73. Jin UH, Suh SJ, Chang HW, et al. Tanshinone IIA from Salvia miltiorrhiza BUNGE inhibits human aortic smooth muscle cell migration and MMP-9 activity through AKT signaling pathway. *J Cell Biochem.* 2008;104(1):15–26.
74. Morla AO, Mogford JE. Control of smooth muscle cell proliferation and phenotype by integrin signaling through focal adhesion kinase. *Biochem Biophys Res Commun.* 2000;272(1):298–302.
75. Karsdal MA, Larsen L, Engsig MT, et al. Matrix metalloproteinase-dependent activation of latent transforming growth factor-beta controls the conversion of osteoblasts into osteocytes by blocking osteoblast apoptosis. *J Biol Chem.* 2002;277(46): 44061–44067.
76. Mao D, Zhang Y, Lu H, Zhang H. Molecular basis underlying inhibition of metastasis of gastric cancer by anti-VEGFa treatment. *Tumour Biol.* 2014;35(8):8217–8223.

77. Zhang H, Chalothorn D, Jackson LF, Lee DC, Faber JE. Transactivation of epidermal growth factor receptor mediates catecholamine-induced growth of vascular smooth muscle. *Circ Res*. 2004;95(10):989–997.
78. Hollborn M, Stathopoulos C, Steffen A, Wiedemann P, Kohen L, Bringmann A. Positive feedback regulation between MMP-9 and VEGF in human RPE cells. *Invest Ophthalmol Vis Sci*. 2007;48(9):4360–4367.
79. Cui Y, Sun YW, Lin HS, et al. Platelet-derived growth factor-BB induces matrix metalloproteinase-2 expression and rat vascular smooth muscle cell migration via ROCK and ERK/p38 MAPK pathways. *Mol Cell Biochem*. 2014;393(1–2):255–263.
80. Rao VH, Kansal V, Stoupa S, Agrawal DK. MMP-1 and MMP-9 regulate epidermal growth factor-dependent collagen loss in human carotid plaque smooth muscle cells. *Physiol Rep*. 2014;2(2):e00224
81. Chew DK, Conte MS, Khalil RA. Matrix metalloproteinase-specific inhibition of Ca2 + entry mechanisms of vascular contraction. *J Vasc Surg*. 2004;40(5):1001–1010.
82. Raffetto JD, Barros YV, Wells AK, Khalil RA. MMP-2 induced vein relaxation via inhibition of [Ca2+]e-dependent mechanisms of venous smooth muscle contraction. Role of RGD peptides. *J Surg Res*. 2010;159(2):755–764.
83. Waitkus-Edwards KR, Martinez-Lemus LA, Wu X, et al. alpha(4)beta(1) Integrin activation of L-type calcium channels in vascular smooth muscle causes arteriole vasoconstriction. *Circ Res*. 2002;90(4):473–480.
84. Raffetto JD, Ross RL, Khalil RA. Matrix metalloproteinase 2-induced venous dilation via hyperpolarization and activation of K+ channels: relevance to varicose vein formation. *J Vasc Surg*. 2007;45(2):373–380.
85. Feletou M, Vanhoutte PM. Endothelium-derived hyperpolarizing factor: where are we now? *Arterioscler Thromb Vasc Biol*. 2006;26(6):1215–1225.
86. Boire A, Covic L, Agarwal A, Jacques S, Sherifi S, Kuliopulos A. PAR1 is a matrix metalloprotease-1 receptor that promotes invasion and tumorigenesis of breast cancer cells. *Cell*. 2005;120(3):303–313.
87. McNamara CA, Sarembock IJ, Gimple LW, Fenton 2nd JW, Coughlin SR, Owens GK. Thrombin stimulates proliferation of cultured rat aortic smooth muscle cells by a proteolytically activated receptor. *J Clin Invest*. 1993;91(1):94–98.
88. Coughlin SR. Thrombin signalling and protease-activated receptors. *Nature*. 2000;407(6801):258–264.
89. Garcia JG, Patterson C, Bahler C, Aschner J, Hart CM, English D. Thrombin receptor activating peptides induce Ca2+ mobilization, barrier dysfunction, prostaglandin synthesis, and platelet-derived growth factor mRNA expression in cultured endothelium. *J Cell Physiol*. 1993;156(3):541–549.
90. Lee HY, You HJ, Won JY, et al. Forkhead factor, FOXO3a, induces apoptosis of endothelial cells through activation of matrix metalloproteinases. *Arterioscler Thromb Vasc Biol*. 2008;28(2):302–308.
91. Lattimer CR, Kalodiki E, Kafeza M, Azzam M, Geroulakos G. Quantifying the degree graduated elastic compression stockings enhance venous emptying. *Eur J Vasc Endovasc Surg*. 2014;47(1):75–80.
92. Lattimer CR, Azzam M, Kalodiki E, Makris GC, Geroulakos G. Compression stockings significantly improve hemodynamic performance in post-thrombotic syndrome irrespective of class or length. *J Vasc Surg*. 2013;58(1):158–165.
93. Raffetto JD, Khalil RA. Matrix metalloproteinases and their inhibitors in vascular remodeling and vascular disease. *Biochem Pharmacol*. 2008;75(2):346–359.
94. Le Devehat C, Khodabandehlou T, Vimeux M, Kempf C. Evaluation of haemorheological and microcirculatory disturbances in chronic venous insufficiency: activity of Daflon 500 mg. *Int J Microcirc Clin Exp*. 1997;17(suppl 1):27–33.

95. Cesarone MR, Belcaro G, Pellegrini L, et al. Venoruton vs Daflon: evaluation of effects on quality of life in chronic venous insufficiency. *Angiology*. 2006;57(2):131–138.
96. Raffetto JD, Khalil RA. Ca(2+)-dependent contraction by the saponoside escin in rat vena cava: implications in venotonic treatment of varicose veins. *J Vasc Surg*. 2011;54(2):489–496.
97. Raffetto JD, Eberhardt RT, Dean SM, Ligi D, Mannello F. Pharmacologic treatment to improve venous leg ulcer healing. *J Vasc Surg Venous Lymphat Disord*. 2016;4(3):371–374.
98. Mann MW. Sclerotherapy: it is back and better. *Clin Plast Surg*. 2011;38(3):475–487. vii.
99. King JT, O'Byrne M, Vasquez M, Wright D. Treatment of truncal incompetence and varicose veins with a single administration of a new polidocanol endovenous microfoam preparation improves symptoms and appearance. *Eur J Vasc Endovasc Surg*. 2015;50(6):784–793.
100. Proebstle TM, Lehr HA, Kargl A, et al. Endovenous treatment of the greater saphenous vein with a 940-nm diode laser: thrombotic occlusion after endoluminal thermal damage by laser-generated steam bubbles. *J Vasc Surg*. 2002;35(4):729–736.
101. Merchant RF, Pichot O, Myers KA. Four-year follow-up on endovascular radiofrequency obliteration of great saphenous reflux. *Dermatol Surg*. 2005;31(2):129–134.
102. Min RJ, Khilnani N, Zimmet SE. Endovenous laser treatment of saphenous vein reflux: long-term results. *J Vasc Interv Radiol*. 2003;14(8):991–996.
103. Aremu MA, Mahendran B, Butcher W, et al. Prospective randomized controlled trial: conventional versus powered phlebectomy. *J Vasc Surg*. 2004;39(1):88–94.
104. Bootun R, Lane TR, Davies AH. The advent of non-thermal, non-tumescent techniques for treatment of varicose veins. *Phlebology*. 2016;31(1):5–14.
105. Tekin AI, Tuncer ON, Memetoglu ME, et al. Nonthermal, nontumescent endovenous treatment of varicose veins. *Ann Vasc Surg*. 2016;36:231–235.
106. Bode W, Fernandez-Catalan C, Grams F, et al. Insights into MMP–TIMP interactions. *Ann N Y Acad Sci*. 1999;878:73–91.
107. Nagase H, Visse R, Murphy G. Structure and function of matrix metalloproteinases and TIMPs. *Cardiovasc Res*. 2006;69(3):562–573.
108. Baker AH, Edwards DR, Murphy G. Metalloproteinase inhibitors: biological actions and therapeutic opportunities. *J Cell Sci*. 2002;115(pt 19):3719–3727.
109. Batra J, Robinson J, Soares AS, Fields AP, Radisky DC, Radisky ES. Matrix metalloproteinase-10 (MMP-10) interaction with tissue inhibitors of metalloproteinases TIMP-1 and TIMP-2: binding studies and crystal structure. *J Biol Chem*. 2012;287(19):15935–15946.
110. Meng Q, Malinovskii V, Huang W, et al. Residue 2 of TIMP-1 is a major determinant of affinity and specificity for matrix metalloproteinases but effects of substitutions do not correlate with those of the corresponding P1′ residue of substrate. *J Biol Chem*. 1999;274(15):10184–10189.
111. Aravind B, Saunders B, Navin T, et al. Inhibitory effect of TIMP influences the morphology of varicose veins. *Eur J Vasc Endovasc Surg*. 2010;40(6):754–765.
112. Badier-Commander C, Couvelard A, Henin D, Verbeuren T, Michel JB, Jacob MP. Smooth muscle cell modulation and cytokine overproduction in varicose veins. An in situ study. *J Pathol*. 2001;193(3):398–407.
113. Caimi G, Ferrara F, Montana M, et al. Behaviour of the plasma concentration of gelatinases and their tissue inhibitors in subjects with venous leg ulcers. *Clin Hemorheol Microcirc*. 2015;60(3):309–316.
114. Li PC, Pan CH, Sheu MJ, Wu CC, Ma WF, Wu CH. Deep sea water prevents balloon angioplasty-induced hyperplasia through MMP-2: an in vitro and in vivo study. *PLoS One*. 2014;9(5):e96927

115. Benjamin MM, Khalil RA. Matrix metalloproteinase inhibitors as investigative tools in the pathogenesis and management of vascular disease. *EXS*. 2012;103:209–279.
116. Hu J, Van den Steen PE, Sang QX, Opdenakker G. Matrix metalloproteinase inhibitors as therapy for inflammatory and vascular diseases. *Nat Rev Drug Discov*. 2007;6(6):480–498.
117. Jacobsen JA, Major Jourden JL, Miller MT, Cohen SM. To bind zinc or not to bind zinc: an examination of innovative approaches to improved metalloproteinase inhibition. *Biochim Biophys Acta*. 2010;1803(1):72–94.
118. Scozzafava A, Supuran CT. Carbonic anhydrase and matrix metalloproteinase inhibitors: sulfonylated amino acid hydroxamates with MMP inhibitory properties act as efficient inhibitors of CA isozymes I, II, and IV, and N-hydroxysulfonamides inhibit both these zinc enzymes. *J Med Chem*. 2000;43(20):3677–3687.
119. Pochetti G, Gavuzzo E, Campestre C, et al. Structural insight into the stereoselective inhibition of MMP-8 by enantiomeric sulfonamide phosphonates. *J Med Chem*. 2006;49(3):923–931.
120. Wojtowicz-Praga SM, Dickson RB, Hawkins MJ. Matrix metalloproteinase inhibitors. *Invest New Drugs*. 1997;15(1):61–75.
121. Skiles JW, Gonnella NC, Jeng AY. The design, structure, and therapeutic application of matrix metalloproteinase inhibitors. *Curr Med Chem*. 2001;8(4):425–474.
122. Puerta DT, Lewis JA, Cohen SM. New beginnings for matrix metalloproteinase inhibitors: identification of high-affinity zinc-binding groups. *J Am Chem Soc*. 2004;126(27):8388–8389.
123. Bernardo MM, Brown S, Li ZH, Fridman R, Mobashery S. Design, synthesis, and characterization of potent, slow-binding inhibitors that are selective for gelatinases. *J Biol Chem*. 2002;277(13):11201–11207.
124. Johnson AR, Pavlovsky AG, Ortwine DF, et al. Discovery and characterization of a novel inhibitor of matrix metalloprotease-13 that reduces cartilage damage in vivo without joint fibroplasia side effects. *J Biol Chem*. 2007;282(38):27781–27791.
125. Chetty C, Bhoopathi P, Joseph P, Chittivelu S, Rao JS, Lakka S. Adenovirus-mediated small interfering RNA against matrix metalloproteinase-2 suppresses tumor growth and lung metastasis in mice. *Mol Cancer Ther*. 2006;5(9):2289–2299.
126. Dorecka M, Francuz T, Garczorz W, Siemianowicz K, Romaniuk W. The influence of elastin degradation products, glucose and atorvastatin on metalloproteinase-1, -2, -9 and tissue inhibitor of metalloproteinases-1, -2, -3 expression in human retinal pigment epithelial cells. *Acta Biochim Pol*. 2014;61(2):265–270.
127. Luan Z, Chase AJ, Newby AC. Statins inhibit secretion of metalloproteinases-1, -2, -3, and -9 from vascular smooth muscle cells and macrophages. *Arterioscler Thromb Vasc Biol*. 2003;23(5):769–775.
128. Ichihara S, Noda A, Nagata K, et al. Pravastatin increases survival and suppresses an increase in myocardial matrix metalloproteinase activity in a rat model of heart failure. *Cardiovasc Res*. 2006;69(3):726–735.
129. Chen Q, Jin M, Yang F, Zhu J, Xiao Q, Zhang L. Matrix metalloproteinases: inflammatory regulators of cell behaviors in vascular formation and remodeling. *Mediators Inflamm*. 2013;2013:928315.
130. Serra R, Gallelli L, Buffone G, et al. Doxycycline speeds up healing of chronic venous ulcers. *Int Wound J*. 2015;12(2):179–184.
131. Renkiewicz R, Qiu L, Lesch C, et al. Broad-spectrum matrix metalloproteinase inhibitor marimastat-induced musculoskeletal side effects in rats. *Arthritis Rheum*. 2003;48(6):1742–1749.
132. Li J, Rush 3rd TS, Li W, et al. Synthesis and SAR of highly selective MMP-13 inhibitors. *Bioorg Med Chem Lett*. 2005;15(22):4961–4966.

INDEX

Note: Page numbers followed by "*f*" indicate figures, and "*t*" indicate tables.

W

X

Z

Kolibri 360 GmbH
Ferdinand-Jühlke-Straße 7
99095 Erfurt, Deutschland
produktsicherheit@kolibri360.de